Medical Image Synthesis

Image synthesis across and within medical imaging modalities is an active area of research with broad applications in radiology and radiation oncology. This book covers the principles and methods of medical image synthesis, along with state-of-the-art research.

First, various traditional non-learning-based, traditional machine-learning-based, and recent deep-learning-based medical image synthesis methods are reviewed. Second, specific applications of different inter- and intra-modality image synthesis tasks and of synthetic image-aided segmentation and registration are introduced and summarized, listing and highlighting the proposed methods, study designs, and reported performances with the related clinical applications of representative studies. Third, the clinical usages of medical image synthesis, such as treatment planning and image-guided adaptive radiotherapy, are discussed. Last, the limitations and current challenges of various medical synthesis applications are explored, along with future trends and potential solutions to solve these difficulties.

The benefits of medical image synthesis have sparked growing interest in a number of advanced clinical applications, such as magnetic resonance imaging (MRI)-only radiation therapy treatment planning and positron emission tomography (PET)/MRI scanning. This book will be a comprehensive and exciting resource for undergraduates, graduates, researchers, and practitioners.

Xiaofeng Yang is an Associate Professor and Vice Chair in Medical Physics Research in the Department of Radiation Oncology at Emory University, where he completed his postdoctoral and medical physics residency training. He holds B.S., M.S., and Ph.D. degrees in Biomedical Engineering from Xi'an Jiaotong University and completed his Ph.D. training and thesis at Emory University. Dr. Yang is also an adjunct faculty member in the medical physics department at Georgia Institute of Technology, as well as in the biomedical informatics department and the Wallace H. Coulter Department of Biomedical Engineering at Emory University and Georgia Institute of Technology. As a board-certified medical physicist with expertise in image-guided radiotherapy, artificial intelligence (AI), multimodality medical imaging, and medical image analysis, Dr. Yang leads the Deep Biomedical Imaging Laboratory. His lab is focused on developing novel AI-aided analytical and computational tools to enhance the role of quantitative imaging in cancer treatment and to improve the accuracy and precision of radiation therapy. His research has been funded by the NIH, DOD, and industrial funding agencies. Dr. Yang has published over 180 peer-reviewed journal papers and has received numerous scientific awards, including awards from SPIE Medical Imaging, AAPM, ASTRO, and SNMMI. In 2020, he received the John Laughlin Young Scientist Award from the American Association of Physicists in Medicine. Dr. Yang currently serves as an Associate Editor for Medical Physics and the Journal of Applied Clinical Medical Physics.

Imaging in Medical Diagnosis and Therapy

Series Editors: Bruce R. Thomadsen, David W. Jordan

For more information about this series, please visit: https://www.routledge.com/Imaging-in-Medical-Diagnosis-and-Therapy/book-series/CRCIMAINMED

Medical Image Synthesis
Methods and Clinical Applications

Edited by
Xiaofeng Yang

CRC Press
Taylor & Francis Group
Boca Raton New York London

CRC Press is an imprint of the
Taylor & Francis Group, an **Informa** business

Front cover image: vs148/Shutterstock.

First edition published 2024
by CRC Press
2385 NW Executive Center Drive, Suite 320, Boca Raton FL 33431

and by CRC Press
4 Park Square, Milton Park, Abingdon, Oxon, OX14 4RN

CRC Press is an imprint of Taylor & Francis Group, LLC

ISBN: 978-1-032-13388-1 (hbk)
ISBN: 978-1-032-15284-4 (pbk)
ISBN: 978-1-003-24345-8 (ebk)

DOI: 10.1201/9781003243458

Typeset in Palatino
by SPi Technologies India Pvt Ltd (Straive)

Contents

Section I   Methods and Principles

Section II   Applications of Inter-Modality Image Synthesis

Section III Applications of Intra-Modality Image Synthesis

Section IV Other Applications of Medical Image Synthesis

Section V   Clinic Usage of Medical Image Synthesis

Section VI Perspectives

Contributors

Jing Cai
Hong Kong Polytechnic University
Kowloon, Hong Kong

Jin Chen
University of Kentucky
Lexington, Kentucky

Ying Chen
Faculty of Information Technology
Beijing University of Technology
Beijing, China

Zhiming Cheng
School of Automation
Hangzhou Dianzi University
Hangzhou, China

Xianjin Dai
Department of Radiation Oncology
Stanford University
California, USA

Ruogu Fang
J. Crayton Pruitt Family Department of
    Biomedical Engineering
and
Center for Cognitive Aging and Memory
McKnight Brain Institute
and
Department of Electrical and Computer
    Engineering
Herbert Wertheim College of Engineering
University of Florida
Florida, USA

Yabo Fu
Memorial Sloan Kettering Cancer Center
New York, USA

Kuang Gong
Gordon Center for Medical Imaging
Massachusetts General Hospital
Harvard Medical School
Massachusetts, USA

Kai Huang
Department of Biomedical Sciences
The University of Texas Health Science
    Center at Houston
Texas, USA

Yu-Hua Huang
Hong Kong Polytechnic University
Kowloon, Hong Kong

Se-In Jang
Gordon Center for Medical
    Imaging
Massachusetts General Hospital
Harvard Medical School
Massachusetts, USA

Yang Lei
Department of Radiation Oncology
and
Winship Cancer Institute
Emory University
Georgia, USA

Taoran Li
Department of Radiation Oncology
University of Pennsylvania
Pennsylvania, USA

Wen Li
Hong Kong Polytechnic University
Hong Kong

Hely Lin
J. Crayton Pruitt Family Department of
    Biomedical Engineering
University of Florida
Florida, USA

Dan Nguyen
Medical Artificial Intelligence and
    Automation (MAIA) Laboratory
Department of Radiation Oncology
University of Texas Southwestern Medical
    Center
Texas, USA

Richard L.J. Qiu
Department of Radiation Oncology
and
Winship Cancer Institute
Emory University
Georgia, USA

Liangqiong Qu
Department of Statistics and Actuarial
    Science
and
The Institute of Data Science
The University of Hong Kong
Hong Kong

Ge Ren
Hong Kong Polytechnic University
Kowloon, Hong Kong

Yang Sheng
Department of Radiation Oncology
Duke University Medical Center
North Carolina, USA

Zhen Tian
Department of Radiation Oncology
The University of Chicago
Illinois, USA

Jing Wang
Department of Radiation Oncology
and
Winship Cancer Institute
Emory University
Georgia, USA

Tonghe Wang
Department of Medical Physics
Memorial Sloan Kettering
Cancer Center
New York, USA

Jackie Wu
Department of Radiation Oncology
Duke University Medical Center
North Carolina, USA

Yao Xiao
The University of Texas MD Anderson
    Cancer Center
Houston, TX, USA

Lei Xing
Department of Radiation Oncology
Stanford University
California, USA

Xiaofeng Yang
Department of Radiation Oncology
and
Winship Cancer Institute
Emory University
Georgia, USA

Lequan Yu
Department of Statistics and Actuarial
    Science
The University of Hong Kong
Hong Kong

Hao Zhang
Department of Medical Physics
Memorial Sloan Kettering Cancer Center
New York, USA

Jie Zhang
University of Kentucky
Lexington, Kentucky

Xiaodan Zhang
Faculty of Information Technology
Beijing University of Technology
Beijing, China

Yongqin Zhang
School of Information Science and
    Technology
Northwest University
Xi'an, China

Zhicheng Zhang
JancsiTech
Hangzhou, China

Wei Zhao
Beihang University
Beijing, China

Yuyin Zhou
Department of Computer Science and
    Engineering
University of California
California, USA

Jiarui Zhu
Hong Kong Polytechnic University
Kowloon, Hong Kong

Lingting Zhu
Department of Statistics and Actuarial
    Science
The University of Hong Kong
Hong Kong

Introduction

Recently, medical image synthesis became a vital aspect of both medical imaging research and clinic study in radiology and radiation oncology fields. Within the medical imaging and medical physics communities, promising work on inter-modality synthesis (e.g., MRI-based synthetic CT/dual energy (DE) CT, CBCT-based synthetic CT, CBCT/CT-based synthetic MRI, CBCT/CT-based stopping power map estimation, CT-based ventilation image, PET-based synthetic CT and image-based dose plan prediction) and intra-modality (cross-modality) synthesis (e.g., low-count PET-based full-count PET image estimation, PET/SPECT attenuation correction, intra-multiparametric MRI transformation, ultrasound/CT/MRI high-resolution image generation, 2D-based 3D volumetric image generation, MRI inhomogeneity correction, low-dose CT, MRI/CT/US denoising and metal artifact reduction) is being performed in PET/SPECT attenuation correction, MRI-based treatment planning, CBCT-guided adaptive radiotherapy, image segmentation, multimodality image registration, high-quality image generation, low-dose PET or CT generation, fast MRI imaging, real-time motion tracking, and numerous other areas. Therefore, we believe that it is time to combine the mostly used and well-known methods – to prepare a book dedicated to the topic of medical image synthesis. The aim of this book is to provide principles and methods, and state-of-the-art research, for undergraduates, graduates, researchers and practitioners. In this book, first, various traditional non-learning-based, traditional machine-learning-based and recent deep-learning-based medical image synthesis methods are reviewed. Second, specific applications of different inter-modality and intra-modality image synthesis tasks and of synthetic image-aided segmentation and registration are introduced and summarized. Third, the clinic usage of medical image synthesis, such as treatment planning and image-guided adaptive radiotherapy, is discussed. Last, the limitations and current challenges of various medical synthesis applications are explored. The future trends or potential solution to solve these difficulties are discussed.

Section I

Methods and Principles

1

Non-Deep-Learning-Based Medical Image Synthesis Methods

Jing Wang and Xiaofeng Yang
Emory University, Atlanta, GA, USA

CONTENTS

1.1 Introduction

Image synthesis across multimodality has been actively studied and widely adopted in radiology and radiation oncology. For diagnosis and treatment purposes, it is quite common to involve medical imaging of several modalities, e.g., computed tomography (CT), magnetic resonance imaging (MRI), and positron emission tomography (PET). However, high-quality image acquisition can be too expensive, time-consuming, or laborsome and thus become infeasible during clinic workflow. In some cases, the radiation constraints and image registration complexity will also restrict the direct use of some imaging modalities [1]. To overcome such obstacles, image synthesis techniques have been introduced to generate difficult images from easy-to-obtain images (intermodality synthesis) or synthesize high-quality images from low-quality inputs (intramodality synthesis).

Before the era of deep learning, image synthesis was conventionally performed with non-learning-based or traditional machine-learning-based methods, including single/multi-atlas-based, random forest, support vector machine (SVM), and other techniques [2].

Such models are considered more difficult when the two involved imaging modalities carry very different information, e.g., between CT and MRI where MRI intensity information is not uniquely related to electron density like the CT Hounsfield unit (HU) value does [3]. In fact, MRI-based synthetic CT (sCT) or sometimes called pseudo–CT (pCT), has been a major topic in such studies since if we can obtain sCTs from standard clinical MRI sequences for treatment planning and patient positioning verification, then an MRI-only radiotherapy (RT) workflow can be implemented to greatly simplify the treatment process and reduce received doses, with a better lesion target and organ at risk (OAR) delineation provided by MRI [2].

In this chapter, we will review the published articles focusing on utilizing traditional computer-aided image synthesis before the development of modern deep learning-based techniques, which mainly include non-learning-based and traditional machine-learning-based methods. The remainder of the chapter is organized as follows: Section 1.2 provides an overview of non-learning-based methods, mainly the atlas-based algorithms. Section 1.3 introduces works utilizing traditional machine-learning-based methods. In Section 1.4 we will discuss the achievements and limitations for these conventional image synthesis techniques.

1.2 Overview of Non-learning-based Methods

Traditionally, one of the dominating applications of image synthesis in medicine is to generate sCTs from MRI sequences since CTs provide accurate patient positioning information and electron density distribution for RT treatment planning but suffer from low soft tissue contrast and extra radiation exposure; while clinical MRIs provide high soft tissue contrast for lesion targeting but don't contain electron density information. Thus, in the early days bulk density override methods were utilized to convert MRIs to sCTs intuitively, e.g., overriding the whole MRI with water-equivalent electron density for RT treatment planning [4–10]. Such early methods simply assume a homogeneous electron density across the volume and the treatment planning with sCTs could lead to a dose discrepancy of greater than 2% compared with the ground truth heterogeneous CTs [2]. Later, more sophisticated overriding methods were developed by assigning different electron density numbers or HU values to different classes on the MRI [11–20]. Though the more specific overriding methods achieved better accuracies, they often require manual contouring to segment different zones (air, bone, soft tissues, etc.) on MRIs.

Compared with the bulk density overriding methods, atlas-based methods are generally considered to achieve improved results with a fully automated workflow and are the most important non-learning-based type of algorithms for generating sCTs from MRIs, which can be used for treatment planning and PET attenuation correction for PET/MR systems. In atlas-based methods, numerous co-registered pre-acquired MRI-CT image pairs form a reference database. The incoming MR image will be registered or warped to the reference MRIs, and the warping deformation will be applied to corresponding reference CTs, which will be brought up to generate the final sCT of the incoming MRI. An example workflow of atlas-based sCT synthesis is shown in Figure 1.1 [21]. Considering the number of atlases CTs used to output the final sCT, there exist single-atlas-based methods and multiple-atlas-based methods.

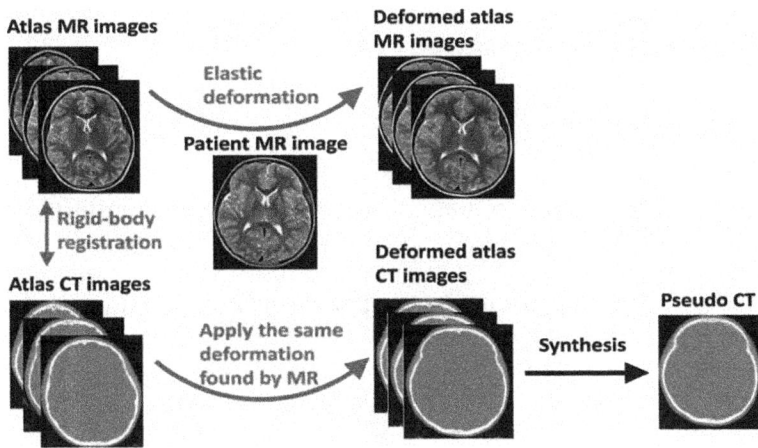

FIGURE 1.1
The workflow of generating sCT images from atlas images. The same warping deformation was applied to both atlas MRIs and the corresponding atlas CTs. The final sCT was a combination of the warped atlas CTs. (Reprint permission from [21].)

1.2.1 Single-Atlas-Based Method

Using a single atlas to create the sCTs is simple and straightforward. In a PET/MR study, Schreibmann et al. [22] used a multimodality optical flow deformable model to register a selected CT atlas template to the incoming patient MR images, by first performing a rigid registration to align the patient template with the patient MR images, and then using a BSpline deformable transform [23] to resolve the warping field between CT and MRIs, and finally applying a Hermosillo algorithm [24, 25] to refine the registration results. Thus, an sCT was created for PET attenuation correction. They compared the sCT attenuation with the true CT attenuation PET results for 17 clinical brain tumor cases, and a high agreement was found through the histogram analysis.

1.2.2 Multi-atlas-based Method

Compared to the single-atlas method, the multi-atlas-based method is more commonly used and considered to perform superior. Burgos et al. [26, 27] used a multi-atlas-based algorithm to synthesize CT from T1w MRI for PET attenuation correction. An atlas database was composed of affinely aligned CT and MRI pairs. To synthesize CT images for the target MRI, the atlas MRIs in the database were registered to the target MRI, and the same transformation was applied to corresponding atlas CTs. The convolution-based fast local normalized correlation coefficient (LNCC) [28] and the local normalized sum of square differences (NSSD), were used as two local similarity measures between the target MRI and transformed atlas MRIs. The ranks of local similarity measures were then converted to spatially varying weights. Then the sCT was generated by averaging the transformed atlas CTs with those weights. Finally, the HU values in sCT were converted to linear attenuation coefficients in cm^{-1} for PET image correction.

Uh et al. compared several atlas-based methods for MRI-CT synthesis of 14 pediatric patients with brain tumors. They used three types of atlas groups containing a single atlas, six atlases, or twelve atlases. For the six atlases group, they employed a simple arithmetic mean process or pattern recognition with Gaussian process (PRGP) [29] to combine the warped CT atlases into the final output, while for the 12 atlases group, they only used the PRGP merging process. Their results suggest the average root-mean-square (RMSD) between pseudo and real CTs improves with more atlases, and the PRGP combination is better than the simple arithmetic mean process.

Sjölund et al. [30] proposed a novel method to combine the multiple registered CT atlases to form a final sCT, which iteratively estimates the final single sCT from the voxel-wise mean of warped reference CTs. They claimed their iterative mean registration estimation made significant improvement compared to a simple mean-estimator of the multiple CT atlases or the single best-warped atlas. Degen et al. [12] addressed the challenge of multimodal image registration with the self-similarity context as a matching criterion, to match CT atlases to patient anatomy. Multimodal local atlas fusion is then performed to jointly combine the registered atlases. The method compared the dissimilarity between the two modalities at each location and only fuse those atlases morphologically similar enough to the incoming MRI. Such a method could extend the MR-CT synthesis technique from brain studies with strict alignment of CT-MRI pairs to whole-body scans without well-aligned CT-MRI pairs. Burgos et al. [14] tried to jointly solve organ segmentation and pCT synthesis by a multi-atlas method. The mean absolute error (MAE) and mean error (ME) were 45.7 ± 4.6 HU and −1.6 ± 7.7 HU, respectively, between sCTs and true CTs.

1.2.3 Patch-based Atlas Method

Though it is intuitive to treat the MRI and CT images as a whole to extract the anatomy information, some studies extend to the patch-based atlas method in the sCT fusion/generation phase. Hofmann et al. [29] built an atlas database of co-registered CT-MR pairs from 17 patients, and then warped the atlas CTs by MRI registration between the atlas and patient MR images. After the prior atlas registration, they used a pattern recognition technique that searched neighboring MR patches in the warped atlases and then generate sCT patches by Gaussian process regression, integrating with the mean value of prior registered CT atlases. By testing the method on three independent patients, the mean percentage difference between the MRI-based attenuation correction and the CT-based PET attenuation is 3.2% +/− 2.5%.

Besides synthesizing CTs from MRI, Roy et al. proposed a patch-based atlas method for generating T2-weighted (T2w) MRI from T1-weighted (T1w) MR images [31]. As an atlas-based method, they also created atlases with co-registered T1w and T2w images. From the atlas image pairs, then multiple subsampled patches were obtained from these image pairs to form numerous atlas patches. For each patch on the targeting T2w, a few similar-looking atlas patches were used to generate the synthetic T2w patches. Similar T1w atlas patches of the neighborhood of that targeting patch with similarity metrics were collected. Then the warped T2w reference atlases corresponding to these picked T1w atlases were combined to obtain the synthetic T2w patches, which were merged to form the final output T2w image. The peak signal-to-noise ratio (PSNR) is 28.68 vs. 25.62, and median cerebral spinal fluid (CSF) intensities are 2.17 vs. 1.91 for high vs. low resolution images, respectively, indicating a better performance on high-resolution datasets.

1.3 Overview of Traditional Machine-Learning-based Methods – Voxel-based Techniques

Though the atlas-based method is probably the most popular technique for image synthesis before the emergence of deep learning, there are other trials using traditional machine learning models such as random forest (RF) to investigate the image synthesis tasks at the voxel level.

Zhao et al. [32] proposed a bi-directional MR-CT synthesis framework using machine learning. The multiple co-registered MR-CT atlas pairs first performed a supervoxel over-segmentation process using the simple linear iterative clustering (SLIC) method, followed by a k-means clustering algorithm for tissue characterization into z-fields. Then for each field (tissue class), there are two RF regressors for sCTs from T1w or for T1w synthesis from CTs. Besides, two RF classifiers were trained to extract features and generate the z labels on both modalities, so that the image synthesis could be feasible by applying the bi-directional RF regressor of each tissue class. In an experiment involving six MR-CT atlas pairs, the structural similarity (SSIM) for sCT is 0.63–0.73, and the SSIM for synthetic MR is 0.7–0.8; while the PSNR for sCT is 27–30, and the PSNR for synthetic MR is 20–28 (Figure 1.2).

Besides the commonly discussed MR-CT synthesis problem, Jog et al. proposed a computationally efficient machine learning method to generate MR images of a desired modality from other input MR modalities, namely, the Regression Ensembles with Patch Learning for Image Contrast Agreement or REPLICA [33]. They extract multi-resolution features

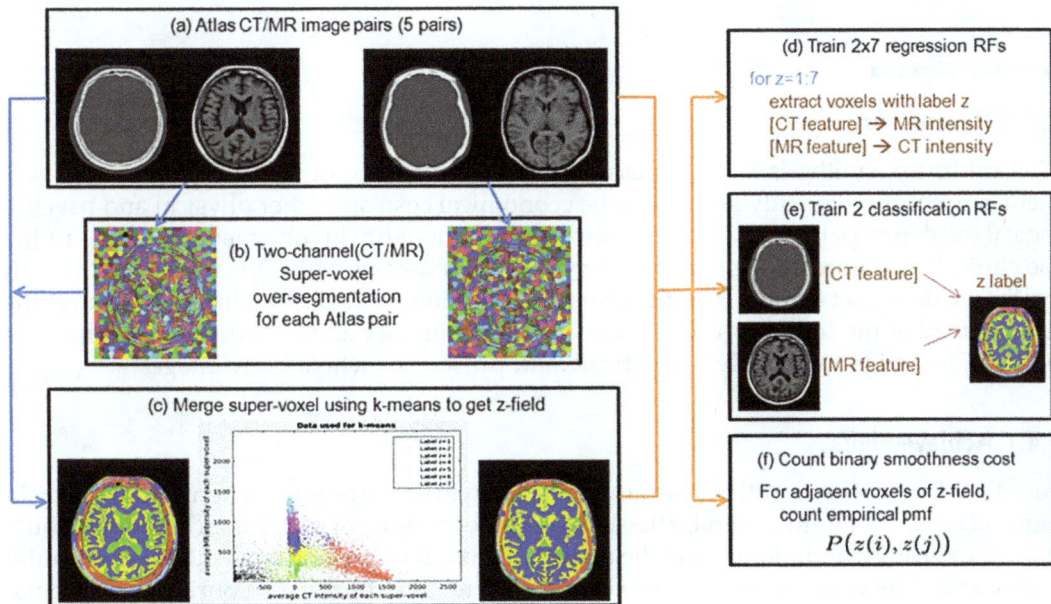

FIGURE 1.2
(a) example of CT-MR atlas pairs; (b) result of SLIC over-segmentation; (c) examples of z-fields after k-means clustering; (d) training of RF regressors; (e) RF classifiers trained to estimate z-fields from single modalities; and (f) computation of pairwise potentials for a Markov Random Field (MRF). (Reprint permission from [32].)

based on voxel intensities and high-resolution context descriptors to feed into an RF training model. The model learns to convert the input MR sequences to the target MR modality images at the voxel level. In an experiment of generating T2w MR images from 32 magnetization-prepared gradient echo (MPRAGE) scans, their method outperformed FUSION (a multi-atlas registration and intensity fusion method [34]) or MIMECS [34, 35] with a PSNR of 50.73, universal quality index (UQI) of 0.89, and an SSIM of 0.87. In another experiment of synthesizing 125 FLuid Attenuated Inversion Recovery (FLAIR) images of 84 patients from T1w, T2w, and P_D-weighted (PDw) sequences, their method achieved a PSNR of 21.73, UQI of 0.84, and SSIM of 0.81. They also pointed out that FUSION is an atlas-based model so the lesion intensity cannot be precisely synthesized if those intensities were not available in the atlases.

Bowles et al. [36] proposed an interesting application of synthesizing pseudo-healthy FLAIR images from an incoming T1w MRI. By subtracting this pseudo-healthy FLAIR from the patient's true FLAIR, they could obtain the difference voxel intensities corresponding to the probability of a pathological lesion, and train a support vector machine (SVM) to generate a probability map. The critical component is again the pseudo-healthy FLAIR synthesis from the input T1w images. They used a regression model to learn the voxel level relationship between healthy FLAIR and T1w, which directly mapped the T1w voxels to synthetic FLAIR images. The lesion detection of the proposed method is satisfactory, with a dice similarity coefficient (DSC) of 0.703, average symmetric surface distance (ASSD) of 1.23, Hausdorff distance (HD) of 38.6, and intraclass correlation (ICC) of 0.985.

We summarize the reviewed work for image synthesis in Table 1.1, including the study details, utilized methods, and obtained results.

1.4 Discussion

Medical image synthesis is an increasingly important topic in today's clinical improvement, since it could greatly save the time, economical cost, and other physical and psychological burden of patients by reducing the imaging modality involvement and simplifying the clinical workflow. Traditional image synthesis focuses on MR-CT synthesis or MR-MR synthesis, to extract anatomy or functional information from existing images, without the need of further medical scans. Most studies are preliminary and included small cohorts of patients, investigating mostly brain/head, and prostate or whole-body images.

1.4.1 Achievements

For MR-CT synthesis, the atlas-based methods seem to be the most commonly adopted technique [21, 30, 39–43]. In general, atlas-based methods perform better than the simple bulk density overriding techniques, and the atlas-based models can be carried out in a fully automated workflow while the density overriding often requires human contouring or segmentation of the MR images. Moreover, using multiple atlases outperforms the single-atlas method since more information can be tailored from the multi-atlas database. The major steps of the atlas-based method are the registration to align atlas MR-CT pairs to form co-registered atlas pairs, the registration of incoming MRI to atlas MRIs, and the fusion of warped CT atlases to generate final sCTs. The first two steps are highly dependent on the accuracy of rigid or non-rigid registration between multimodal images, and that is beyond the scope of this chapter.

TABLE 1.1

Summary of the Reviewed Work for Image Synthesis in Sections 2.1–2.3 and 3

Author	Year	Tasks	Location	Method	Dataset	Results
Hofmann et al. [29]	2008	sCT for PET/MRI correction	Brain	Patch-based atlas	20 patients	PET quantification with a mean error of 3.2% +/− 2.5%
Schreibmann et al. [22]	2010	sCT for PET/MRI correction	Brain	Single-atlas-based	17 brain tumor cases	mean voxel-to-voxel difference < 2 HU
Burgos et al. [26]	2013	sCT for PET/MRI correction	Brain	Multi-atlas-based	28 subjects	Average mean absolute residual (MAR) between sCTs and ground truth is 73–108 in HU for different CT modalities
Burgos et al. [27]	2014	sCT for PET/MRI correction	Brain	Multi-atlas-based	18 + 41 brain cases for parameter optimization and validation	The average MAE is 113 HU for full head and 240 HU for skull
Uh et al. [21]	2014	sCT from MRI	Brain	Single/multi-atlas-based	14 pediatric patients	More atlases and novel combination methods yielded better results
Sjölund et al. [30]	2015	sCT from MRI	Whole head	Multi-atlas-based	10 patients	Best mean absolute errors (HU) is 114.5 ± 20.5
Roy et al. [31]	2016	MRI synthesis from other sequences	Whole head	Patch-based atlas	44 patients	PSNR is 28.68 vs. 25.62 dB, and median CSF intensities are 2.17 vs. 1.91 for high- vs. low-resolution images
Degen et al. [37]	2016	sCT from MRI	Whole body	Multi-atlas-based	18 3D CT atlases	The dice overlap of segmentation labels between MRI and CT is greatly enhance after fusion
Bowles et al. [36]	2017	MRI synthesis from other sequences	Brain	SVM, Regression model	127 subjects	Lesion detection: DSC of 0.703, ASSD of 1.23, HD of 38.6, and ICC of 0.985
Burgos et al. [38]	2017	sCT from MRI	Prostate	Multi-atlas-based	15 subjects	MAE of 45.7 ± 4.6 HU, and ME of −1.6 ± 7.7 HU
Zhao et al. [32]	2017	Bi-directional MR/CT synthesis	Whole head	RF	6 MR/CT atlas pairs	SSIM for sCT is 0.63–0.73, and the SSIM for synthetic MR is 0.7–0.8; while the PSNR for sCT is 27–30, and the PSNR for synthetic MR is 20–28
Jog et al. [33]	2017	MRI synthesis from other sequences	Brain	RF	21 and 84 patients for T2w and FLAIR synthesis	A PSNR of 50.73, UQI of 0.89, and a SSIM of 0.87 for T2w MRI synthesis; a PSNR of 21.73, UQI of 0.84, and SSIM of 0.81 for FLAIR synthesis

Mean absolute error (MAE), mean absolute residual (MAR), mean error (ME), peak signal-to-noise ratio (PSNR), structural similarity (SSIM), universal quality index (UQI), mean surface distance (MSD), Hausdorff distance (HD), dice similarity coefficient (DSC), root-mean-square distance (RMSD), averaged symmetric surface distance (ASSD), intraclass correlation (ICC), and cerebral spinal fluid (CSF).

The last step, however, is unique in image synthesis and the novelty of many sCT studies focused on proposing devised fusion strategy. Common practices are taking the weighted average of warped atlas CTs as the sCTs or employing a patch-based method to fuse the final output. Similar principles also apply to atlas-based MR-MR synthesis.

Another commonly used type of implementation for CT-MR or MR-MR synthesis is a voxel-based method [15, 44–46]. The main idea is to obtain the voxel or supervoxel correlation between multimodal images via regression or machine learning models. Such techniques are claimed to outperform the atlas-based fusion methods [33]. In the scope of synthesizing CTs from MRIs, apart from the standard MRI sequences alone [47–53], there exist techniques utilizing ultra-short echo time (UTE) sequences in which the bone structures are more easily differentiated from air. However, taking these UTE sequences is costly in time and resources and not included in standard clinical workflow, so we didn't elaborate on those UTE work in this chapter.

Besides the most frequently seen CT and MR images, the CT synthesis from other modalities such as transrectal ultrasound was also investigated, which could potentially be useful in brachytherapy [54]. In the future, image synthesis tasks involving more imaging modalities can be expected, especially with the fast-developing deep-learning techniques.

1.4.2 Limitations

Though the atlas-based methods have achieved promising accuracy in image synthesis for both CT and MR images and have been considered as a robust method especially on unexpected or unusual cases [1], some limitations exist in such techniques. The first limitation is inherent within the method itself that the final synthetic images come from the co-registered atlas database. That explains why a single atlas can lead to noticeable deviance [40] due to its insufficiency of representing patient variations. Increasing the number of atlases in the database may relieve the problem, but could still perform poorly for atypical patients, e.g., patients with large tumors or surgical cavities [21]. Besides, increasing the atlases numbers can burden the cost of the already complex registration process in both the atlas MR-CT co-registration phase and the incoming MRI registration phase. Moreover, the effect of adding atlases is still not fully investigated. The optimal number of atlases needed for various tasks can depend on the similarity between atlases and the incoming images [42], while some studies reported limited benefits of adding atlases beyond 15 patients [55]. For machine learning techniques, they are often used with patch-based voxel level methods involving non-standard MRI sequences such as UTE that elongated patient stays, while for those used with standard MRI sequences, the need to manually contour bone and air may limit their use [2].

1.5 Conclusion

We have overviewed the state-of-the-art progress of conventional medical image synthesis before the blossom of deep learning. The main topics are MR-CT synthesis or MR-MR synthesis, and there are three major pathways to implement the tasks: the bulk density overriding, the atlas-based synthesis, and the traditional machine-learning-based methods. Though these techniques are relatively simple compared to the latest deep learning models, they have proven to be successful in generating desired images without the need for additional modality scans, and thus help reduce patient stays and lower radiation exposure.

References

1. Wang, T., et al., A review on medical imaging synthesis using deep learning and its clinical applications. *Journal of Applied Clinical Medical Physics*, 2021. **22**(1): pp. 11–36.
2. Johnstone, E., et al., Systematic review of synthetic computed tomography generation methodologies for use in magnetic resonance imaging–only radiation therapy. *International Journal of Radiation Oncology* Biology* Physics*, 2018. **100**(1): pp. 199–217.
3. Karlsson, M., et al., Dedicated magnetic resonance imaging in the radiotherapy clinic. *International Journal of Radiation Oncology* Biology* Physics*, 2009. **74**(2): pp. 644–651.
4. Schad, L.R., et al., Radiosurgical treatment planning of brain metastases based on a fast, three-dimensional MR imaging technique. *Magnetic Resonance Imaging*, 1994. **12**(5): pp. 811–819.
5. Ramsey, C.R. and A.L. Oliver, Magnetic resonance imaging based digitally reconstructed radiographs, virtual simulation, and three-dimensional treatment planning for brain neoplasms. *Medical Physics*, 1998. **25**(10): pp. 1928–1934.
6. Prabhakar, R., et al., Feasibility of using MRI alone for 3D radiation treatment planning in brain tumors. *Japanese Journal of Clinical Oncology*, 2007. **37**(6): pp. 405–411.
7. Wang, C., et al., MRI-based treatment planning with electron density information mapped from CT images: a preliminary study. *Technology in Cancer Research & Treatment*, 2008. **7**(5): pp. 341–347.
8. Weber, D., et al., Open low-field magnetic resonance imaging for target definition, dose calculations and set-up verification during three-dimensional CRT for glioblastoma multiforme. *Clinical Oncology*, 2008. **20**(2): pp. 157–167.
9. Chen, L., et al., MRI-based treatment planning for radiotherapy: dosimetric verification for prostate IMRT. *International Journal of Radiation Oncology* Biology* Physics*, 2004. **60**(2): pp. 636–647.
10. Chen, L., et al., Dosimetric evaluation of MRI-based treatment planning for prostate cancer. *Physics in Medicine & Biology*, 2004. **49**(22): p. 5157.
11. Eilertsen, K., et al., A simulation of MRI based dose calculations on the basis of radiotherapy planning CT images. *Acta Oncologica*, 2008. **47**(7): pp. 1294–1302.
12. Karotki, A., et al., Comparison of bulk electron density and voxel-based electron density treatment planning. *Journal of Applied Clinical Medical Physics*, 2011. **12**(4): pp. 97–104.
13. Korsholm, M.E., L.W. Waring, and J.M. Edmund, A criterion for the reliable use of MRI-only radiotherapy. *Radiation Oncology*, 2014. **9**(1): pp. 1–7.
14. Chin, A.L., et al., Feasibility and limitations of bulk density assignment in MRI for head and neck IMRT treatment planning. *Journal of Applied Clinical Medical Physics*, 2014. **15**(5): pp. 100–111.
15. Doemer, A., et al., Evaluating organ delineation, dose calculation and daily localization in an open-MRI simulation workflow for prostate cancer patients. *Radiation Oncology*, 2015. **10**(1): pp. 1–9.
16. Jonsson, J.H., et al., Treatment planning using MRI data: an analysis of the dose calculation accuracy for different treatment regions. *Radiation Oncology*, 2010. **5**(1): pp. 1–8.
17. Kristensen, B.H., et al., Dosimetric and geometric evaluation of an open low-field magnetic resonance simulator for radiotherapy treatment planning of brain tumours. *Radiotherapy and Oncology*, 2008. **87**(1): pp. 100–109.
18. Lambert, J., et al., MRI-guided prostate radiation therapy planning: Investigation of dosimetric accuracy of MRI-based dose planning. *Radiotherapy and Oncology*, 2011. **98**(3): pp. 330–334.
19. Stanescu, T., et al., A study on the magnetic resonance imaging (MRI)-based radiation treatment planning of intracranial lesions. *Physics in Medicine & Biology*, 2008. **53**(13): p. 3579.
20. Stanescu, T., et al., 3T MR-based treatment planning for radiotherapy of brain lesions. *Radiology and Oncology*, 2006. **40**(2): pp. 125–132.
21. Uh, J., et al., MRI-based treatment planning with pseudo CT generated through atlas registration. *Medical Physics*, 2014. **41**(5): p. 051711.

22. Schreibmann, E., et al., MR-based attenuation correction for hybrid PET-MR brain imaging systems using deformable image registration. *Medical Physics*, 2010. **37**(5): pp. 2101–2109.

23. Mattes, D., et al., PET-CT image registration in the chest using free-form deformations. *IEEE Transactions on Medical Imaging*, 2003. **22**(1): pp. 120–128.

24. De Craene, M., et al., Incorporating metric flows and sparse Jacobian transformations in ITK. *Insight Journal*, 2006. http://hdl.handle.net/1926/183

25. Cuadra, M.B., et al., Atlas-based segmentation of pathological MR brain images using a model of lesion growth. *IEEE Transactions on Medical Imaging*, 2004. **23**(10): pp. 1301–1314.

26. Burgos, N., et al., Attenuation correction synthesis for hybrid PET-MR scanners. In *International Conference on Medical Image Computing and Computer-Assisted Intervention*. 2013. Springer.

27. Burgos, N., et al., Attenuation correction synthesis for hybrid PET-MR scanners: application to brain studies. *IEEE Transactions on Medical Imaging*, 2014. **33**(12): pp. 2332–2341.

28. Cachier, P., et al., Iconic feature based nonrigid registration: the PASHA algorithm. *Computer Vision and Image Understanding*, 2003. **89**(2–3): pp. 272–298.

29. Hofmann, M., et al., MRI-based attenuation correction for PET/MRI: a novel approach combining pattern recognition and atlas registration. *Journal of Nuclear Medicine*, 2008. **49**(11): pp. 1875–1883.

30. Sjölund, J., et al., Generating patient specific pseudo-CT of the head from MR using atlas-based regression. *Physics in Medicine & Biology*, 2015. **60**(2): p. 825.

31. Roy, S., et al., Patch based synthesis of whole head MR images: Application to EPI distortion correction. In *International Workshop on Simulation and Synthesis in Medical Imaging*. 2016. Springer.

32. Zhao, C., et al., A supervoxel based random forest synthesis framework for bidirectional MR/CT synthesis. In *International Workshop on Simulation and Synthesis in Medical Imaging*. 2017. Springer.

33. Jog, A., et al., Random forest regression for magnetic resonance image synthesis. *Medical Image Analysis*, 2017. **35**: pp. 475–488.

34. Roy, S., A. Carass, and J.L. Prince, Magnetic resonance image example-based contrast synthesis. *IEEE Transactions on Medical Imaging*, 2013. **32**(12): pp. 2348–2363.

35. Roy, S., A. Carass, and J. Prince, A compressed sensing approach for MR tissue contrast synthesis. In *Biennial International Conference on Information Processing in Medical Imaging*. 2011. Springer.

36. Bowles, C., et al., Brain lesion segmentation through image synthesis and outlier detection. *NeuroImage: Clinical*, 2017. **16**: pp. 643–658.

37. Degen, J., and M.P. Heinrich, Multi-atlas based pseudo-CT synthesis using multimodal image registration and local atlas fusion strategies. In *Proceedings of the IEEE Conference on Computer Vision and Pattern Recognition Workshops*. 2016.

38. Burgos, N., et al., Iterative framework for the joint segmentation and CT synthesis of MR images: application to MRI-only radiotherapy treatment planning. *Physics in Medicine & Biology*, 2017. **62**(11): p. 4237.

39. Dowling, J.A., et al., An atlas-based electron density mapping method for magnetic resonance imaging (MRI)-alone treatment planning and adaptive MRI-based prostate radiation therapy. *International Journal of Radiation Oncology* Biology* Physics*, 2012. **83**(1): pp. e5–e11.

40. Demol, B., et al., Dosimetric characterization of MRI-only treatment planning for brain tumors in atlas-based pseudo-CT images generated from standard T1-weighted MR images. *Medical Physics*, 2016. **43**(12): pp. 6557–6568.

41. Andreasen, D., et al., Patch-based generation of a pseudo CT from conventional MRI sequences for MRI-only radiotherapy of the brain. *Medical Physics*, 2015. **42**(4): pp. 1596–1605.

42. Andreasen, D., K. Van Leemput, and J.M. Edmund, A patch-based pseudo-CT approach for MRI-only radiotherapy in the pelvis. *Medical Physics*, 2016. **43**(8Part1): pp. 4742–4752.

43. Edmund, J.M., et al., Cone beam computed tomography guided treatment delivery and planning verification for magnetic resonance imaging only radiotherapy of the brain. *Acta Oncologica*, 2015. **54**(9): pp. 1496–1500.

44. Dowling, J.A., et al., Automatic substitute computed tomography generation and contouring for magnetic resonance imaging (MRI)-alone external beam radiation therapy from standard MRI sequences. *International Journal of Radiation Oncology* Biology* Physics*, 2015. **93**(5): pp. 1144–1153.

45. Jonsson, J.H., et al., Treatment planning of intracranial targets on MRI derived substitute CT data. *Radiotherapy and Oncology*, 2013. **108**(1): pp. 118–122.

46. Johansson, A., et al., Voxel-wise uncertainty in CT substitute derived from MRI. *Medical Physics*, 2012. **39**(6Part1): pp. 3283–3290.

47. Kapanen, M., and M. Tenhunen, T1/T2*-weighted MRI provides clinically relevant pseudo-CT density data for the pelvic bones in MRI-only based radiotherapy treatment planning. *Acta Oncologica*, 2013. **52**(3): pp. 612–618.

48. Korhonen, J., et al., A dual model HU conversion from MRI intensity values within and outside of bone segment for MRI-based radiotherapy treatment planning of prostate cancer. *Medical Physics*, 2014. **41**(1): p. 011704.

49. Korhonen, J., et al., Absorbed doses behind bones with MR image-based dose calculations for radiotherapy treatment planning. *Medical Physics*, 2013. **40**(1): p. 011701.

50. Korhonen, J., et al., Influence of MRI-based bone outline definition errors on external radiotherapy dose calculation accuracy in heterogeneous pseudo-CT images of prostate cancer patients. *Acta Oncologica*, 2014. **53**(8): pp. 1100–1106.

51. Koivula, L., L. Wee, and J. Korhonen, Feasibility of MRI-only treatment planning for proton therapy in brain and prostate cancers: dose calculation accuracy in substitute CT images. *Medical Physics*, 2016. **43**(8Part1): pp. 4634–4642.

52. Kim, J., et al., Implementation of a novel algorithm for generating synthetic CT images from magnetic resonance imaging data sets for prostate cancer radiation therapy. *International Journal of Radiation Oncology* Biology* Physics*, 2015. **91**(1): pp. 39–47.

53. Yu, H., et al., Toward magnetic resonance–only simulation: segmentation of bone in MR for radiation therapy verification of the head. *International Journal of Radiation Oncology* Biology* Physics*, 2014. **89**(3): pp. 649–657.

54. Satheesh, B.A., et al., Pseudo-CT image synthesis from ultrasound images for potential use in Brachytherapy treatment planning-initial results. In *TENCON 2021–2021 IEEE Region 10 Conference (TENCON)*. 2021. IEEE.

55. Siversson, C., et al., MRI only prostate radiotherapy planning using the statistical decomposition algorithm. *Medical Physics*, 2015. **42**(10): pp. 6090–6097.

2

Deep-Learning-Based Medical Image Synthesis Methods

Yang Lei and Richard L.J. Qiu
Emory University, Atlanta, GA, USA

Tonghe Wang
Memorial Sloan Kettering Cancer Center, New York, NY, USA

Xiaofeng Yang
Emory University, Atlanta, GA, USA

CONTENTS

DOI: 10.1201/9781003243458-3

2.1 Introduction

Image synthesis involves the generation of artificial or simulated images in a specific target image modality or target domain using input images from a different image modality [1]. The objective of a synthesis task is to use synthetic images to replace the physical patient during the imaging procedure. This is motivated by a variety of factors, such as the infeasibility of a specific image acquisition, the added cost and labor of the imaging procedure, the ionizing radiation exposure to patients by some of those imaging procedures, or the introduction of uncertainties via performing image registration across different modalities. Over the past decade, image synthesis research has been introduced considerable attention in radiation oncology, radiology, and biology fields [2]. Potential clinical applications of image synthesis include radiation treatment using magnetic resonance imaging (MRI) only [3–10], positron emission tomography (PET) attenuation correction (AC) and image quality enhancement [11, 12], stopping power estimation[13–15], medical image quality improvement [23–32], medical image reconstruction [33, 34], synthetic image-aided auto-delineation [16–22], super-resolution image estimation [35–37], and others [38–42].

Image synthesis technology has been a hot topic of investigation for the past decade. Conventional machine learning-based image synthesis methods often rely on prior knowledge that is explicitly and manually designed for converting images from one modality to another [43–50]. The robustness and accuracy of these methods are based on the way of prior knowledge definition and investigation [51–53]. In recent years, deep learning-based methods have dominated the field of image synthesis [54]. Contrary to standard machine learning, deep learning does not depend on manually extracted features ruled by prior knowledge [55–57]. It employs neural networks (NN) or convolutional neural networks (CNN) with hidden layers of neurons or convolutional kernels to learn features automatically. These methods mostly follow a general workflow that adopts a data-driven methodology to map image intensities.

The typical work includes a learning/training stage, where the network-based model learns the correlation between the arrival one and the object one, and an estimation stage, where the learned model generates the synthetic target based on an input. Deep learning-based methods have several advantages over conventional machine learning-based methods. They exhibit higher generalizability since the same network architecture can be applied to several sets of image modalities with the smallest adaptations. This makes it possible to fast map different imagery modalities which is clinically meaningful. Despite the requirement for substantial attempts in data gathering and mining during the training stage, the prediction of an image usually only takes a few seconds. Due to these advantages, there has been a significant increase in the clinical research interest of medical images used in radiotherapy.

In this chapter, we aim to:

- Summarize the latest network architecture designs of deep learning-based medical image synthesis.
- Summarize the latest deep learning-based medical image synthesis applications.
- Emphasize significant involvements and recognize current contests.

2.2 Literature Searching

The focus of this review is confined to deep learning-based methods designed for synthetic image estimation. Medical image synthesis applications include studies on multimodality MRI synthesis, proton stopping power estimation, image quality improvement/enhancement, super/high-resolution visualization, MRI-only-based radiotherapy, inter-modality image registration, segmentation, PET AC, and data augmentation for image analysis.

Searches of peer-reviewed journal articles and conference proceedings references were conducted on PubMed by means of the conditions in title or abstract as of December 2022: ("estimate*" OR "reconstruct*" OR "trans*" OR "syn*" OR "restore*" OR "corr*" OR "generate*") AND "deep learning" AND "generative" OR "adversarial" OR "convolution*" OR "neural" ("MR" OR "MRI" OR "CT" OR "PET" OR "Medical" OR "Biomedical" etc.).

2.3 Network Architecture

The frameworks of the reviewed investigations are roughly categorized into four categories: NN, CNN, FCN, and GAN. These categories of methods are not entirely distinct from one another; instead, they demonstrate incremental increases in architectural complexity. After introducing the four categories of network design, we will then summarize feasible loss functions that can be used to train these networks.

2.3.1 NN

A NN is a network or circuit of biological neurons, or, in a modern sense, an artificial NN, composed of artificial neurons or nodes. In medical image synthesis, a most popular NN design is the autoencoder (AE) network.

AE and its variants have been extensively explored in the literature and remain widely employed in the field of analysis of medical images [58]. AE typically involves a NN encoding layer, which represents the image via latent layers with minimum representation error. The aim of this layer is to re-establish the arrival data within a low-dimensional latent space. AE is able to discover relevant patterns in the data by restricting the dimension of latent space vector representation.

To avoid potential overfitting issues caused by AE learning an identity function, recently, improvement of AE has been studied. A widely-known enhancement of AE is the stacked autoencoder (SAE), which is constructed by utilizing stacking operators. SAEs are composed of multiple AEs arranged in stacked layers. The yield of an individual layer is coupled with the input of the subsequent layers [59]. SAE serves a greedy layer-wise learning to obtain optimal parameters. The advantage of SAE is its ability to capture the significant hierarchical features of the input data [59].

Another variant of the AE is the denoising autoencoder (DAEs), which enables the creation of improved higher-level feature representations [60]. DAEs avoid learning insignificant solutions for the trained model, such as estimating a denoised output from its noisy input [61]. The stacked denoising autoencoder (SDAE) is a learning model that uses the capabilities of DAE [62].

By incorporating a sparsity constraint, it can to overcome the weakness of AEs potentially, which typically exhibit a modest number of neurons. By introducing many neurons to exhibit a low average output via the sparse AE, the network ensures that many neurons remain inactive for most of the time [63].

SAE requires layer-wise learning, which can be time-consuming and tedious due to the fully connected layers used in its construction. Li et al. did the earliest attempt to train a convolutional AE (CAE) in an end-to-end manner without pre-learning [64]. Guo et al. recommended that CAEs are beneficial for learning features for images, preserving local structures, and avoiding alteration of feature space [65]. Wang et al. recommended a hybrid model of transfer learning and CAE for automated chest screening [66].

2.3.2 CNN

CNN is a variant of NNs. The core of CNN is the convolution layer, which is aimed at information mining [67]. The convolution layer generates information maps based on its optimized convolutional kernels. A pooling layer conducts a summarizing of the information by taking either the maximum or average value within the specified neighboring region, thereby reducing the structural resolution of arrival information. The rectified linear unit (ReLU) is the commonly employed activation function in CNNs [68]. The fully connected layer establishes connections between each neuron in the preceding layer and every neuron in the succeeding layer such that the information can be transferred from the beginning to the end layer. Then, the end layer predicts the probability from the previous information.

During training, the CNN architecture's learnable parameters are employed to predict the objective categorization of the training data. The loss function, typically cross-entropy (CE) loss, is then computed, and the weights are updated via the gradient descent (GD) method through back-propagation. Adam gradient descent (AGD) and Stochastic gradient descent (SGD) are among the greatest widespread methods used for optimization.

Lecun et al. earlier developed a CNN model, called LeNet, for hand-written numeral identification [69]. LeNet is compiled of a set of convolutional layers, fully connected layers, and pooling layers. The progress in the computer unit and the sufficient data made CNN training possible. AlexNet was introduced by Krizhevsky et al. They won the ILSVRC-2012 Image Classification Competition [70] with a greatly lesser misclassification rate than the second rank [71]. Subsequently, CNNs have attracted prevalent consideration, and their alternatives have been established, leading to state-of-the-art performance in various image-processing tasks. ZFNet was developed by Zeiler and Fergus. It is developed as an enhancement to increase the execution of AlexNet. This study [72] showed that shallow networks could realize the boundary, intensity, and structural information of an image. They also demonstrated that deeper networks can achieve better performance. The primary enhancement in ZFNet is the utilization of a deconvolution network for visualizing the information maps.

Simonyan and Zisserman introduced the Visual Geometry Group (VGG) to investigate the capabilities of network models with intense layers [73]. The focal advance of VGG is an utter assessment of the network's performance when increasing its layers' depth. The study showed that a significant improvement on the prior-art configurations can be achieved by increasing the depth to 16–19 layers. GoogLeNet was introduced to expand the network structure [74]. Through the integration of the proposed inception module, GoogLeNet emerged victorious in the ImageNet Large-Scale Visual Recognition Challenge

2014 (ILSVRC14), an image classification and detection competition. The inception module aids the CNN model in providing more descriptive representations of the entered information, all the while augmenting the size of the network model.

These new technologies allow the network to have increased size which improves the performance. Nevertheless, the merely deeper network would result in overfitting issues. In order to reduce the complexity of optimizing a deep model and solve the overfitting consequence produced by adding more layers, He et al. introduced a residual network (ResNet) [75]. ResNet, primarily comprised of residual blocks, has been proven capable of surpassing a 100-layer obstacle and even reaching 1,000 layers.

Taking inspiration from the ResNet, Huang et al. later introduced a densely connected convolutional network (DenseNet) by establishing connections between individual layer to every other layer [76]. In contrast to ResNet, which focuses on learning the fundamental change between the input data and the output data, DenseNet [77] intended to merge multiple frequencies of information from the preceding and flow layers using dense block [78].

2.3.3 FCN

Most CNNs take an image or patch as input and then estimate a voxel-wise value, which is correlated to the central of the patch. The first proposal for a CNN was introduced by Shelhamer et al., in which the fully connected layers of previous CNN are substituted with convolution layers. The new network is referred to as a fully convolutional network (FCN) since all the layers of the network are convolutional layers. Due to significant enhancements in deconvolution kernels used for up-sampling the feature map, an FCN enables the model to achieve dense voxel-wise prediction for the entire full-size volume, rather than patch-wise classification, as seen in traditional CNNs [79]. FCN allows segmenting the entire image in a single forward pass, combining high-resolution information with low-resolution information, and then passing them to the remaining layers. FCN can improve localization performance and produce more accurate output.

The U-Net architecture, which was originally developed by Ronneberger [80], is a widely-known FCN model used in auto-delineation. The U-Net comprises an encoding set of FCN layers and a decoding set of FCN layers. In between, long skip connections exist between the two sets. These operators supply high-frequency information to the decoding path, allowing the network to address the potential imbalance between the localization accuracy of a delineated organ and the organ boundary specification. This issue arises because using bulky-sized data requires extra downsized operators, which in turn can lower the localization precision. In contrast, minor-sized data are not able to capture unlimited texture information to represent the organ boundary. Later on, V-Net [81] appears to be an improved network based on the U-Net architecture, with a similar structure.

In place of relying solely on single optimization, as done in traditional FCNs, the main objective behind the deep supervision model [82, 83] is to offer intended optimization for middle sets of FCN layers and transmit learned information to the remaining layers. The approach extends optimization to the hidden and shallower layers of FCN, which increases the judicial capability of learned information in distinguishing multiorgans in auto-delineating projects. Lately, attention gates and transformers have been integrated into FCN to advance the execution of recognition [84]. The attention gate has the capability of learning how to diminish unrelated information and emphasize outstanding information that is beneficial for users' tasks.

2.3.4 GAN

GANs have garnered significant attention in medical imaging because of their capacity to generate data without the need for explicit modeling of the probability density function. As compared to CNN, GANs include an additional sub-network, called discriminator network. This approach offers an innovative means of integrating samples with no ground truth data into learning procedures and enforcing sophisticated constancy. This approach demonstrated its utility in numerous cases, including but not limited to image reconstruction [85], image enhancement [86–89], segmentation [90–92], diagnosis [93], data generation [94], and multimodality image estimation [95].

Characteristic GANs are composed of two opposing models, called a generator model and a discriminator model [96]. The generator model is used to produce synthetic samples that are closely approximated to target distribution from a feature representation field. The discriminator model is used to differentiate between the generated sample and the definite sample. The discriminator model forces the generator model to obtain more faithful samples by imposing penalties on idealistic generated samples during the optimization process. The two networks compete. As is summarized by Yi et al., [97], Multiple GAN variants can be categorized into three groups: 1) variants of the discriminator's objective, 2) variants of the generator's objective, and 3) variants of the architecture.

2.3.4.1 Conditional GAN

Traditional GAN was trained without limitations on data generation. Subsequently, it was enhanced by incorporating restricted restrictions to generate samples with required assets, known as conditional generative adversarial network (cGAN). The generator model of cGAN is built via FCN, either an end-to-end manner or a non-end-to-end manner. The former manner can produce equal-sized samples as input, while the latter manner cannot. The end-to-end FCN is commonly structured with a set of encoding layers and a set of decoding layers. The combination of these two sets of layers is able to derive equal-sized output. The encoding set uses convolutional layers with different settings of stride sizes and w/o pooling based on the users' purpose. On the other hand, the decoding set comprises deconvolution layers to achieve end-to-end mapping, several convolution layers, and a final layer for regression. Short skip connections, such as residual blocks [87] or dense blocks [95], can be employed by connecting the two sets of layers. The benefit of a residual block is mining the disagreement between the input and output. The aim of the dense block is the aggregation of the low- and high-frequency information from input. The long residual block creates a skip connection that directs information between long-distance layers, guiding the entire layers to concentrate on the optimization purpose [98]. Certain studies have incorporated attention gates into the long skip connection to trap crucial information instead of expanding the receptive field [17]. The kernels updated in hidden layers can be used to remove irrelevant features. The non-end-to-end FCN, on the other hand, typically only consists of a set of encoding layers, possibly followed by NN layers. The discriminator of the GAN model is implemented similarly to non-end-to-end FCN, expect the last layer culminating in a sigmoid or softmax operator to facilitate recognition or assessment.

Many different variations of the cGAN model have been suggested in bumping into specific anticipated requirements. This review work explores several cGAN models that can be either designed for or adaptable to medical image synthesis, including pix2pix, deep convolutional GAN (DCGAN), and Information Maximizing GAN (InfoGAN).

2.3.4.1.1 DCGAN

DCGAN can yield better and more steady training results when a fully convolutional layer is used as opposed to a fully convolutional layer. This architecture is illustrated in the work of [99]. In the core of the framework, pooling layers were replaced with fractional-stride convolutions, which allowed it to learn from a random input noise vector by using its own spatial up-sampling to generate an image from it. Two important changes were adopted to modify the architecture of early cGAN: batch normalization and leaky ReLU. Batch normalization [100] was employed to address inadequate initialization and avoid model collapse. The model collapse was a significant obstacle for GAN frameworks in earlier research. In all the layers of a discriminator, Leaky ReLU [101] activation was introduced as a replacement for maxout activation [96], leading to an improvement in the resolution of image output [38].

2.3.4.1.2 Pix2pix

The Pix2pix model, which is trained in a supervised manner, is an image-to-image translation model introduced by Isola et al. [102]. The network aims to learn a translation between the arrival data and the learning target and requires the generated data to be tightly close to the distribution of the target. To achieve the goal, the discriminator loss, which assesses the truthfulness of generated data, and the image-wise error (e.g., mean absolute error (MAE)) are employed. The use of image-wise error trains the model under supervised means, making it more fit for imaging synthesis tasks when the ground truth data are given. An illustration of this is the task of MRI-only radiation therapy, where the planning CT and co-registered MRI are provided for learning the model. After training, the synthetic CT (sCT) generated from the input MRI has accurate Hounsfield values (HUs) like a real CT, which are critical for accurate dose calculation in radiotherapy planning.

2.3.4.1.3 InfoGAN

Recently, there are groups of synthesis tasks that show that the trained model should not only rely on image intensity loss but also on histogram similarity: for example, the cone beam CT (CBCT) scatter correction [103, 104] and PET AC [105–108]. Under these requirements, a model may not be properly learned by just utilizing image-wise error as the loss function [109, 110]. InfoGAN is a variation of cGAN that was developed to learn disentangled representations via counting information-theoretic extensions [111]. It does this by maximizing the mutual information, which is an assessment of distribution similarity [112]. InfoGAN has been successful in disentangling written characters from the digit structures on the Modified National Institute of Standards and Technology public dataset [112]. It has also been used for generating volumetric CT from 2D kilovoltage image [113].

2.3.4.2 CycleGAN

Numerous diverse variations of CycleGAN models have been proposed for synthesis tasks recently. This work summarizes the use of these models in medical imaging [87, 95, 114–116]. As mentioned, in cGANs, there are two sub-networks: a discriminator and a generator. These two sub-networks are trained alternatively and competitively. Mismatches might persist between the input and output for specific clinic use, even with successful image registration, leading to potential issues. To tackle this concern, CycleGAN proposes an extra round to enforce the model to approach a one-to-one representation. CycleGAN extends the route of a regular cGAN via compelling an inverse revolution, which involves transforming a synthetic sample back to the input data distribution [117]. This additional constraint further guides and enhances the precision of the learned model.

The CycleGAN is composed of two comprehensive loops: the initial loop performs the transformation from the input data distribution to the output data distribution and then from the output data distribution back to the input data distribution. The subsequent loop conducts a transformation from the output data distribution to the input data distribution and then from the input data distribution back to the output data distribution.

2.3.4.2.1 Res-CycleGAN

CycleGAN with residual blocks has shown promising results for image synthesis tasks when the input and output modalities share similar intensities, for example, CBCT vs. CT [87], low dose PET vs. full dose PET [118], etc. In these models, the generators of CycleGAN utilize residual blocks.

2.3.4.2.2 Dense-CycleGAN

When the input and target images are drastically different from each other, the main struggle in molding a transformation/translation/generation lies in the fact that the position and structure of the input image and output image can differ dramatically between dissimilar samples. To improve the prediction of each voxel within anatomic regions [119], several dense blocks, inspired by densely connected CNN, were introduced to capture multifrequency information [95]. As shown in the generator architecture (Figure 2.4), the dense block can be implemented to achieve this goal.

2.3.4.2.3 UNIT

Multimodal MRI synthesis [120] involves mapping between multiple image domains instead of just one input distribution to the other single output distribution. These types of tasks have been addressed by unsupervised image-to-image translation networks (UNIT). UNIT seeks to learn the shared histogram of samples among diverse distributions. It can be challenging to estimate the cross-distribution from a single modality without additional assumptions. In order to solve this, a shared-latent space hypothesis was established [102].

2.3.4.2.4 BicycleGAN

For multimodality image synthesis, such as mapping MRI to CT, Zhu et al. [115] proposed the BicycleGAN to improve upon the UNIT network. BicycleGAN seeks to map the relationship between output and input images via a GAN-based network [93]. It does this by condensing the uncertainty inherent in the representation into a compact vector that is stochastically collected during testing. The BicycleGAN is trained to estimate the synthetic image based on the arrival image with the target (synthetic) image domain label. To prevent mode collapse, the BicycleGAN utilizes the domain label to minimize the complexity of training the model. BicycleGAN clearly promotes the alignment between the hidden feature representation and the estimated image [38, 39].

2.3.4.2.5 StarGAN

StarGAN is a multimodal image translation variant of CycleGAN [116]. One of the challenges of multimodal estimation is the constrained scalability when dealing with larger than two required output distributions, as it necessitates constructing separate models independently for each set of data distributions. To tackle this restriction, StarGAN was recommended as a solution for multiple modality image synthesis tasks using a single model. This model architecture enables simultaneous prediction of multimodality images within a single network. Additionally, it grants the flexibility to translate an arrival data to whichever anticipated output requirement. StarGAN has been applied to multimodal MRI synthesis [120].

2.3.5 Loss Function

As mentioned earlier, the architecture of GANs depends on the constant amendment of the component networks, which is directly influenced by the definition of their respective optimization.

2.3.5.1 Image Distance Loss

The image distance loss function used in GANs for image synthesis typically consists of two parts: intensity-based loss and similarity-based loss. Examples of intensity-based loss are MAE and mean square error (MSE) [121, 122]. Some other works used l_p-norm ($p \in (1, 2)$) [95]. Structural loss, such as the gradient difference loss (GDL), assesses an overall structural similarity in the structure between the generated image and the learning target image. GDL is calculated by comparing the gradients of the two samples. As compared between two variables A and B, the GDL is defined as:

$$\text{GDL}(A,B) = \sum_{i,j,k} \left\{ \begin{array}{l} \left(\left| A_{i,j,k} - A_{i-1,j,k} \right| - \left| B_{i,j,k} - B_{i-1,j,k} \right| \right)^2 \\ + \left(\left| A_{i,j,k} - A_{i,j-1,k} \right| - \left| B_{i,j,k} - B_{i,j-1,k} \right| \right)^2 \\ + \left(\left| A_{i,j,k} - A_{i,j,k-1} \right| - \left| B_{i,j,k} - B_{i,j,k-1} \right| \right)^2 \end{array} \right\} \tag{2.1}$$

where i, j, and k represent the voxel index of the 3D dimension, respectively. In general, a combination of these two types of loss is used to train GANs for image synthesis tasks. It allows the model to simultaneously optimize for both pixel-level accuracy and overall structure.

2.3.5.2 Histogram Matching Loss

In order to ensure similarity in histogram distribution between the synthetic image and learning target, Lei et al. introduced a histogram matching loss, termed MaxInfo loss [113]. This term measures the mutual dependency between the probability distributions of the synthetic and ground truth images

$$\text{MaxInfo}(A,B) = \sum_{i,j,k} P(A_{i,j,k}, B_{i,j,k}) \log \frac{P(A_{i,j,k}, B_{i,j,k})}{P(A_{i,j,k}) \cdot P(B_{i,j,k})} \qquad (2.2)$$

$P(A, B)$ is a joint probability between variables A and B. $P(X)$ means the marginal probability of the variable X.

2.3.5.3 Perceptual Loss

The quality of the output synthetic image can be negatively affected by discrepancies in anatomical structures between the input data and the learning object samples. These discrepancies can lead to blurred structure/edge boundaries [113]. Using only image distance loss functions, such as MAE and GDE, GAN-based models will not be able to fabricate harsh boundary edges due to the ambiguity introduced by the mismatch between the input and the learning target during training.

To improve boundary contrast and sharpness, perceptual loss is often utilized [123–126]. This loss term measures the perceptual information difference between the generated image and the learning target by using hierarchical features via a feature pyramid network (FPN) architecture [86, 130–133]. For instance, in a lung sCT project, by using thoracic CT during optimization, the FPN can be pretrained [127, 128] with lung contours [91, 129, 130]. FPN, denoted by F_s, extracts multiple hierarchical features from the learning target (A) and the generated one (B), respectively, that is, $f_A = \bigcup_{i=1}^{N} F_s^i(A)$ and $f_B = \bigcup_{i=1}^{N} F_s^i(B)$. N denotes the number of layers for hierarchical feature extraction. The perceptual loss is termed as the Euclidean distance measured between f_A and f_B, and is calculated as:

$$L_p(f_A, f_B) = \sum_{i=1}^{N} \frac{\omega_i}{C_i \cdot H_i \cdot W_i \cdot D_i} \left\| F_s^i(A) - F_s^i(B) \right\|_2^2, \qquad (2.3)$$

where C_i denotes the number of feature map channels at the ith pyramid level. H_i, W_i, and D_i denotes the height, width, and depth of that feature map. ω_i is a balancing parameter for the feature level i. The weight for higher pyramid levels is often increased (set as $\omega_i = p^{i-1}$ with $\in(1, 2)$) [113] to account for the coarser semantic feature at those levels.

2.3.5.4 Discriminator Loss

CE losses are often used to train the discriminator [86]. The objective of the discriminator is to assess the validity of the arrival synthetic one. So, it should be able to discriminate/regard whether the arrival one is real or fake. Given the source image I_{src}, the generator model G updated from the last iteration, the discriminator D, and the learning target I_{tar}, the discriminator can be obtained by optimizing the following:

$$D = \arg\min_{D} \left\{ CE(D(I_{tar}), 1) + CE(D(G(I_{src})), 0) \right\} \qquad (2.4)$$

In the above equation, "1" characterizes real and "0" characterizes the fake.

2.3.5.5 Adversarial Loss

The generator loss function is typically made up of a combination of different loss terms that serve various purposes. As mentioned earlier, the objective of the generator is to deceive the discriminator. Given the learned discriminator D, source image I_{src}, learning target image I_{tar}, and generator G, the generator loss measured by CE can be expressed as follows:

$$L_{adv} = CE\big(D\big(G\big(I_{src}\big)\big),1\big),\tag{2.5}$$

The synthetic image $G(I_{src})$ would be close to real for the discriminator D upon minimizing the loss term of Equation (2.2).

2.4 Applications

Recently, deep learning-based methods have been successful in a number of medical image synthesis applications. We provide a brief summary of some recent examples of these applications, including the specific imaging modalities and clinical applications. These applications will be further discussed in later chapters.

2.4.1 Multimodality MRI Synthesis

MRI is a widely used medical imaging technique that provides valuable anatomical and functional information for various medical applications [131–134]. By applying different MRI pulse sequences, it is possible to acquire multicontrast images of the same anatomy [135–137], which can provide complementary information for the assessment, diagnosis, and treatment planning of diverse diseases. For example, in a brain MR scan, a T1-weighted (T1) scan presents the distinction between white and grey matters. Both T1-weighted and contrast-enhanced (T1c) scans are able to assess tumor-shape changes with enhanced demarcation around the tumor [138, 139]. The T2-weighted (T2) images show the fluid clearly in relation to the cortical tissue, and fluid-attenuated inversion recovery (FLAIR) images can be used to clearly outline the contour of a lesion [140, 141]. By combining the strengths of these modalities, it is possible to unveil rich underlying tissue information that can aid in diagnosis and treatment management [46, 95, 142–145]. However, the use of different imaging protocols for different patients and institutions can make it difficult to consistently apply the same scan setting to all patients, even with the same disease such as glioblastoma [146]. To overcome this challenge, cross-modal image synthesis has been introduced as a method to generate missing modalities using available ones as input [147] as shown in Figure 2.1.

Recently, both cGAN and CycleGAN models have been explored for MRI multimodality image synthesis [147–154]. Yu et al. [149] enhanced the proficiency of an image-to-image estimation method for a 3D MRI synthesis and a FLAIR image estimation from the T1 modality. The contest between the generator and the discriminator models resulted in an attainable Nash equilibrium. Dar et al. [150] focused on single-modal MRI estimation between the T1 and T2 scans using the cGAN model. The model was trained under

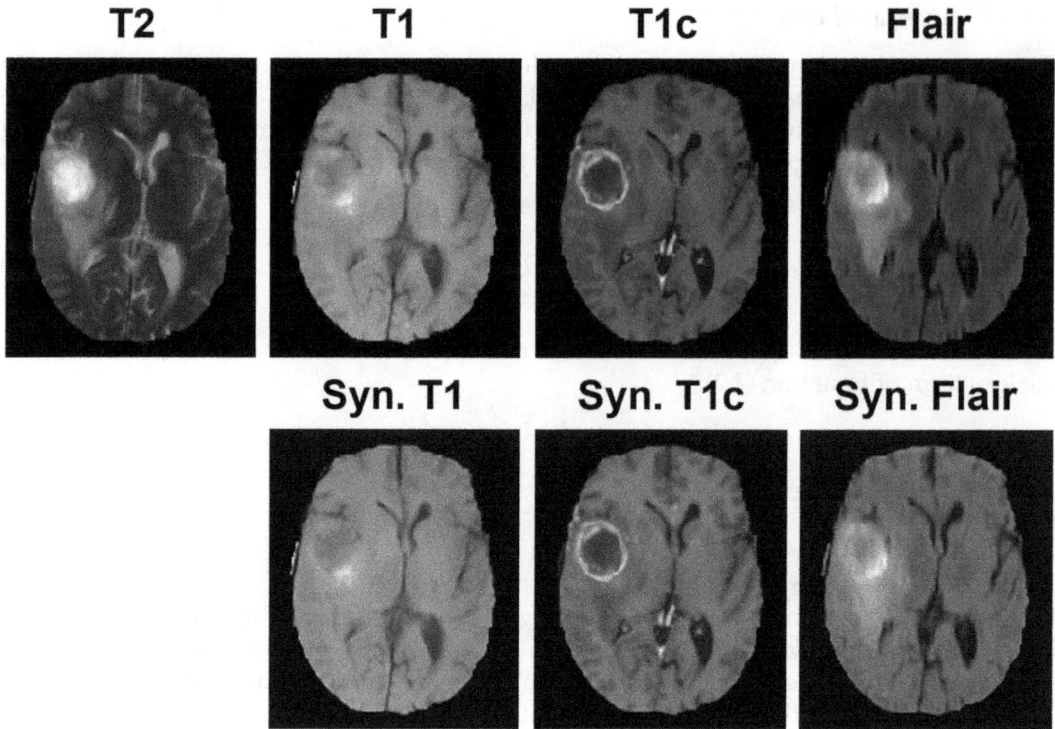

FIGURE 2.1
Modality image synthesis results. (From [52])

voxel-wise and perceptual losses via a well-registered dataset and under cycle loss via unregistered data set. The evaluation was conducted on brain MRI scan from glioma patients and healthy cases, demonstrating improved synthesis imaging precision. These models are commonly augmented to capture the distinctive correlation between arrival and required modalities.

Olut et al. [148] developed a cGAN-based approach for T1- and T2-based MRI synthesis. Joyce et al. [155] anticipated an encoder–decoder model to generate FLAIR from T1, T2, and diffusion MRIs. Chartsias et al. [156] developed a CNN that takes all available MRI modalities to generate its missing modalities. An accessible GAN-based model was proposed by Lei et.al., [151] to allow for the flexible selection of uninformed subsets of multi-modalities as source data and estimate the corresponding required image modality. GANs [152, 154] are also explored for multimodal MRI synthesis.

2.4.2 MRI-only Radiation Therapy Treatment Planning

MR image provides more excellent soft-tissue contrast than that of CT, which makes it useful for improving organ-at-risk (OAR) auto-delineation for radiotherapy [157–162]. However, the dose calculation algorithms used in radiation therapy treatment planning systems count on the electron density distribution that is produced from planning CT. Therefore, MRI is often aligned to CT and behaved in conjunction with the CT images for radiation planning [163]. In order to overcome this issue, methods have been developed recently that can create electron density image from MRI, a process known as sCT generation [164–166], as shown in Figure 2.2.

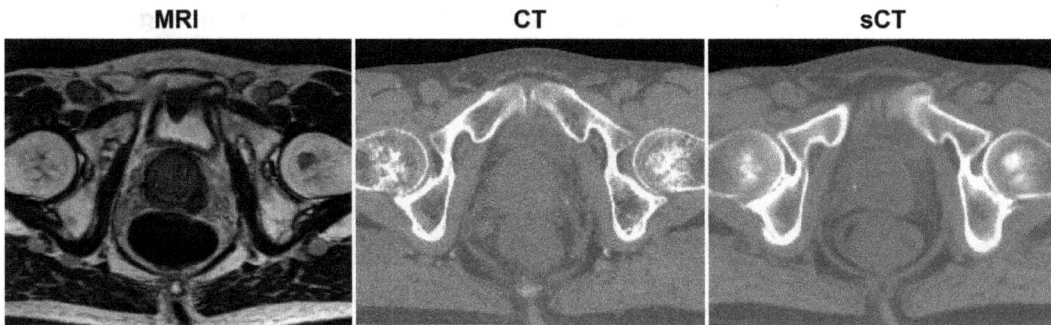

FIGURE 2.2
Results of synthesizing CT using MRI as the input modality. (From [95])

CGAN has been used to improve the quality of the generated sCT images by introducing an additional discriminator to distinguish the sCT from real CT, compared to the previously deep learning-based methods [167]. However, GAN-based methods typically require training pairs of MRI and CT images to be perfectly registered, which can be difficult to reach due to the high levels of accuracy needed for image synthesis [168, 169]. If there are some residual local mismatches between the MRI and CT training data, that is, soft-tissue misalignment after bone-based rigid registration, cGAN-based methods would produce a degenerative network, decreasing their accuracy. Wolterink et al. demonstrated that training with co-registered MRI/CT is not mandatory for the CycleGAN-based sCT estimation task [168].

2.4.3 CBCT Improvement/Enhancement

The incorporation of CBCT on top of medical linear accelerators (LINACs) has granted the use of daily image guidance radiation therapy (IGRT). This procedure enhances the reproducibility of patient setup [170]. While CBCT is a valuable means for IGRT, the physical appearances, namely a large scatter-to-primary proportion, result in image artifacts, such as streaking, shading, cupping, and reduced image contrast [171]. All these causes impede accessible CBCT, restricting the full operation of the knowledge supplied by frequent imaging [172, 173]. Despite these shortfalls, the use of CBCT has fronted to grown curiosity in adaptive radiotherapy (ART), where dose could be calculated daily via the patient's setup on the treatment couch [174, 175]. ART could help alleviate patient setup faults and accommodate patient's daily variation, such as weight loss or inflammation [176]. Eliminating the uncertainty would result in reduced margins on target volumes and better sparing of OARs, hypothetically resulting in advanced target doses with fewer side effects [177].

CycleGAN models have recently been employed for CBCT image quality improvement because of their efficient capacity to count the residual between different data domains, especially when the fundamental target and the OARs' boundaries are similar, even in cases where the transforming is nonlinear [87]. A CBCT correction result is shown in Figure 2.3. ResNet sub-network structures can be employed in CycleGAN framework to force learn to minimize the difference between the CBCT (input) and the planning CT (learning target) [178–180]. Liang et al. also employed a CycleGAN which achieved good performance at the HN site [181].

CBCT	Corrected CBCT	Planning CT

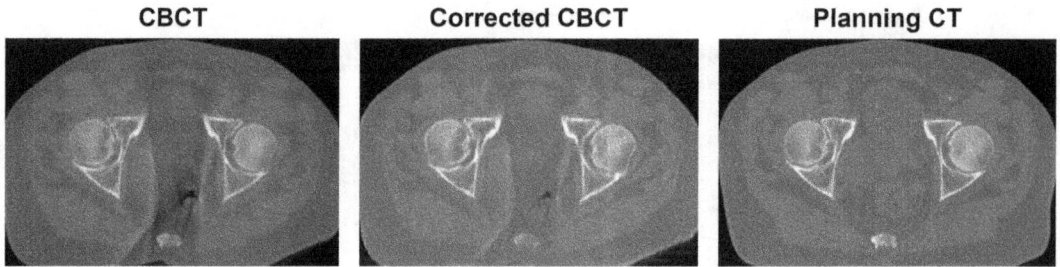

FIGURE 2.3
Results of CBCT improvement. (From [87])

2.4.4 Low-count PET and PET Attenuation Correction

Image synthesis methods have been developed to promote PET AC and low-count PET reconstruction. The use of GANs has been proposed for directly estimating AC PET from non-AC PET (NAC PET). Dong et al. utilized a CycleGAN model to accomplish whole-body PET AC, marking the first instance of its application in this context [86], showing that their method was reliable through the inclusion of successive imaging to assess the arrival PET density adjustments with time on their AC PET and ground truth (as shown in Figure 2.4).

Low-count PET has a wide range of applications in pediatric PET scans and radiotherapy reaction assessments due to its improvement in terms of appropriate motion discipline and decreased patient dose. Nevertheless, a low-count PET scan can lead to heightened image noise and decreased contrast-to-noise ratio (CNR). Reconstructing a standard- or full-count PET from a low-count PET is not able to be accomplished through straightforward post-processing operations, like denoizing, due to the fact that reducing the radiation dose alters the underlying biological and metabolic processes, resulting in not only noise but also changes in local uptake values [183]. Additionally, even with the same tracer injection dose, the uptake distribution and signal level can vary significantly among patients. To address these challenges, a CycleGAN model is employed for low-count PET reconstruction [118]. It directly operates on low-count PET data and estimates full-count PET images.

While CycleGAN demonstrates the practicality of transforming low-count PET to full-count PET, several researchers have explored the use of combined PET/MR images as arrival

NAC	Corrected NAC	AC

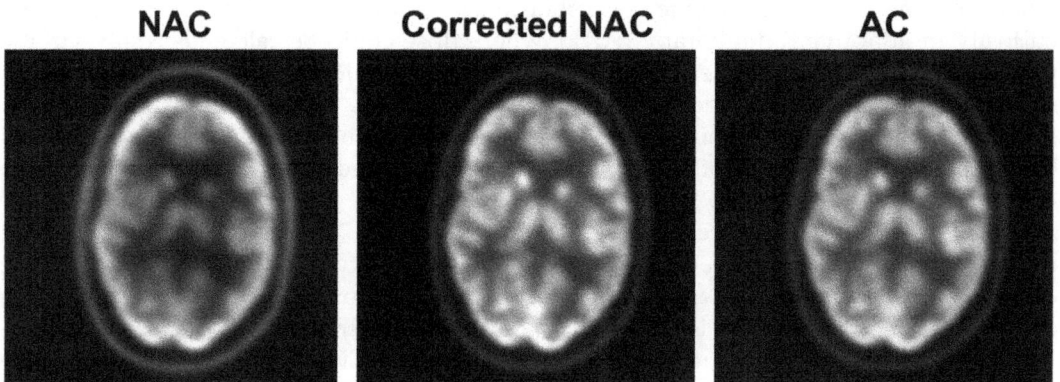

FIGURE 2.4
Results of a corrected PET image using NAC PET as input. (From [182])

data to auxiliary enhance the quality of output, particularly once MRI data is accessible. The inclusion of MR scan, which provide anatomical knowledge, can significantly enhance the network's functioning compared to using only PET images. As pointed out in a clinical study of uptake status, Chen et al. achieved 83% exactness via PET-only and 89% exactness via combining PET and MRI together (PET+MRI) [184]. The PET+MRI results are impressive for revealing the primary anatomical patterns, which likely contributed to the improved accuracy. The significance of MRIs was confirmed in a research conducted by Xiang et al., where they observed a major improvement in the peak signal-to-noise ratio (PSNR) [185]. MRIs provided crucial structural information that contributed to the estimation of high-quality PET, despite the significant visual differences between structural MRIs and PETs.

2.5 Summary and Discussion

GANs gained popularity in the field of image synthesis in recent years. These networks, including cGAN- and CycleGANs, have been shown to produce more faithful estimated samples with superior distribution resemblance to authentic data and competent quantitative metrics than traditional model-based methods. In comparison to traditional machine learning approaches, the GAN models offer greater generalization. The same network and architecture are able to be directed to altered sets of imaging modalities under marginal adjustments, enabling straightforward expansion of their efforts to various imaging modalities.

However, there are still some challenges that need to be addressed in the use of GANs for medical image synthesis. One challenge is the need for paired datasets, which can be difficult to collect and require high accuracy in image registration. CycleGAN can relax this requirement to some extent by allowing the use of unpaired datasets, but this may not always improve numerical performance.

Another challenge is the inconsistency of GAN performance when input images exhibit significant differences from their training datasets. This can be a concern in clinical settings where unusual cases, such as patients with implants or abnormal anatomy, may be encountered. Despite these challenges, research in medical image synthesis using GANs is expected to continue to advance in the coming years.

Disclosures

The authors declare no conflicts of interest.

References

[1] X. Yang et al., "MRI-based synthetic CT for radiation treatment of prostate cancer," *International Journal of Radiation Oncology • Biology • Physics ASTRO*, 2018, vol. 102, no. 3, pp. S193–S194, doi: 10.1016/j.ijrobp.2018.07.086

[2] T. Wang, Y. Lei, Y. Fu, W. Curran, T. Liu, and X. Yang, "Medical imaging synthesis using deep learning and its clinical applications: A review," *arXiv: Medical Physics*, vol. doi: arXiv:2004.10322, 2020.

[3] Y. Liu et al., "MRI-based treatment planning for proton radiotherapy: Dosimetric validation of a deep learning-based liver synthetic CT generation method," (in eng), *Physics in Medicine and Biology*, vol. 64, no. 14, p. 145015, Jul 16 2019, doi: 10.1088/1361-6560/ab25bc

[4] Y. Liu et al., "MRI-based treatment planning for liver stereotactic body radiotherapy: Validation of a deep learning-based synthetic CT generation method," (in eng), *The British Journal of Radiology*, vol. 92, no. 1100, p. 20190067, Aug 2019, doi: 10.1259/bjr.20190067

[5] Y. Liu et al., "Evaluation of a deep learning-based pelvic synthetic CT generation technique for MRI-based prostate proton treatment planning," (in eng), *Physics in Medicine and Biology*, vol. 64, no. 20, p. 205022, Oct 21 2019, doi: 10.1088/1361-6560/ab41af

[6] G. Shafai-Erfani et al., "MRI-based proton treatment planning for base of skull tumors," *International Journal of Particle Therapy*, vol. 6, no. 2, pp. 12–25, Fall 2019, doi: 10.14338/IJPT-19-00062.1

[7] Y. Lei et al., "MRI-based pseudo CT synthesis using anatomical signature and alternating random forest with iterative refinement model," (in eng), *Journal of Medical Imaging (Bellingham)*, vol. 5, no. 4, p. 043504, Oct 2018, doi: 10.1117/1.JMI.5.4.043504

[8] Y. Lei et al., "Pseudo CT estimation using patch-based joint dictionary learning," presented at the *IEEE Eng Med Biol Soc*, Jul, 2018.

[9] X. Yang et al., "Pseudo CT estimation from MRI using patch-based random forest," in *Medical Imaging 2017: Image Processing*, 2017, vol. 10133, no. 101332Q.

[10] X. Yang et al., "Patient-specific synthetic CT generation for MRI-only prostate radiotherapy treatment planning," *Medical Physics*, 2018, vol. 45, no. 6, pp. E701–E702.

[11] X. Yang et al., "MRI-based attenuation correction for brain PET/MRI based on anatomic signature and machine learning," (in eng), *Physics in Medicine and Biology*, vol. 64, no. 2, p. 025001, Jan 7 2019, doi: 10.1088/1361-6560/aaf5e0

[12] X. Yang and B. Fei, "Multiscale segmentation of the skull in MR images for MRI-based attenuation correction of combined MR/PET," *Journal of the American Medical Informatics Association*, vol. 20, no. 6, pp. 1037–1045, Nov–Dec 2013, doi: 10.1136/amiajnl-2012-001544

[13] T. Wang et al., "Learning-based stopping power mapping on dual energy CT for proton radiation therapy," *arXiv: Medical Physics*, vol. doi: arXiv:2005.12908, 2020.

[14] S. Charyyev et al., "Learning-based synthetic dual energy CT imaging from single energy CT for stopping power ratio calculation in proton radiation therapy," *arXiv: Medical Physics*, vol. doi: arXiv:2005.12908 2020.

[15] S. Charyyev et al., "High quality proton portal imaging using deep learning for proton radiation therapy: A phantom study," *Biomedical Physics & Engineering Express*, vol. 6, no. 3, p. 035029, Apr 27 2020, doi: 10.1088/2057-1976/ab8a74

[16] X. Dai et al., "Synthetic MRI-aided head-and-neck organs-at-risk auto-delineation for CBCT-guided adaptive radiotherapy," *arXiv: Medical Physics*, vol. doi: arXiv:2010.04275, 2020.

[17] X. Dong et al., "Synthetic MRI-aided multi-organ segmentation on male pelvic CT using cycle consistent deep attention network," (in eng), *Radiotherapy and Oncology*, vol. 141, pp. 192–199, Dec 2019, doi: 10.1016/j.radonc.2019.09.028

[18] Y. Lei et al., "CT prostate segmentation based on synthetic MRI-aided deep attention fully convolution network," *Medical Physics*, vol. 47, no. 2, pp. 530–540, Feb 2020, doi: 10.1002/mp.13933

[19] Y. Lei et al., "CBCT-Based Synthetic MRI Generation for CBCT-Guided Adaptive Radiotherapy," in *Artificial Intelligence in Radiation Therapy*, Cham, D. Nguyen, L. Xing, and S. Jiang, Eds., 2019: Springer International Publishing, pp. 154–161.

[20] Y. Lei et al., "Male pelvic multi-organ segmentation aided by CBCT-based synthetic MRI," (in eng), *Physics in Medicine and Biology*, vol. 65, no. 3, p. 035013, Feb 4 2020, doi: 10.1088/1361-6560/ab63bb

[21] Y. Liu et al., "Head and neck multi-organ auto-segmentation on CT images aided by synthetic MRI," *Medical Physics*, vol. 47, no. 9, pp. 4294–4302, Sep 2020, doi: 10.1002/mp.14378

[22] Y. Fu et al., "Pelvic multi-organ segmentation on cone-beam CT for prostate adaptive radio-therapy," *Medical Physics*, vol. 47, no. 8, pp. 3415–3422, Aug 2020, doi: 10.1002/mp.14196

[23] J. M. Wolterink, T. Leiner, M. A. Viergever, and I. Išgum, "Generative adversarial networks for noise reduction in low-dose CT," *IEEE Transactions on Medical Imaging*, vol. 36, no. 12, pp. 2536–2545, 2017, doi: 10.1109/TMI.2017.2708987

[24] Y. Lei et al., "A denoising algorithm for CT image using low-rank sparse coding," in *Medical Imaging 2018: Image Processing*, 2018, vol. 10574, no. 105741P.

[25] T. Wang et al., "Deep learning-based image quality improvement for low-dose computed tomography simulation in radiation therapy," (in eng), *Journal of Medical Imaging (Bellingham)*, vol. 6, no. 4, p. 043504, Oct 2019, doi: 10.1117/1.JMI.6.4.043504

[26] X. Dai et al., "Intensity non-uniformity correction in MR imaging using residual cycle genera-tive adversarial network," *Physics in Medicine and Biology*, vol. 65, no. 21, p. 215025, Nov 27 2020, doi: 10.1088/1361-6560/abb31f

[27] T. Wang et al., "Dosimetric study on learning-based cone-beam CT correction in adaptive radia-tion therapy," (in eng), *Medical Dosimetry*, vol. 44, no. 4, pp. e71–e79, Winter 2019, doi: 10.1016/j.meddos.2019.03.001

[28] Y. Lei et al., "Improving image quality of cone-beam CT using alternating regression forest," in *Medical Imaging 2018: Image Processing*, 2018, vol. 10573, no. 1057345.

[29] X. Yang et al., "Attenuation and scatter correction for whole-body PET using 3D generative adversarial networks," in *Journal of Nuclear Medicine*, 2019, vol. 60, no. S1, p. 174.

[30] S. Charyyev et al., "Synthetic dual energy CT images from single energy CT image for proton radiotherapy," *Medical Physics*, 2020, vol. 47, no. 6, pp. E378–E379.

[31] S. Charyyev et al., "Learning-based synthetic dual energy CT imaging from single energy CT for stopping power ratio calculation in proton radiation therapy," *The British Journal of Radiology*, vol. 95, no. 1129, p. 20210644, Jan 1 2022, doi: 10.1259/bjr.20210644

[32] Y. Liu et al., "A deep-learning-based intensity inhomogeneity correction for MR imaging," *Medical Physics*, 2019, vol. 46, no. 6, pp. E386–E386.

[33] G. Yang et al., "DAGAN: Deep de-aliasing generative adversarial networks for fast compressed sensing MRI reconstruction," *IEEE Transactions on Medical Imaging*, vol. 37, no. 6, pp. 1310–1321, 2018, doi: 10.1109/TMI.2017.2785879

[34] J. Harms et al., "Cone-beam CT-derived relative stopping power map generation via deep learning for proton radiotherapy," *Medical Physics*, vol. 47, no. 9, pp. 4416–4427, Sep 2020, doi: 10.1002/mp.14347

[35] Y. Lei et al., "High-resolution CT image retrieval using sparse convolutional neural network," in *Medical Imaging 2018: Image Processing*, 2018, vol. 10573, no. 105733F.

[36] H. Xie et al., "Synthesizing high-resolution magnetic resonance imaging using parallel cycle-consistent generative adversarial networks for fast magnetic resonance imaging," *Medical Physics*, vol. 49, no. 1, pp. 357–369, 2022/01/01 2022, doi: 10.1002/mp.15380

[37] Y. Lei et al., "High-resolution ultrasound imaging reconstruction using deep attention genera-tive adversarial networks," *Medical Physics*, 2019, vol. 46, no. 6, pp. E491–E491.

[38] Y. Lei, R. L. J. Qiu, T. Wang, W. J. Curran, T. Liu, and X. Yang, "Chapter 7 - Generative Adversarial Networks for Medical Image Synthesis," in *Biomedical Image Synthesis and Simulation*, N. Burgos and D. Svoboda Eds.: Academic Press, 2022, pp. 105–128.

[39] Y. Lei, R. L. Qiu, T. Wang, W. Curran, T. Liu, and X. Yang, "Generative adversarial network for image synthesis," *ArXiv*, vol. doi: arXiv:2012.15446, 2020.

[40] H. Xie et al., "Generation of contrast-enhanced CT with residual cycle-consistent genera-tive adversarial network (Res-CycleGAN)," in *Medical Imaging 2021: Image Processing*, 2021, vol. 11595, no. 1159540 SPIE. [Online]. Available: 10.1117/12.2581056

[41] Y. Lei et al., "Deep learning-based 3D image generation using a single 2D projection image," in *Medical Imaging 2021: Image Processing*, 2021, vol. 11596, no. 115961V SPIE. [Online]. Available: 10.1117/12.2580796

[42] X. Dai et al., "Deep learning-based volumetric image generation from projection imaging for prostate radiotherapy," in *Medical Imaging 2021: Image Processing*, 2021, vol. 11598, no. 115981R SPIE. [Online]. Available: 10.1117/12.2581053

[43] D. Andreasen, K. Van Leemput, R. H. Hansen, J. A. L. Andersen, and J. M. Edmund, "Patch-based generation of a pseudo CT from conventional MRI sequences for MRI-only radiotherapy of the brain," (in eng), *Medical Physics*, vol. 42, no. 4, pp. 1596–1605, Apr 2015, doi: 10.1118/1.4914158

[44] S. Aouadi et al., "Sparse patch-based method applied to MRI-only radiotherapy planning," *Physica Medica*, vol. 32, no. 3, p. 309, 2016/09/01/ 2016, doi: 10.1016/j.ejmp.2016.07.173

[45] A. Torrado-Carvajal et al., "Fast patch-based pseudo-CT synthesis from T1-weighted MR images for PET/MR attenuation correction in brain studies," (in eng), *Journal of Nuclear Medicine*, vol. 57, no. 1, pp. 136–143, Jan 2016, doi: 10.2967/jnumed.115.156299

[46] Y. Lei et al., "Magnetic resonance imaging-based pseudo computed tomography using anatomic signature and joint dictionary learning," (in eng), *Journal of Medical Imaging (Bellingham)*, vol. 5, no. 3, p. 034001, Jul 2018, doi: 10.1117/1.JMI.5.3.034001

[47] T. Huynh et al., "Estimating CT Image from MRI Data Using Structured Random Forest and Auto-Context Model," (in eng), *IEEE Transactions on Medical Imaging*, vol. 35, no. 1, pp. 174–183, Jan 2016, doi: 10.1109/Tmi.2015.2461533

[48] D. Andreasen, J. M. Edmund, V. Zografos, B. H. Menze, and K. Van Leemput, "Computed tomography synthesis from magnetic resonance images in the pelvis using multiple random forests and auto-context features," (in eng), *Proceedings of SPIE*, vol. 9784, no. 978417, 2016, doi: Artn 978417. 10.1117/12.2216924

[49] X. Yang et al., "A learning-based approach to derive electron density from anatomical MRI for radiation therapy treatment planning," in *International Journal of Radiation Oncology • Biology • Physics ASTRO*, 2017, vol. 99, no. 2, pp. S173–S174, doi: 10.1016/j.ijrobp.2017.06.437

[50] G. Shafai-Erfani et al., "Dose evaluation of MRI-based synthetic CT generated using a machine learning method for prostate cancer radiotherapy," (in eng), *Medical Dosimetry*, vol. 44, no. 4, pp. e64–e70, Winter 2019, doi: 10.1016/j.meddos.2019.01.002

[51] T. Wang et al., "Machine learning in quantitative PET: A review of attenuation correction and low-count image reconstruction methods," *Physica Medica*, vol. 76, pp. 294–306, Aug 2020, doi: 10.1016/j.ejmp.2020.07.028

[52] X. Dai et al., "Multimodal MRI synthesis using unified generative adversarial networks," *Medical Physics*, 10.1002/mp.14539 vol. 47, no. 12, pp. 6343–6354, Dec 2020, doi: 10.1002/mp.14539

[53] C.-W. Chang et al., "Deep learning-based fast volumetric image generation for image-guided proton FLASH radiotherapy," *arXiv preprint arXiv:2210.00971*, 2022.

[54] Y. Lei et al., "Deep learning in multi-organ segmentation," *ArXiv*, vol. doi: abs/2001.10619, 2020.

[55] S. Momin et al., "Knowledge-based radiation treatment planning: A data-driven method survey," *arXiv: Medical Physics*, vol. doi: arXiv:2009.07388, 2020.

[56] Y. Fu, Y. Lei, T. Wang, W. J. Curran, T. Liu, and X. Yang, "Deep learning in medical image registration: A review," *Physics in Medicine and Biology*, vol. 65, no. 20, p. 20TR01, Oct 22 2020, doi: 10.1088/1361-6560/ab843e

[57] T. Wang et al., "Synthesizing high-resolution CT from low-resolution CT using self-learning," in *Medical Imaging 2021: Image Processing*, 2021, vol. 11595, no. 115952N SPIE. [Online]. Available: 10.1117/12.2581080

[58] K. Raza and N. K. J. A. Singh, "A tour of unsupervised deep learning for medical image analysis," *ArXiv*, vol. abs/1812.07715, 2018.

[59] H. C. Shin, M. R. Orton, D. J. Collins, S. J. Doran, and M. O. Leach, "Stacked autoencoders for unsupervised feature learning and multiple organ detection in a pilot study using 4D patient data," (in eng), *IEEE Transactions on Pattern Analysis and Machine Intelligence*, vol. 35, no. 8, pp. 1930–1943, Aug 2013, doi: 10.1109/Tpami.2012.277

[60] P. Vincent, H. Larochelle, I. Lajoie, Y. Bengio, and P.-A. J. J. M. L. R. Manzagol, "Stacked denoising autoencoders: Learning useful representations in a deep network with a local denoising criterion," *Journal of Machine Learning Research*, vol. 11, pp. 3371–3408, 2010.

[61] V. Alex, K. Vaidhya, S. Thirunavukkarasu, C. Kesavadas, and G. Krishnamurthi, "Semisupervised learning using denoising autoencoders for brain lesion detection and segmentation," (in eng), *Journal of Medical Imaging (Bellingham, Wash)*, vol. 4, no. 4, p. 041311, Oct 2017, doi: 10.1117/1.Jmi.4.4.041311

[62] K. Vaidhya, S. Thirunavukkarasu, A. Varghese, and G. Krishnamurthi, "Multi-modal brain tumor segmentation using stacked denoising autoencoders," in *Brainles@MICCAI*, 2015.

[63] S. F. Qadri, Z. Zhao, D. Ai, M. Ahmad, and Y. Wang, "Vertebrae segmentation via stacked sparse autoencoder from computed tomography images," *Eleventh International Conference on Digital Image Processing (ICDIP 2019)*. SPIE, 2019.

[64] F. Li, H. Qiao, B. Zhang, and X. Xi, "Discriminatively boosted image clustering with fully convolutional auto-encoders," *Pattern Recognition*, vol. 83, pp. 161–173, 2017.

[65] X. Guo, X. Liu, E. Zhu, and J. Yin, "Deep Clustering with Convolutional Autoencoders," in *Neural Information Processing*, Cham, D. Liu, S. Xie, Y. Li, D. Zhao, and E.-S. M. El-Alfy, Eds., 2017: Springer International Publishing, pp. 373–382.

[66] C. Wang, A. Elazab, F. Jia, J. Wu, and Q. Hu, "Automated chest screening based on a hybrid model of transfer learning and convolutional sparse denoising autoencoder," (in eng), *Biomedical Engineering Online*, vol. 17, no. 1, p. 63, May 23 2018, doi: 10.1186/s12938-018-0496-2

[67] T. Zhou, S. Ruan, and S. Canu, "A review: Deep learning for medical image segmentation using multi-modality fusion," *Array*, vol. 3–4, p. 100004, 2019/09/01/ 2019, doi: 10.1016/j.array.2019.100004

[68] K. M. He, X. Y. Zhang, S. Q. Ren, and J. Sun, "Delving deep into rectifiers: Surpassing human-level performance on ImageNet classification," (in eng), *Proceedings of the IEEE Conference on Computer Vision and Pattern Recognition*, pp. 1026–1034, 2015, doi: 10.1109/Iccv.2015.123

[69] Y. Lecun, L. Bottou, Y. Bengio, and P. Haffner, "Gradient-based learning applied to document recognition," *Proceedings of the IEEE*, vol. 86, no. 11, pp. 2278–2324, 1998, doi: 10.1109/5.726791

[70] O. Russakovsky et al., "ImageNet large scale visual recognition challenge," *International Journal of Computer Vision*, vol. 115, no. 3, pp. 211–252, 2015/12/01 2015, doi: 10.1007/s11263-015-0816-y

[71] A. Krizhevsky, I. Sutskever, and G. E. Hinton, "ImageNet classification with deep convolutional neural networks," (in eng), *Communications of the ACM*, vol. 60, no. 6, pp. 84–90, Jun 2012, doi: 10.1145/3065386

[72] M. D. Zeiler and R. Fergus, "Visualizing and Understanding Convolutional Networks," in *Computer Vision – ECCV 2014*, Cham, D. Fleet, T. Pajdla, B. Schiele, and T. Tuytelaars, Eds., 2014: Springer International Publishing, pp. 818–833.

[73] K. Simonyan and A. Zisserman, "Very deep convolutional networks for large-scale image recognition," *CoRR*, vol. abs/1409.1556, 2014.

[74] C. Szegedy et al., "Going deeper with convolutions," in *2015 IEEE Conference on Computer Vision and Pattern Recognition (CVPR)*, 7–12 June 2015, pp. 1–9, doi: 10.1109/CVPR.2015.7298594

[75] K. He, X. Zhang, S. Ren, and J. Sun, "Deep Residual Learning for Image Recognition," in *2016 IEEE Conference on Computer Vision and Pattern Recognition (CVPR)*, 27–30 June 2016, pp. 770–778, doi: 10.1109/CVPR.2016.90

[76] G. Huang, Z. Liu, L. V. D. Maaten, and K. Q. Weinberger, "Densely connected convolutional networks," in *2017 IEEE Conference on Computer Vision and Pattern Recognition (CVPR)*, 21–26 July 2017 2017, pp. 2261–2269, doi: 10.1109/CVPR.2017.243

[77] T. Wang et al., "Dosimetric evaluation of MRI-based synthetic CT for stereotactic radiosurgery of brain cancer," *Medical Physics*, 2018, vol. 45, no. 6, pp. E701–E701.

[78] H. Xie et al., "Magnetic resonance imaging contrast enhancement synthesis using cascade networks with local supervision," *Medical Physics*, vol. 49, no. 5, pp. 3278–3287, 2022.

[79] E. Shelhamer, J. Long, and T. Darrell, "Fully convolutional networks for semantic segmentation," (in eng), *IEEE Transactions on Pattern Analysis and Machine Intelligence*, vol. 39, no. 4, pp. 640–651, Apr 2017, doi: 10.1109/Tpami.2016.2572683

[80] O. Ronneberger, P. Fischer, and T. Brox, "U-Net: Convolutional networks for biomedical image segmentation," (in eng), *Medical Image Computing and Computer-Assisted Intervention*, vol. 9351, pp. 234–241, 2015, doi: 10.1007/978-3-319-24574-4_28

[81] F. Milletari, N. Navab, and S.-A. Ahmadi, "V-Net: Fully Convolutional neural networks for volumetric medical image segmentation," in *Fourth International Conference on 3D Vision*, pp. 565–571, 2016.

[82] Y. Lei et al., "Ultrasound prostate segmentation based on multidirectional deeply supervised V-Net," (in eng), *Medical Physics*, vol. 46, no. 7, pp. 3194–3206, Jul 2019, doi: 10.1002/mp.13577

[83] B. Wang et al., "Deeply supervised 3D fully convolutional networks with group dilated convolution for automatic MRI prostate segmentation," (in eng), *Medical Physics*, vol. 46, no. 4, pp. 1707–1718, Apr 2019, doi: 10.1002/mp.13416

[84] J. Schlemper et al., "Attention gated networks: Learning to leverage salient regions in medical images," *Medical Image Analysis*, vol. 53, pp. 197–207, Apr 2019, doi: 10.1016/j.media.2019.01.012

[85] X. Ying, H. Guo, K. Ma, J. Y. Wu, Z. Weng, and Y. Zheng, "X2CT-GAN: Reconstructing CT from biplanar X-rays with generative adversarial networks," in *CVPR*, 2019.

[86] X. Dong et al., "Deep learning-based attenuation correction in the absence of structural information for whole-body PET imaging," (in eng), *Physics in Medicine and Biology*, Dec 23 2019, doi: 10.1088/1361-6560/ab652c

[87] J. Harms et al., "Paired cycle-GAN-based image correction for quantitative cone-beam computed tomography," (in eng), *Medical Physics*, vol. 46, no. 9, pp. 3998–4009, Sep 2019, doi: 10.1002/mp.13656

[88] R. Liu et al., "Synthetic dual-energy CT for MRI-only based proton therapy treatment planning using label-GAN," (in eng), *Physics in Medicine and Biology*, vol. 66, no. 6, p. 065014, Mar 9 2021, doi: 10.1088/1361-6560/abe736

[89] R. Liu et al., "Synthetic dual-energy CT for MRI-based proton therapy treatment planning using label-GAN," *Physics in Medicine and Biology*, vol. 66, no. 6, Mar 8 2021, doi: 10.1088/1361-6560/abe736

[90] W. Dai, N. Dong, Z. Wang, X. Liang, H. Zhang, and E. P. Xing, "SCAN: Structure correcting adversarial network for organ segmentation in chest X-rays," in *DLMIA/ML-CDS@MICCAI*, 2017.

[91] X. Dong et al., "Automatic multiorgan segmentation in thorax CT images using U-net-GAN," (in eng), *Medical Physics*, vol. 46, no. 5, pp. 2157–2168, May 2019, doi: 10.1002/mp.13458

[92] X. Yang et al., "Synthetic MRI-aided multi-organ CT segmentation for head and neck radiotherapy treatment planning," *International Journal of Radiation Oncology, Biology, Physics*, vol. 108, no. 3, p. e341, 2020.

[93] Q. Zhang, H. Wang, H. Lu, D. Won, and S. W. Yoon, "Medical image synthesis with generative adversarial networks for tissue recognition," in *2018 IEEE International Conference on Healthcare Informatics (ICHI)*, 4–7 June 2018, pp. 199–207, doi: 10.1109/ICHI.2018.00030

[94] C. Han, K. Murao, S. I. Satoh, and H. Nakayama, "Learning more with less: GAN-based medical image augmentation," *ArXiv*, vol. abs/1904.00838, 2019.

[95] Y. Lei et al., "MRI-only based synthetic CT generation using dense cycle consistent generative adversarial networks," (in eng), *Medical Physics*, vol. 46, no. 8, pp. 3565–3581, Aug 2019, doi: 10.1002/mp.13617

[96] I. J. Goodfellow et al., "Generative Adversarial Nets," in *NIPS*, 2014.

[97] X. Yi, E. Walia, and P. Babyn, "Generative adversarial network in medical imaging: A review," *Medical Image Analysis*, vol. 58, p. 101552, 2019.

[98] D. Nie et al., "Medical image synthesis with deep convolutional adversarial networks," *IEEE Transactions on Biomedical Engineering*, vol. 65, no. 12, pp. 2720–2730, Dec 2018, doi: 10.1109/TBME.2018.2814538

[99] M. Mehralian and B. Karasfi, "RDCGAN: Unsupervised representation learning with regularized deep convolutional generative adversarial networks," in *2018 9th Conference on Artificial Intelligence and Robotics and 2nd Asia-Pacific International Symposium*, 10–10 Dec 2018, pp. 31–38, doi: 10.1109/AIAR.2018.8769811

[100] S. Ioffe and C. Szegedy, "Batch normalization: Accelerating deep network training by reducing internal covariate shift," *presented at the Proceedings of the 32nd International Conference on International Conference on Machine Learning*, Vol. 37, Lille, France, 2015.

[101] A. L. Maas, A. Y. Hannun, and A. Y. Ng, "Rectifier nonlinearities improve neural network acoustic models," in *ICML Workshop on Deep Learning for Audio, Speech and Language Processing*, 2013.

[102] P. Isola, J. Zhu, T. Zhou, and A. A. Efros, "Image-to-image translation with conditional adversarial networks," in *2017 IEEE Conference on Computer Vision and Pattern Recognition (CVPR)*, 21–26 July 2017, pp. 5967–5976, doi: 10.1109/CVPR.2017.632

[103] X. Yang et al., "CBCT-guided prostate adaptive radiotherapy with CBCT-based synthetic MRI and CT," in *International Journal of Radiation Oncology • Biology • Physics ASTRO*, 2019, vol. 105, no. 1, p. S250, doi: 10.1016/j.ijrobp.2019.06.372

[104] R. Qiu et al., "Daily coverage assessment using deep learning generated synthetic CT for lung SBRT patients," *Medical Physics*, 2022, vol. 49, no. 6: WILEY 111 RIVER ST, HOBOKEN 07030-5774, NJ USA, pp. E730–E731.

[105] T. Wang, Y. Lei, Y. Fu, W. Curran, T. Liu, and X. Yang, "Machine learning in quantitative PET imaging," *ArXiv*, vol. doi: abs/2001.06597, 2020.

[106] X. Yang et al., "PET attenuation correction using MRI-aided two-stream pyramid attention network," *Journal of Nuclear Medicine*, vol. 61, no. supplement 1, p. 110, 2020. [Online]. Available: http://jnm.snmjournals.org/content/61/supplement_1/110.abstract

[107] X. Yang et al., "Whole-body PET estimation from ultra-short scan durations using 3D cycle-consistent generative adversarial networks," *Journal of Nuclear Medicine*, 2019, vol. 60, no. supplement 1, p. 247.

[108] X. Yang et al., "CT-aided low-count whole-body PET imaging using cross-modality attention pyramid network," *Journal of Nuclear Medicine*, vol. 61, no. supplement 1, p. 1416, 2020. [Online]. Available: http://jnm.snmjournals.org/content/61/supplement_1/1416.abstract

[109] X. Dong et al., "Synthetic CT generation from non-attenuation corrected PET images for whole-body PET imaging," (in eng), *Physics in Medicine and Biology*, vol. 64, no. 21, p. 215016, Nov 4 2019, doi: 10.1088/1361-6560/ab4eb7

[110] X. Dong et al., "Whole-body PET estimation from low count statistics using deep convolutional neural networks," *Medical Physics*, 2019, vol. 46, no. 6: WILEY 111 RIVER ST, HOBOKEN 07030-5774, NJ USA, pp. E193–E193.

[111] X. Dong et al., "Jack Krohmer Junior investigator competition winner: Deep learning-based self attenuation correction for whole-body PET imaging," *Medical Physics*, 2019, vol. 46, no. 6: WILEY 111 RIVER ST, HOBOKEN 07030-5774, NJ USA, pp. E192–E192.

[112] X. Chen, Y. Duan, R. Houthooft, J. Schulman, I. Sutskever, and P. Abbeel, "InfoGAN: Interpretable representation learning by information maximizing generative adversarial nets," *presented at the Proceedings of the 30th International Conference on Neural Information Processing Systems*, Barcelona, Spain, 2016.

[113] Y. Lei et al., "Deep learning-based real-time volumetric imaging for lung stereotactic body radiation therapy: A proof of concept study," *Physics in Medicine & Biology*, vol. 65, no. 23, p. 235003, 2020/12/18 2020, doi: 10.1088/1361-6560/abc303

[114] M.-Y. Liu, T. Breuel, and J. Kautz, "Unsupervised image-to-image translation networks," *ArXiv*, vol. abs/1703.00848, 2017.

[115] J.-Y. Zhu et al., "Toward multimodal image-to-image translation," *ArXiv*, vol. abs/1711.11586, 2017.

[116] Y. Choi, M. Choi, M. Kim, J. Ha, S. Kim, and J. Choo, "StarGAN: Unified generative adversarial networks for multi-domain image-to-image translation," in *2018 IEEE/CVF Conference on Computer Vision and Pattern Recognition*, 18–23 June 2018, pp. 8789–8797, doi: 10.1109/CVPR.2018.00916

[117] Y. Liu et al., "CBCT-based synthetic CT generation using deep-attention cycleGAN for pancreatic adaptive radiotherapy," (in eng), *Medical Physics*, vol. 47, no. 6, pp. 2472–2483, Jun 2020, doi: 10.1002/mp.14121

[118] Y. Lei et al., "Whole-body PET estimation from low count statistics using cycle-consistent generative adversarial networks," (in eng), *Physics in Medicine and Biology*, vol. 64, no. 21, p. 215017, Nov 4 2019, doi: 10.1088/1361-6560/ab4891

[119] X. Dai et al., "Automated delineation of head and neck organs at risk using synthetic MRI-aided mask scoring regional convolutional neural network," *Medical Physics*, vol. 48, no. 10, pp. 5862–5873, Oct 2021, doi: 10.1002/mp.15146

[120] X. Dai et al., "Multimodal MRI synthesis using unified generative adversarial networks," *Medical Physics*, 2020/10/14 2020, doi: 10.1002/mp.14539

[121] T. Wang et al., "A review on medical imaging synthesis using deep learning and its clinical applications," *Journal of Applied Clinical Medical Physics*, 2020, doi: 10.1002/acm2.13121

[122] Y. Lei et al., "MRI-based synthetic CT generation using deep convolutional neural network," in *Medical Imaging 2019: Image Processing*, 2019, vol. 10949, no. 109492T.

[123] Y. Liu et al., "Synthetic contrast MR image generation using deep learning," *Medical Physics*, 2021, vol. 48, no. 6: WILEY 111 RIVER ST, HOBOKEN 07030-5774, NJ USA.

[124] T. Wang et al., "Synthetic dual energy CT imaging from single energy CT using deep attention neural network," in *Medical Imaging 2021: Image Processing*, 2021, vol. 11595, no. 1159547 SPIE. [Online]. Available: 10.1117/12.2580966

[125] G. Shafai-Erfani et al., "MRI-based prostate proton radiotherapy using deep-learning-based synthetic CT," *Medical Physics*, 2019, vol. 46, no. 6: WILEY 111 RIVER ST, HOBOKEN 07030-5774, NJ USA, pp. E476–E477.

[126] G. Shafai-Erfani et al., "A novel method of generating synthetic CT from MRI for proton treatment planning for base of skull tumors," *Medical Physics*, 2019, vol. 46, no. 6: WILEY 111 RIVER ST, HOBOKEN 07030-5774, NJ USA, pp. E474–E474.

[127] X. Yang et al., "Med-A-nets: Segmentation of multiple organs in chest CT image with deep adversarial networks," *Medical Physics*, 2018, vol. 45, no. 6: WILEY 111 RIVER ST, HOBOKEN 07030-5774, NJ USA, pp. E154–E154.

[128] Y. Lei et al., "Learning-based CBCT correction using alternating random forest based on auto-context model," (in eng), *Medical Physics*, vol. 46, no. 2, pp. 601–618, Feb 2019, doi: 10.1002/mp.13295

[129] J. Yang et al., "Autosegmentation for thoracic radiation treatment planning: A grand challenge at AAPM 2017," (in eng), *Medical Physics*, vol. 45, no. 10, pp. 4568–4581, Oct 2018, doi: 10.1002/mp.13141

[130] X. Dong et al., "Chest multiorgan segmentation of CT images with U-net-GAN," *Medical Physics*, 2019, vol. 46, no. 6: WILEY 111 RIVER ST, HOBOKEN 07030-5774, NJ USA, pp. E371–E371.

[131] S. Ogawa, T.-M. Lee, A. R. Kay, and D. W. Tank, "Brain magnetic resonance imaging with contrast dependent on blood oxygenation," *Proceedings of the National Academy of Sciences*, vol. 87, no. 24, pp. 9868–9872, 1990.

[132] S. W. Young, "Nuclear magnetic resonance imaging: Basic principles," United States: N. p., 1985. Web. https://www.osti.gov/biblio/5722201

[133] D. B. Plewes and W. Kucharczyk, "Physics of MRI: A primer," *Journal of Magnetic Resonance Imaging*, vol. 35, no. 5, pp. 1038–1054, 2012.

[134] J. Jeong et al., "Machine-learning based classification of glioblastoma using delta-radiomic features derived from dynamic susceptibility contrast enhanced magnetic resonance images: Introduction," (in eng), *Quantitative Imaging in Medicine and Surgery*, vol. 9, no. 7, pp. 1201–1213, Jul 2019, doi: 10.21037/qims.2019.07.01

[135] X. Dai et al., "Synthetic MRI-aided delineation of organs at risk in head-and-neck radiotherapy," *Medical Physics*, 2021, vol. 48, no. 6. WE-D-TRACK 6-5 https://w4.aapm.org/meetings/2021AM/programInfo/programAbs.php?sid=9186&aid=58317

[136] X. Dai et al., "Synthetic MRI-aided multi-organ segmentation in head-and-neck cone beam CT," in *Medical Imaging 2021: Image-Guided Procedures, Robotic Interventions, and Modeling*, 2021, vol. 11598, no. 115981M SPIE. [Online]. Available: 10.1117/12.2581128

[137] Y. Lei et al., "MRI classification using semantic random forest with auto-context model," (in eng), *Quantitative Imaging in Medicine and Surgery*, vol. 11, no. 12, pp. 4753–4766, 2021, doi: 10.21037/qims-20-1114

[138] X. Dong et al., "Air, bone and soft-tissue segmentation on 3D brain MRI using semantic classification random forest with auto-context model," *ArXiv*, vol. doi: arXiv:1911.09264, 2019.

[139] Y. Gao et al., "MRI-based material mass density and relative stopping power estimation via deep learning for proton therapy," *arXiv preprint arXiv:2210.05804*, 2022.

[140] H. Lu, L. M. Nagae-Poetscher, X. Golay, D. Lin, M. Pomper, and P. C. van Zijl, "Routine clinical brain MRI sequences for use at 3.0 Tesla," *Journal of Magnetic Resonance Imaging*, vol. 22, no. 1, pp. 13–22, Jul 2005, doi: 10.1002/jmri.20356

[141] R. Bitar et al., "MR pulse sequences: What every radiologist wants to know but is afraid to ask," *Radiographics*, vol. 26, no. 2, pp. 513–537, 2006.

[142] T. Wang et al., "MRI-based treatment planning for brain stereotactic radiosurgery: Dosimetric validation of a learning-based pseudo-CT generation method," (in eng), *Medical Dosimetry*, vol. 44, no. 3, pp. 199–204, Autumn 2019, doi: 10.1016/j.meddos.2018.06.008

[143] T. Wang et al., "Multiparametric MRI-guided dose boost to dominant intraprostatic lesions in CT-based High-dose-rate prostate brachytherapy," (in eng), *The British Journal of Radiology*, vol. 92, no. 1097, p. 20190089, May 2019, doi: 10.1259/bjr.20190089

[144] Y. Lei et al., "MRI-based synthetic CT generation using semantic random forest with iterative refinement," (in eng), *Physics in Medicine and Biology*, vol. 64, no. 8, p. 085001, Apr 5 2019, doi: 10.1088/1361-6560/ab0b66

[145] Y. Fu, Y. Lei, T. Wang, W. J. Curran, T. J. Liu, and X. J. A. Yang, "Deep learning in medical image registration: A review," *ArXiv*, vol. abs/1912.12318, 2019.

[146] T. Wang, Y. Lei, W. Curran, T. Liu, and X. Yang, "Contrast-enhanced MRI synthesis from non-contrast MRI using attention CycleGAN," in *Medical Imaging 2021: Biomedical Applications in Molecular, Structural, and Functional Imaging*, 2021, vol. 11600, no. 116001L SPIE. [Online]. Available: 10.1117/12.2581064

[147] K. Armanious et al., "MedGAN: Medical image translation using GANs," *Computerized Medical Imaging and Graphics*, vol. 79, p. 101684, Jan 2020, doi: 10.1016/j.compmedimag.2019.101684

[148] S. Olut, Y. H. Sahin, U. Demir, and G. Unal, "Generative Adversarial Training for MRA Image Synthesis Using Multi-contrast MRI," in Islem Rekik, Gozde Unal, Ehsan Adeli and Sang Hyun Park (eds) *Predictive Intelligence in Medicine*, (Lecture Notes in Computer Science, 2018, Chapter 18, pp. 147–154.

[149] B. Yu, L. Zhou, L. Wang, J. Fripp, and P. Bourgeat, "3D cGAN based cross-modality MR image synthesis for brain tumor segmentation," in *2018 IEEE 15th International Symposium on Biomedical Imaging (ISBI 2018)*, 2018: IEEE, pp. 626–630.

[150] S. U. Dar, M. Yurt, L. Karacan, A. Erdem, E. Erdem, and T. Cukur, "Image synthesis in multi-contrast MRI with conditional generative adversarial networks," *IEEE Transactions on Medical Imaging*, vol. 38, no. 10, pp. 2375–2388, Oct 2019, doi: 10.1109/TMI.2019.2901750

[151] Y. Lei et al., "Brain MRI classification based on machine learning framework with auto-context model," in *Medical Imaging 2019: Biomedical Applications in Molecular, Structural, and Functional Imaging*, 2019, vol. 10953: SPIE, doi: 10.1117/12.2512555

[152] A. Sharma and G. Hamarneh, "Missing MRI pulse sequence synthesis using multi-modal generative adversarial network," *IEEE Transactions on Medical Imaging*, vol. 39, no. 4, pp. 1170–1183, 2019.

[153] B. Yu, L. Zhou, L. Wang, Y. Shi, J. Fripp, and P. Bourgeat, "Ea-GANs: Edge-aware generative adversarial networks for cross-modality MR image synthesis," (in eng), *IEEE Transactions on Medical Imaging*, vol. 38, no. 7, pp. 1750–1762, Jul 2019, doi: 10.1109/tmi.2019.2895894

[154] M. Yurt, S. U. H. Dar, A. Erdem, E. Erdem, and T. Çukur, "mustGAN: Multi-stream generative adversarial networks for MR image synthesis," *arXiv preprint arXiv:1909.11504*, 2019.

[155] T. Joyce, A. Chartsias, and S. A. Tsaftaris, "Robust multi-modal MR image synthesis," in *International Conference on Medical Image Computing and Computer-Assisted Intervention*, 2017: Springer, pp. 347–355.

[156] A. Chartsias, T. Joyce, M. V. Giuffrida, and S. A. Tsaftaris, "Multimodal MR synthesis via modality-invariant latent representation," (in eng), *IEEE Transactions on Medical Imaging*, vol. 37, no. 3, pp. 803–814, Mar 2018, doi: 10.1109/tmi.2017.2764326

[157] C. F. Njeh, "Tumor delineation: The weakest link in the search for accuracy in radiotherapy," *Journal of Medical Physics*, vol. 33, no. 4, pp. 136–140, Oct 2008, doi: 10.4103/0971-6203.44472

[158] S. Devic, "MRI simulation for radiotherapy treatment planning," *Medical Physics*, vol. 39, no. 11, pp. 6701–6711, Nov 2012, doi: 10.1118/1.4758068

[159] M. A. Schmidt and G. S. Payne, "Radiotherapy planning using MRI," (in eng), *Physics in Medicine and Biology*, vol. 60, no. 22, pp. R323–R361, Nov 21 2015, doi: 10.1088/0031-9155/60/22/R323

[160] T. Wang et al., "Lung tumor segmentation of PET/CT using dual pyramid mask R-CNN," in *Medical Imaging 2021: Image Processing*, 2021, vol. 11596, no. 1159632 SPIE. [Online]. Available: 10.1117/12.2580987

[161] T. Wang et al., "Prostate and tumor segmentation on PET/CT using dual mask R-CNN," in *Medical Imaging 2021: Image Processing*, 2021, vol. 11600, no. 116000S SPIE. [Online]. Available: 10.1117/12.2580970

[162] L. A. Matkovic et al., "Prostate and dominant intraprostatic lesion segmentation on PET/CT using cascaded regional-net," *Physics in Medicine & Biology*, vol. 66, no. 24, p. 245006, 2021/12/07 2021, doi: 10.1088/1361-6560/ac3c13

[163] U. A. van der Heide, A. C. Houweling, G. Groenendaal, R. G. Beets-Tan, and P. Lambin, "Functional MRI for radiotherapy dose painting," *Magnetic Resonance Imaging*, vol. 30, no. 9, pp. 1216–1223, Nov 2012, doi: 10.1016/j.mri.2012.04.010

[164] R. G. Price, J. P. Kim, W. L. Zheng, I. J. Chetty, and C. Glide-Hurst, "Image guided radiation therapy using synthetic computed tomography images in brain cancer," (in eng), *International Journal of Radiation Oncology*, vol. 95, no. 4, pp. 1281–1289, Jul 15 2016, doi: 10.1016/j.ijrobp.2016.03.002

[165] J. M. Edmund and T. Nyholm, "A review of substitute CT generation for MRI-only radiation therapy," *Radiation Oncology*, vol. 12, no. 1, p. 28, Jan 26 2017, doi: 10.1186/s13014-016-0747-y

[166] E. Johnstone et al., "Systematic review of synthetic computed tomography generation methodologies for use in magnetic resonance imaging-only radiation therapy," (in eng), *International Journal of Radiation Oncology*, vol. 100, no. 1, pp. 199–217, Jan 1 2018, doi: 10.1016/j.ijrobp.2017.08.043

[167] D. Nie et al., "Medical image synthesis with deep convolutional adversarial networks," *IEEE Transactions on Biomedical Engineering*, Mar 9 2018, doi: 10.1109/TBME.2018.2814538

[168] J. M. Wolterink, A. M. Dinkla, M. H. F. Savenije, P. R. Seevinck, C. A. T. van den Berg, and I. Išgum, "Deep MR to CT Synthesis Using Unpaired Data," in *Simulation and Synthesis in Medical Imaging*, Cham, S. A. Tsaftaris, A. Gooya, A. F. Frangi, and J. L. Prince, Eds., 2017: Springer International Publishing, pp. 14–23.

[169] X. Yang et al., "MRI-based synthetic CT for MRI-only proton radiotherapy treatment planning," *Medical Physics*, 2018, vol. 45, no. 6, pp. E360–E360.

[170] Y. Liu et al., "CBCT-based synthetic CT using deep learning for pancreatic adaptive radiotherapy," *Medical Physics*, 2019, vol. 46, no. 6: WILEY 111 RIVER ST, HOBOKEN 07030-5774, NJ USA, pp. E131–E132.

[171] Y. Liu et al., "Deep-learning-based synthetic-CT generation method for CBCT-guided proton therapy," *Medical Physics*, 2019, vol. 46, no. 6, pp. E475–E475.

[172] L. Shi, T. Tsui, J. Wei, and L. Zhu, "Fast shading correction for cone beam CT in radiation therapy via sparse sampling on planning CT," *Medical Physics*, vol. 44, no. 5, pp. 1796–1808, 2017.

[173] Y. Xu et al., "A practical cone-beam CT scatter correction method with optimized Monte Carlo simulations for image-guided radiation therapy," *Physics in Medicine and Biology*, vol. 60, no. 9, p. 3567, 2015.

[174] Y. Lei et al., "Accurate CBCT prostate segmentation aided by CBCT-based synthetic MRI," *Medical Physics*, 2019, vol. 46, no. 6, pp. E132–E132.

[175] Y. Lei et al., "Thoracic CBCT-based synthetic CT for lung stereotactic body radiation therapy," *Medical Physics*, 2021, vol. 48, no. 6. MO-IePD-TRACK 3-7. https://w4.aapm.org/meetings/2021AM/programInfo/programAbs.php?sid=9281&aid=58445

[176] J. Janopaul-Naylor et al., "Synthetic CT-aided online CBCT multi-organ segmentation for CBCT-guided adaptive radiotherapy of pancreatic cancer," *International Journal of Radiation Oncology, Biology, Physics*, vol. 108, no. 3, pp. S7–S8, 2020, doi: 10.1016/j.ijrobp.2020.07.2080

[177] S. Acharya et al., "Online magnetic resonance image guided adaptive radiation therapy: First clinical applications," *International Journal of Radiation Oncology, Biology, Physics*, vol. 94, no. 2, pp. 394–403, 2016, doi: 10.1016/j.ijrobp.2015.10.015

[178] R. L. Qiu et al., "Chest CBCT-based synthetic CT using cycle-consistent adversarial network with histogram matching," in *Medical Imaging 2021: Image Processing*, 2021, vol. 11596, no. 115961Z SPIE. [Online]. Available: 10.1117/12.2581094

[179] X. Dai et al., "Synthetic CT-aided multiorgan segmentation for CBCT-guided adaptive pancreatic radiotherapy," *Medical Physics*, vol. 48, no. 11, pp. 7063–7073, Nov 2021, doi: 10.1002/mp.15264

[180] X. Dai et al., "Synthetic CT-based multi-organ segmentation in cone beam CT for adaptive pancreatic radiotherapy," in *Medical Imaging 2021: Image Processing*, 2021, vol. 11596, no. 1159623 SPIE. [Online]. Available: 10.1117/12.2581132

[181] X. Liang et al., "Generating synthesized computed tomography (CT) from cone-beam computed tomography (CBCT) using CycleGAN for adaptive radiation therapy," *arXiv preprint arXiv:.13350*, 2018.

[182] X. Dong et al., "Deep learning-based attenuation correction in the absence of structural information for whole-body positron emission tomography imaging," (in eng), *Physics in Medicine and Biology*, vol. 65, no. 5, p. 055011, Mar 2 2020, doi: 10.1088/1361-6560/ab652c

[183] A. Le et al., "Multi-level canonical correlation analysis for standard-dose PET image estimation," (in eng), *IEEE Transactions on Image Processing*, vol. 25, no. 7, pp. 3303–3315, 2016, doi: 10.1109/TIP.2016.2567072

[184] K. T. Chen et al., "Ultra-low-dose (18)F-florbetaben amyloid PET imaging using deep learning with multi-contrast MRI inputs," (in eng), *Radiology*, vol. 290, no. 3, pp. 649–656, Mar 2019, doi: 10.1148/radiol.2018180940

[185] L. Xiang, Y. Qiao, D. Nie, L. An, Q. Wang, and D. Shen, "Deep auto-context convolutional neural networks for standard-dose PET image estimation from low-dose PET/MRI," (in eng), *Neurocomputing*, vol. 267, pp. 406–416, Dec 6 2017, doi: 10.1016/j.neucom.2017.06.048

Section II

Applications of Inter-Modality Image Synthesis

3

MRI-Based Image Synthesis

Tonghe Wang
Memorial Sloan Kettering Cancer Center, New York, NY, USA

Xiaofeng Yang
Emory University, Atlanta, GA, USA

CONTENTS

3.1 Introduction

Image synthesis between different medical imaging modalities/protocols is an active research field with a great clinical interest in radiation oncology and radiology. It aims to facilitate a specific clinical workflow by bypassing or replacing a certain imaging procedure when the acquisition is infeasible, costs additional time/labor/expense, has ionizing radiation exposure, or introduces uncertainty from image registration between different modalities. The proposed benefit has raised increasing interest in a number of potential clinical applications such as magnetic resonance imaging (MRI)-only radiation therapy treatment planning, positron emission tomography (PET)/MRI scanning, etc.

Image synthesis with its potential applications has been investigated for decades. The conventional methods usually rely on models with explicit human-defined rules about the conversion of images from one modality to the other. These models are usually application-specific depending on the unique characteristics of the involved pair of imaging modalities, thus can be diverse in methodologies and complexities. It is also hard to build such a model when the two imaging modalities provide distinct information, such as anatomic imaging and functional imaging. This is partially why the majority of these studies are limited to image synthesis between computed tomography (CT) images from MRI.[1] These methods usually require case-by-case parameter tuning for optimal performance.

Owing to the widespread success of machine learning in the computer vision field in recent years, the latest breakthrough in artificial intelligence has been integrated into medical image synthesis. In addition to CT-MRI synthesis, image synthesis in other imaging modalities such as PET and cone-beam CT (CBCT) is now viable. As a result, more and more applications could benefit from the recent advancements in image synthesis techniques.[2–4] Deep learning, as a large subset of machine learning and artificial intelligence, has dominated this field in the past several years. Deep learning utilizes a neural network with many layers containing a huge number of neurons to extract useful features from images. Various networks and architectures have been proposed for better performance on different tasks. Deep learning-based image synthesis methods usually share a common framework that uses a data-driven approach for image intensity mapping. The workflow usually consists of a training stage for the network to learn the mapping between the input and its target, and a predication stage to synthesize the target from an input. Compared with conventional model-based methods, deep learning-based methods are more generalizable since the same network and architecture for a pair of image modalities can be applied to different pairs of image modalities with minimal adjustment. This allows rapid expansion of applications using a similar methodology to a variety of imaging modalities that are clinically desired for image synthesis. The performance of the deep learning-based methods largely depends on the representativeness of the training datasets rather than case-specific parameters. Although network training may require lots of effort in collecting and cleaning training datasets, the prediction usually takes only a few seconds. Due to these advantages, deep learning-based methods have attracted great research and clinical interest in medical imaging and radiation therapy.

MRI-based image synthesis is one of the first investigations for medical image synthesis, and remains the most studied topic in this field. Inspired by its success, many applications aiming at synthesizing from and to other imaging modalities have been actively investigated. The main clinical motivation of MR-based synthesis is to replace a certain image modality with MR acquisition.[5] Such image modality usually involves ionizing radiation, e.g., CT, which is preferred to avoid or mitigate in terms of its potential side effect from radiation dose. In this chapter, we would like to focus on MRI-based synthetic CT. The results of the synthetic CT in current studies are still considerably different from real CT, which prevents it from direct diagnostic usage. However, many studies showed its feasibility for non- or indirect diagnostic purposes, such as treatment planning for radiation therapy and PET attenuation correction.

Current radiation therapy workflow requires both MRI and CT performed on patients for treatment planning since MR images provide good soft-tissue contrast that is essential for tumor and organs-at-risk (OARs) delineation,[6] while CT images are used to derive electron density maps for dose calculation as well as reference images for pre-treatment positioning. The contours of the tumor and OARs are delineated on MR images and then propagated to CT images by image registration for treatment planning and dose evaluation. However, performing both imaging modalities would have additional cost and time for the patient and also introduces systematic positioning errors up to 2 mm during the CT-MRI image registration process.[7-9] Moreover, CT scan also introduces non-negligible ionization dose to patients,[10] especially those requiring re-simulation. Thus, it is highly desirable to skip CT scans with a solely MRI-based treatment planning workflow. Emerging MR-Linac technology also motivates the exclusive use of MRI in radiotherapy.[11,12] MR cannot directly replace CT in current radiotherapy since the signal of MR images is from hydrogen nucleus, thus cannot provide material attenuation coefficients for electron density calibration and subsequent dose calculation.

Replacing CT with MR is also preferred in current PET imaging. CT is widely combined with PET in order to perform both imaging exams serially on the same table. The CT images are used to derive the 511 keV linear attenuation coefficient map to model photon attenuation by a piecewise linear scaling algorithm.[13,14] The linear attenuation coefficient map is then used to correct the PET images for the attenuated annihilation photons in the patient body in order to achieve satisfactory image quality. MR has been proposed to be incorporated with PET as a promising alternative to existing PET/CT systems for its advantages of excellent soft-tissue contrast and radiation dose-free, with a similar challenge as in radiation therapy that MR images cannot be directly used to derive the 511 keV attenuation coefficients for attenuation correction process. Therefore, MR-to-CT image synthesis could be used in PET/MR systems for photon attenuation correction.

There is no one-to-one relationship between MR voxel intensity versus CT Hounsfield unit (HU) values, thus it leads to a huge difference in image appearance and contrast between them, which makes intensity-based calibration methods infeasible. For example, air is dark and bone is bright on CT, while both are dark on MR. Conventional methods proposed in the literature either segment the MR images into several groups of materials and then assign corresponding CT HU numbers,[15-20] or register the MR images with an atlas with known CT HU numbers.[21-23] These methods heavily rely on the performance of segmentation and registration, which is very challenging due to the indistinguishable air/bone boundary and large inter-patient variation.

Tables 3.1 and 3.2 list the studies that synthesized CT images from MR for radiation therapy and PET attenuation correction, respectively. Figure 3.1 shows a pie chart of the number of articles categorized by the applications. For synthetic CT in radiation therapy, the mean average error (MAE) is the most commonly and consistently used metrics by which almost every study reported the image quality of its synthetic CT. For synthetic CT in PET AC, the quality of PET with attenuation corrected by synthetic CT is more evaluated than the synthetic CT itself. For studies that presented several variants of methods, we listed the one with the best MAE for radiation therapy, or the best PET quality for PET AC.

3.1.1 Synthetic CT Quality

Most studies have the MAE of the synthetic CT within the patient's body around 40 to 70 HU. Studies with better results show the MAE comparable to the typical uncertainties observed in real CT scanning. For example, the MAE of soft tissue reported in refs[24-30] is less than 40 HU. On the other hand, the MAE of bone or air is usually larger than 100 HU. The relatively inferior performance on bone and air is within expectation due to their ambiguous contrast on MR images. Another potential reason can be the mis-registration between the CT and MR images in the patient datasets. Such misalignment, which mostly happens on the bone, not only causes the intensity mapping error during the training process, but also results in the overestimation of error in the evaluation study since the error from mis-registration was counted as a synthesis error. There are two studies reporting much higher MAE of the rectum (~70 HU) than soft tissue.[28,31] It may be attributed to its mismatch of the rectum on CT and MR due to different filling statuses. Moreover, since the number of bone pixels are much less than those of soft tissue in the patient's body, the training process tends to map pixels to low HU region in the prediction stage. Potential solutions can include assigning higher loss weights on bone or adding bone-only images for training.[24]

Learning-based methods showed better performance than conventional methods in synthetic CT accuracy in multiple studies, which demonstrated the advantage of the data-driven approaches over model-based methods.[27,31-33] For example, synthetic CT by the atlas-based method was generally noisier and prone to error of registration, which led to higher image error than learning-based methods. However, atlas-based methods were also shown to be more robust than learning-based methods to image quality variation in some cases.[31] One of the limitations of the learning-based method is that its performance can be a catastrophic failure when applied to datasets that are very different from the training

TABLE 3.1

Summary of Studies on MR-based Synthetic CT for Radiation Therapy[59]

Network	MR Parameters	Site, and # of Patients in Training/Testing	Key Findings in Image Quality	Key Findings in Dosimetry	Author, Year
U-net	1.5T T1w without contrast	Brain: 18, 6-fold cross validation	MAE (HU): 84.8 ± 17.3	N/A*	Han, 2017 [32]
GAN	N/A	Brain: 16 Pelvis: 22	MAE (HU): 92.5 ± 13.9	N/A	Nie et al., 2018 [33]
CNN	T1w	Brain: 16, leave-one-out Pelvis: 22, leave-one-out	MAE (HU): 85.4 ± 9.24 (brain) 42.4 ± 5.1 (Pelvis)	N/A	Xiang et al., 2018 [60]
CNN	1.5T T1w	Brain: 52, 2-fold cross validation	MAE (HU): 67 ± 11	Dose difference <1%	Dinkla et al., 2018 [30]
U-net	3T T2w	Pelvis: 39, 4-fold cross validation	MAE (HU): 32.7 ± 7.9	Dose difference <1%	Arabi et al., 2018 [31]
U-net	3T T2w	Pelvis: 36 training/15 testing	MAE (HU): 29.96 ± 4.87	Dose difference of max dose in PTV <1.01%	Chen et al., 2018 [27]
GAN	1T post-Gadolinium T1w	Brain: 15, 5-fold cross validation	MAE (HU): 89.3 ± 10.3	N/A	Emami et al., 2018 [26]
GAN	Dixon in-phase, fat and water	Pelvis: 91 (59 prostate+18 rectal+14 cervical cancer), 32 (prostate) training/59 (rest) testing	MAE (HU): 65 ± 10 (Prostate) 56 ± 5 (Rectum) 59 ± 6 (Cervix)	Dose difference < 1.6%	Maspero et al., 2018 [61]
U-net	3T in-phase Dixon T2w	Head and neck: 22 training/12 testing	MAE (HU): 75 ± 9	Mead dose difference −0.03% ± 0.05% overall, −0.07% ± 0.22% in >90% of prescription dose volume	Dinkla et al., 2019 [25]
U-net	1.5T T1w without contrast	Pelvis: 20, 5-fold cross validation	MAE (HU): 40.5 ± 5.4 (2D) 37.6 ± 5.1 (3D)	N/A	Fu et al., 2019 [24]
U-net	3T in-phase Dixon T1w	Brain: 47 training/13 testing	MAE (HU): 17.6 ± 3.4	Mean target dose difference 2.3 ± 0.1%	Gupta et al., 2019 [29]
GAN	1.5T post-Gadolinium T1w	Brain: 77, 70% training/12% validation/18% testing	MAE (HU): 47.2 ± 11.0	Mean dose volume histogram (DVH) metrics difference <1%	Kazemifar et al., 2019 [44]
GAN	3T T2w	Pelvis: 39, training/testing: 25/14, 25/14, 25/11	MAE (HU): 34.1 ± 7.5	PTV V95% difference < 0.6%	Largent et al., 2019 [28]

Network	MRI input	Patients / validation	MAE	Dose / evaluation metric	Reference
CycleGAN	Brain: T1w, Pelvis: T2w	Brain: 24, Pelvis: 20, Leave-one-out cross validation	MAE (HU): 55.7 ± 9.4 (Brain), 50.8 ± 15.5 (Pelvis)	N/A	Lei et al., 2019[62]
U-net	1.5T T1w	Brain: 30 training/10 testing	MAE (HU): 75 ± 23	PTV V95% difference 0.27% ± 0.79%	Liu et al., 2019[63]
CycleGAN	3T/1.5T T1w	Liver: 21, leave-one-out cross validation	MAE (HU): 72.87 ± 18.16	Mean DVH metrics difference <1% for both photon and proton plans	Liu et al., 2019[64] and Liu et al., 2019[50]
CycleGAN	1.5T T2w	Pelvis: 17, leave-one-out cross validation	MAE (HU): 51.32 ± 16.91	Mean DVH metrics difference <1% (Proton plan)	Liu et al., 2019[50]
U-net	1.5T T1w	Brain: 57 training/28 validation/4 testing	MAE (HU): (82, 147)[+]	Gamma passing rate: >95% at (1%, 1mm) for photon plan, >90% at (2%, 2mm) for proton plan	Neppl et al., 2019[65]
GAN	0.35T T1w	Breast: 48 training/12 testing	MAE (HU): 16.1 ± 3.5	PTV D95 difference<1%	Olberg et al., 2019[66]
CycleGAN	1.5T T1w	Brain: 50	MAE (HU): 54.55 ± 6.81	PTV D95 difference<0.5% (proton plan)	Shafai-Erfani et al., 2019[51]
U-net	1.5T T2w	Head and neck: 23 training/10 testing	MAE (HU): 131 ± 24	N/A	Wang et al., 2019[67]
U-net	3T T1w Dixon	Pelvis: 27, 3-fold cross validation	MAE (HU): (33, 40)	N/A	Florkow et al., 2020[40]
GAN	T1w+T2w+ FLAIR	Brain: 15	MAE (HU): 108.1 ± 24.0	DVH metrics difference < 1%	Koike et al., 2020[68]
GAN	T1w+T2w+ Contrast-enhanced T1w+ Contrast-enhanced T1w Dixon water	Head and neck: 30 training/15 testing	MAE (HU): 69.98 ± 12.02	Mean average dose difference <1%	Qi et al., 2020[39]
GAN	1.5T Pre-contrast T1w+post-contrast T1w+T2w	Head and neck: 32, 8-fold cross validation	MAE (HU): 75.7 ± 14.6	N/A	Tie et al., 2020[41]
GAN	1.5T and 3T T2w from three scanners	Pelvis: 11 training from two scanner/8 testing from one scanner	MAE (HU): 48.5 ± 6	Maximum dose difference in target = 1.3%	Boni et al., 2020

* N/A: not available, i.e., not explicitly indicated in the publication.
+ Numbers in parentheses indicate minimum and maximum values.

TABLE 3.2

Summary of Studies on MR-based Synthetic CT for PET Attenuation Correction[59]

Network	MR Parameters	Site, and # of Patients in Training/Testing	Key Findings in Synthetic CT Quality	Key Findings in PET Quality	Author, Year
U-net	Dixon and ZTE	Brain: 14, leave-two-out	MAE (%): 12.62 ± 1.46	Absolute bias <3% among 8 VOIs	Gong et al., 2018[43]
U-net (Encoder-decoder)	3T UTE	Brain: 30 pre-training/6 training/8 testing	N/A*	Bias (%): −0.8 ± 0.8 to 1.1 ± 1.3 among 23 VOIs	Jiang et al., 2018[57]
U-net	3T Dixon and ZTE	Pelvis:26, 10 training/16 testing	Mean error (HU): −12 ± 78	RMSE (%): 2.68 among 30 bone lesions, 4.07 among 60 soft-tissue lesions	Leynes et al., 2018[42]
U-net (Encoder-decoder)	1.5T T1w	Brain: 30 training/10 testing	N/A	Bias (%): −3.2 ± 1.3 to 0.4 ± 0.8	Liu et al., 2018[36]
U-net	1.5T T1w	Brain: 44 training/11 validation/11 testing	Global Bias (%): −1.06 ± 0.81	Global Bias(%): −0.49 ± 1.7 for 11C-WAY-100635 −1.52 ± 0.73 for 11C-DASB	Spuhler et al., 2019[69]
U-net	Dixon-VIBE	Pelvis: 28 pairs from 19 patients, 4-fold cross validation	MAE (%): 2.36 ± 3.15	Bias (%): 0.27 ± 2.59 for fat −0.03 ± 2.98 for soft tissue −0.95 ± 5.09 for bone	Torrado-Carvajal et al., 2019[56]
U-net	ZTE	Brain: 23 training/47 testing	N/A	Bias (%): −1.8 ± 1.9 to 1.7 ± 2.6 among 70 VOIs	Blanc-Durand et al., 2019[37]
U-net	UTE	Brain: 79 (pediatric), 4-fold cross validation	N/A	Bias (%): −0.2 to 0.5 in 95% CI	Ladefoged et al., 2019[38]
GAN	3T T1w	Brain: 40, 2-fold cross validation	MAE (HU): 302 ± 79 (bone)	Absolute bias < 4% among 63 VOIs	Arabi et al., 2019[55]

* N/A: not available, i.e., not explicitly indicated in the publication.

datasets. The difference may come from abnormal anatomy, and degraded image quality due to severe artifacts and noise. Atlas-based methods, on the other hand, generate a weighted average of templates from its library, and thus are more predictable in unexpected cases.

Note that the reported results among these studies cannot be fairly compared because of different datasets, training, and testing strategy, etc. Thus, it is difficult to conclude the method with the best performance. Some studies compared methods using the same datasets, which may reveal the advantages and limitations of those methods.

Different types of MR sequences have been adopted for synthetic CT generation. The specific MR sequence used usually depends on the accessibility. The optimal sequence yielding the best performance so far is unknown. T1-weighted and T2-weighted sequences, due to their wide availability, enables learning models to be trained from a relatively large

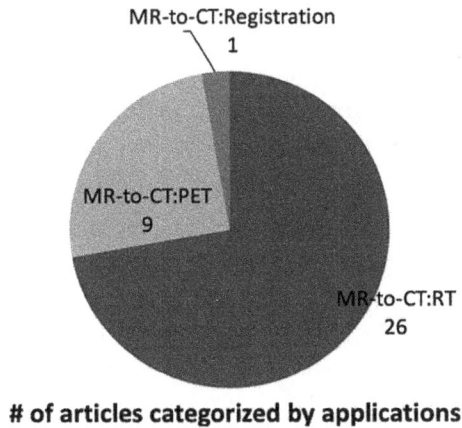

# of articles categorized by applications

FIGURE 3.1
Pie chart of numbers of articles in different categories of applications.

number of datasets with CT and co-registered T1- or T2-weighted MR images. T2-w images may be superior to T1-w since they intrinsically have better geometric accuracy in regions where subject-induced susceptibility is large, such as the nasal cavity, and have fewer chemical shift artifacts at fat and tissue boundaries. However, air and bone have little contrast in either T1- or T2-weighted MR images, which may adversely affect the extraction of the corresponding features in learning-based methods. Two-point Dixon sequence can separate water and fat, and has already been applied in commercial PET/MR scanners with a combination of volume-interpolated breath-hold examination (VIBE) for Dixon-based soft-tissue and air segmentation for PET AC as a clinical standard.[34,35] Its drawback is again the poor contrast of bone, which results in the misclassification of bone as fat. In order to enhance the bone contrast to facilitate the feature extraction in learning-based methods, ultrashort echo time (UTE)– and/or zero echo time (ZTE) MR sequences have been recently used due to its capability to generate positive image contrast from bone.[36] Ladefoged *et al.* and Blanc-Durand *et al.* studied the feasibility of UTE and ZTE MR sequences using U-net in PET/MR AC, repectively.[37,38] However, the advantage of this specialized sequence has not been validated by comparing compared the using of UTE/ZTE and conventional MR sequence under the same deep learning network. Moreover, the UTE/ZTE MR images have little diagnostic value on soft tissue but have a long acquisition time, which may hinder its utility in time-sensitive cases.

Other studies attempted to use multiple MR images with different contrast as input in training and prediction in order to include more additional features in the network. For example, Qi *et al.* proposed to use a 4-channel input that includes T1w, T2w, contrast-enhanced T1w, and contrast-enhanced T1w Dixon water images. Compared with the results from fewer channels, the 4-channel result has a lower MAE.[39] Florkow *et al.* investigated single and multi-channel input using magnitude MR images and Dixon reconstructed water, fat, in-phase, and opposed-phase images obtained from a single T1w multi-echo gradient-echo acquisition.[40] They found that multi-channel input is able to improve synthetic CT generation more than single-channel input. Tie *et al.* used T2-w and pre- and post-contrast T1w MR images in a multi-channel multi-path architecture and showed a significant improvement over multi-channel single-path and single-channel results.[41] An attractive combination is UTE/ZTE and Dixon, which provide a contrast of bone against air and fat against soft tissue, respectively.[42,43] Leynes *et al.* showed that the synthetic CT using ZTE and Dixon MR has less error than that using Dixon alone.[42] Although the image quality improvement has been validated, the cost and benefit of performing additional MR sequences for synthetic CT generation need to be further evaluated in specific applications since it usually requires extra cost and acquisition time.

The image registration between MR and other modalities during training has been an issue. U-net and generative adversarial network (GAN)-based methods are susceptible to registration error if a pixel-to-pixel loss is used. Kazemifar *et al.* proposed a possible solution that uses mutual information as the loss function in the generator of GAN to bypass the registration step in the training.[44] As CycleGAN was originally developed for unpaired image-to-image translation, CycleGAN-based methods feature higher robustness to registration error since it introduces cycle consistence loss to enforce the structural consistency between original one and cycle one, (e.g., force cycle MRI generated from synthetic CT to be the same as original MRI).[45-48]

3.1.2 MR-only Radiation Therapy

For the studies aiming for radiation therapy, many of them evaluated the dosimetry accuracy of synthetic CT by calculating radiation treatment dose using the same treatment plan and comparing it with that of real CT as ground truth. It is shown that the dose difference is about 1%, which is small when compared with the current total uncertainty of dose delivery on patients (5%) during the entire radiation therapy pathway. Compared to the large improvement in image accuracy, the improvement from learning-based methods over conventional methods in dosimetry accuracy on photon radiation therapy is relatively small. The conclusion about the significance of the improvement is also mixed.[27,31] A potential reason is that the dose calculation on photon plans is quite forgiving to image inaccuracy, especially in homogeneous regions such as the brain. For the widely studied volumetric modulated arc therapy (VMAT), the contribution from errors in images to dose also tends to cancel out in an arc. However, the small dosimetric improvement may still be worthwhile in cases such as stereotactic radiosurgery (SRS) and stereotactic body radiation therapy (SBRT) where a large amount of dose is to be delivered into a small volume. In such cases, the dose calculation accuracy could be sensitive to the errors on synthetic CT around the target volume.[3] The recent adoption of non-coplanar beams may also be challenging to MR-based synthetic CT since the beam path length can be sensitive to the prediction error of patient surface due to the beam obliquity, which is worth further investigation.

Studies have also evaluated synthetic CT in the context of proton radiation treatment for prostate, liver, and brain cancer.[49-51] Unlike photons, proton beams deposit dose with a very high-dose gradient at the distal end of the beam. The proton treatment plan thus has highly conformal dose distribution to the target by proton beams coming from several angles. The local HU inaccuracy along the beam path on the planning CT would lead to a shift of the highly conformal high-dose area, which may cause a tumor to be substantially under-dosed or the OARs to be over-dosed.[52] As shown in ref[50] most dose difference of using synthetic CT was at the distal end of the proton beam. As reported by Liu *et al.*,[49,50] the largest and mean absolute range difference is 0.56 cm and 0.19 cm among their 21 liver cancer patients, and 0.75 cm and 0.23 cm among 17 prostate cancer patients.

In addition to dosimetry accuracy for treatment planning, another important aspect for the evaluation of synthetic CT is its geometry fidelity for patient setup. Unfortunately, the studies on synthetic CT positioning accuracy are sparse. Fu *et al.* conducted a patient alignment test by rigidly aligning the synthetic CT and real CT to the CBCT acquired at the first fraction.[24] The translation vector distance and absolute Euler angle difference between the two alignments were found to be less than 0.6mm and 0.5° on average, respectively. Gupta *et al.* adopted a similar study, and found the translation difference was less than 0.7mm in one direction.[29] Apart from alignment with CBCT, the alignment between the derived DRR from synthetic CT and kV image of the patient is also of clinical interest. However, no study on DRR alignment accuracy is found in the reviewed literature. Note that the geometry accuracy of synthetic CT is not only affected by the performance of methods, but also the geometric distortion on MR images caused by magnetic field inhomogeneity as well as subject-induced susceptibility and chemical shift. Methods to mitigate MR distortion are also important in improving synthetic CT accuracy in patient positioning.

3.1.3 PET Attenuation Correction

For the studies aiming for PET attenuation correction, the bias on PET quantification caused by the synthetic CT error has been evaluated. Although it is difficult to specify the tolerance level of quantification error before it affects a clinician's judgment, the general consensus is that quantification errors of 10% or less typically do not affect diagnosis.[53] Based on the average relative bias represented by these studies, almost all of the proposed methods in the studies met this criterion. However, it should be noted that due to the variation among study objects, the bias in some volumes of interest (VOIs) of some patients may exceed 10%.[37,42] It suggests that special attention should be given to the standard deviation of bias as well as its mean when interpreting results since the proposed methods may have poor local performance that would affect some patients. On the other hand, listing or plotting the results of every data point, or at least the range, instead of simply giving a mean ± STD in presenting results, would be more informative in demonstrating the performance of the proposed methods.

Since bone has the highest attenuation capability due to its high density and atomic number,[54] its accuracy on synthetic CT plays a vital role in the final accuracy of the attenuation-corrected PET. Compared with the evaluation for radiation therapy, the bias and geometry accuracy of bone on the synthetic CT is more often evaluated for PET AC. Multiple studies showed that synthetic CT with better bone accuracy tends to generate more accurate PET globally.[37,43,55,56] The more accurate synthetic CT images by learning-based methods than conventional methods also lead to a more accurate PET AC. Such improvements were found to be significant in the reviewed studies. It was shown that PET AC by conventional synthetic CT methods have about 5% bias on average among selected VOIs, while for learning-based methods, the bias was reduced to around 2%.[36,42,43,56,57]

In addition to the two widely studied applications, i.e., radiation treatment simulation and PET AC, using synthetic CT from MR to aid intra-modality image registration has been proven promising. Direct image registration between CT and MR is very challenging due to the distinct image contrast and can be unreliable in deformable registration algorithms where large distortion is allowed. McKenzie *et al.* proposed a CycleGAN-based network to generate synthetic CT, and used the synthetic CT to replace the MR in MR-CT registration for head-and-neck.[58] In this way, the multimodality registration problem is reduced to a mono-modality one. In their study as summarized in Table 3.3, it was found that with the same deformable registration algorithm, the average landmark error decreased from 9.8 ± 3.1 mm in direct MR-CT registration to 6.0 ± 2.1 mm in using synthetic CT as bridge. Similar results were also found in the registration at CT-MR direction.

TABLE 3.3

Summary of Studies on MR-based Synthetic CT for Registration[59]

Network	MR Parameters	Site, and # of Patients in Training/Testing	Key Findings in Synthetic CT Quality	Key Findings in Registration Accuracy	Author, Year
CycleGAN	0.35T	Head and neck: 25, 5-fold cross validation	N/A*	landmark error (mm): 6.0 ± 2.1 (MR-to-CT) 6.6 ± 2.0 (CT-to-MR)	McKenzie *et al.*, 2019[58]

* N/A: not available, i.e., not explicitly indicated in the publication.

3.1.4 Discussion

Recent years have witnessed the trend of deep learning being increasingly used in the application of medical imaging. The latest networks and techniques have been borrowed from the computer vision field and adapted to specific clinical tasks for radiology and radiation oncology. As reviewed in this paper, learning-based image synthesis is an emerging and active field since all of these reviewed studies were published within the last three years. With the development of both artificial intelligence and computing hardware, more learning-based methods are expected to facilitate the clinical workflow with novel applications. Although the reviewed literature show the success of deep learning-based image synthesis in various applications, there are still some common open questions that need to be answered in future studies.

In the implementation of the network, due to the limitations of the graphics processing unit (GPU) memory, some of the deep learning approaches are trained on two-dimensional (2D) slices. Since the loss functions of 2D models do not account for continuity in the third dimension, slice discontinuities can be observed. Some studies trained models in 3D patches to exploit 3D information with even less memory burden[62], while a potential drawback is that the larger scale image features may be hard to extract[25]. Training on three-dimensional (3D) image stacks is expected to achieve a more homogeneous conversion result. Fu *et al.* compared the performance between 2D and 3D models using the same U-net[24]. They found that the 3D model generated synthetic CT with smaller MAE and more accurate bone region. However, to achieve robust performance, the 3D model needs more training data since it has more parameters. A compromised solution is to use multiple adjacent slices that may allow the model to capture more image context information or to train different networks for all three orthogonal 2D planes to allow pseudo-3D information[70].

The reviewed studies showed the advantages of learning-based methods over conventional methods in performance as well as clinical applications. Learning-based methods generally outperform conventional methods in generating more realistic synthetic images with higher similarity to real images and better quantitative metrics. In the implementation, depending on the hardware, training a model usually takes several hours to days for learning-based methods. However, once the model is trained, it can be applied to new patients to generate synthetic images within a few seconds or minutes. Conventional methods vary a lot in specific methodologies and implementations, resulting in a wide range of run times. Iterative methods such as compressed sensing (CS) were shown to be unfavorable for large time and resource consuming.

In the training stage, most of the reviewed studies require paired datasets, i.e., the source image and target image need to have pixel-to-pixel correspondence. This requirement poses difficulties in collecting sufficient eligible datasets, as well as demands high accuracy in image registration. Some networks such as CycleGAN can relax the requirement of the paired datasets to be unpaired datasets, which can be beneficial for clinical applications in enrolling a large number of patient datasets for training.

Although the advantages of learning-based methods have been demonstrated, it should be noted that their performance can be unpredictable when the input images are very different from the training datasets. In most of the reviewed studies, unusual cases are excluded. However, these unusual cases can happen from time to time in clinic, and should be handled with caution. For example, it is not uncommon to see a patient with a hip prosthesis in a pelvis scan. The hip prosthesis creates severe artifacts on both CT and MR images; thus, it can be of clinical interest to see the related effect of its inclusion in training or testing datasets, which has not been studied yet. Similar unusual cases can also be seen in other forms in all imaging modalities and are worth investigating, including all kinds of implants that can introduce artifacts, obese patients that present much higher noise levels on image than average, and patients with anatomical abnormality.

Due to the limitation in the number of available datasets, most studies used N-fold cross-validation or leave-N-out strategy. The small to intermediate number of patients in training/testing is proper for a feasibility study, but is far from enough in evaluating the clinical utility and potential impact. Moreover, the representativeness of training/testing dataset needs special attention in a clinical study. The missing of diverse demographics may reduce the robustness and generality in the performance of the model. Most of the studies trained models using data from a single institution with a single scanner. As replacing/equipping with a new scanner is common in practice, it is interesting to know how the trained model would perform on another scanner of different model or vendor when the image characteristic cannot be exactly matched. Boni *et al.* recently presented a proof-of-concept study that predicted synthetic images of one site using a model trained on another two sites, and demonstrated clinically acceptable synthetic results[71]. Further studies could include datasets from multiple centers and adopted a leave-one-center-out training/test strategy in order to validate the consistency and robustness of the network.

Before being deployed into clinical workflow, there are still a few challenges to be addressed. To account for the potentially unpredictable synthetic images that can be resulted from noncompliance with imaging protocols as training data, or unexpected anatomic structures, additional quality assurance (QA) steps would be essential in clinical practice. The QA procedure would aim to check the consistency of the performance of the model routinely or after upgrade by re-training the network with more patient datasets, as well as to check the synthetic image quality of a patient-specific case.

References

1. Johnstone E, Wyatt JJ, Henry AM, et al. Systematic review of synthetic computed tomography generation methodologies for use in magnetic resonance imaging-only radiation therapy [published online ahead of print 2017/12/20]. *International Journal of Radiation Oncology, Biology, Physics.* 2018;100(1):199–217.
2. Wang T, Lei Y, Manohar N, et al. Dosimetric study on learning-based cone-beam CT correction in adaptive radiation therapy. *Medical Dosimetry.* 2019;44(4):e71–e79.
3. Wang T, Manohar N, Lei Y, et al. MRI-based treatment planning for brain stereotactic radiosurgery: Dosimetric validation of a learning-based pseudo-CT generation method. *Medical Dosimetry.* 2019;44(3):199–204.

4. Yang X, Wang T, Lei Y, et al. MRI-based attenuation correction for brain PET/MRI based on anatomic signature and machine learning. *Physics in Medicine & Biology*. 2019;64(2):025001.

5. Yang X, Lei Y, Shu H-K, et al. Pseudo CT estimation from MRI using patch-based random forest. *Paper presented at*: Medical Imaging 2017: Image Processing 2017.

6. Khoo VS, Joon DL. New developments in MRI for target volume delineation in radiotherapy [published online ahead of print 2006/09/19]. *The British Journal of Radiology*. 2006;79(Spec No 1):S2–15.

7. Nyholm T, Nyberg M, Karlsson MG, Karlsson M. Systematisation of spatial uncertainties for comparison between a MR and a CT-based radiotherapy workflow for prostate treatments. *Radiation Oncology*. 2009;4(1):54.

8. Ulin K, Urie MM, Cherlow JM. Results of a multi-institutional benchmark test for cranial CT/MR image registration. *International Journal of Radiation Oncology, Biology, Physics*. 2010;77(5):1584–1589.

9. van der Heide UA, Houweling AC, Groenendaal G, Beets-Tan RG, Lambin P. Functional MRI for radiotherapy dose painting. *Magnetic Resonance Imaging*. 2012;30(9):1216–1223.

10. Devic S. MRI simulation for radiotherapy treatment planning. *Medical Physics*. 2012;39(11):6701–6711.

11. Lagendijk JJW, Raaymakers BW, Raaijmakers AJE, et al. MRI/linac integration. *Radiotherapy and Oncology*. 2008;86(1):25–29.

12. Fallone BG, Murray B, Rathee S, et al. First MR images obtained during megavoltage photon irradiation from a prototype integrated linac-MR system. *Medical Physics*. 2009;36(6Part1):2084–2088.

13. Kinahan PE, Townsend DW, Beyer T, Sashin D. Attenuation correction for a combined 3D PET/CT scanner [published online ahead of print 1998/11/04]. *Medical Physics*. 1998;25(10):2046–2053.

14. Burger C, Goerres G, Schoenes S, Buck A, Lonn AH, Von Schulthess GK. PET attenuation coefficients from CT images: Experimental evaluation of the transformation of CT into PET 511-keV attenuation coefficients [published online ahead of print 2002/07/12]. *European Journal of Nuclear Medicine and Molecular Imaging*. 2002;29(7):922–927.

15. Lee YK, Bollet M, Charles-Edwards G, et al. Radiotherapy treatment planning of prostate cancer using magnetic resonance imaging alone. *Radiotherapy and Oncology*. 2003;66(2):203–216.

16. Jonsson JH, Karlsson MG, Karlsson M, Nyholm T. Treatment planning using MRI data: An analysis of the dose calculation accuracy for different treatment regions. *Radiation Oncology*. 2010;5(1):62.

17. Lambert J, Greer PB, Menk F, et al. MRI-guided prostate radiation therapy planning: Investigation of dosimetric accuracy of MRI-based dose planning. *Radiotherapy and Oncology*. 2011;98(3):330–334.

18. Kristensen BH, Laursen FJ, Løgager V, Geertsen PF, Krarup-Hansen A. Dosimetric and geometric evaluation of an open low-field magnetic resonance simulator for radiotherapy treatment planning of brain tumours. *Radiotherapy and Oncology*. 2008;87(1):100–109.

19. Johansson A, Karlsson M, Nyholm T. CT substitute derived from MRI sequences with ultrashort echo time. *Medical Physics*. 2011;38(5):2708–2714.

20. Hsu S-H, Cao Y, Huang K, Feng M, Balter JM. Investigation of a method for generating synthetic CT models from MRI scans of the head and neck for radiation therapy. *Physics in Medicine & Biology*. 2013;58(23):8419.

21. Dowling JA, Lambert J, Parker J, et al. An atlas-based electron density mapping method for magnetic resonance imaging (MRI)-alone treatment planning and adaptive MRI-based prostate radiation therapy. *International Journal of Radiation Oncology, Biology, Physics*. 2012;83(1):e5–e11.

22. Uh J, Merchant TE, Li Y, Li X, Hua C. MRI-based treatment planning with pseudo CT generated through atlas registration [published online ahead of print 2014/05/03]. *Medical Physics*. 2014;41(5):051711.

23. Sjölund J, Forsberg D, Andersson M, Knutsson H. Generating patient specific pseudo-CT of the head from MR using atlas-based regression. *Physics in Medicine & Biology*. 2015;60(2):825.

24. Fu J, Yang Y, Singhrao K, et al. Deep learning approaches using 2D and 3D convolutional neural networks for generating male pelvic synthetic computed tomography from magnetic resonance imaging [published online ahead of print 2019/06/21]. *Medical Physics*. 2019;46(9):3788–3798.

25. Dinkla AM, Florkow MC, Maspero M, et al. Dosimetric evaluation of synthetic CT for head and neck radiotherapy generated by a patch-based three-dimensional convolutional neural network [published online ahead of print 2019/06/18]. *Medical Physics*. 2019;46(9):4095–4104.

26. Emami H, Dong M, Nejad-Davarani SP, Glide-Hurst CK. Generating synthetic CTs from magnetic resonance images using generative adversarial networks [published online ahead of print 2018/06/15]. *Medical Physics*. 2018;45(8):3627–3636.

27. Chen S, Qin A, Zhou D, Yan D. Technical Note: U-net-generated synthetic CT images for magnetic resonance imaging-only prostate intensity-modulated radiation therapy treatment planning [published online ahead of print 2018/10/21]. *Medical Physics*. 2018;45(12):5659–5665.

28. Largent A, Barateau A, Nunes JC, et al. Comparison of deep learning-based and patch-based methods for pseudo-CT generation in MRI-based prostate dose planning [published online ahead of print 2019/09/11]. *International Journal of Radiation Oncology, Biology, Physics*. 2019;105(5):1137–1150.

29. Gupta D, Kim M, Vineberg KA, Balter JM. Generation of synthetic CT images from MRI for treatment planning and patient positioning using a 3-channel U-net trained on sagittal images [published online ahead of print 2019/10/15]. *Frontiers in Oncology*. 2019;9:964.

30. Dinkla AM, Wolterink JM, Maspero M, et al. MR-only brain radiation therapy: Dosimetric evaluation of synthetic CTs generated by a dilated convolutional neural network. *International Journal of Radiation Oncology, Biology, Physics*. 2018;102(4):801–812.

31. Arabi H, Dowling JA, Burgos N, et al. Comparative study of algorithms for synthetic CT generation from MRI: Consequences for MRI-guided radiation planning in the pelvic region [published online ahead of print 2018/09/15]. *Medical Physics*. 2018;45(11):5218–5233.

32. Han X. MR-based synthetic CT generation using a deep convolutional neural network method [published online ahead of print 2017/02/14]. *Medical Physics*. 2017;44(4):1408–1419.

33. Nie D, Trullo R, Lian J, et al. Medical image synthesis with deep convolutional adversarial networks [published online ahead of print 2018/07/12]. *IEEE Transactions on Bio-Medical Engineering*. 2018;65(12):2720–2730.

34. Freitag MT, Fenchel M, Baumer P, et al. Improved clinical workflow for simultaneous whole-body PET/MRI using high-resolution CAIPIRINHA-accelerated MR-based attenuation correction [published online ahead of print 2017/11/07]. *European Journal of Radiology*. 2017;96:12–20.

35. Izquierdo-Garcia D, Hansen AE, Förster S, et al. An SPM8-based approach for attenuation correction combining segmentation and nonrigid template formation: Application to simultaneous PET/MR brain imaging. *Journal of Nuclear Medicine*. 2014;55(11):1825–1830.

36. Liu F, Jang H, Kijowski R, Bradshaw T, McMillan AB. Deep learning MR imaging-based attenuation correction for PET/MR imaging [published online ahead of print 2017/09/20]. *Radiology*. 2018;286(2):676–684.

37. Blanc-Durand P, Khalife M, Sgard B, et al. Attenuation correction using 3D deep convolutional neural network for brain 18F-FDG PET/MR: Comparison with Atlas, ZTE and CT based attenuation correction. *PLoS One*. 2019;14(10):e0223141.

38. Ladefoged CN, Marner L, Hindsholm A, Law I, Hojgaard L, Andersen FL. Deep learning based attenuation correction of PET/MRI in pediatric brain tumor patients: Evaluation in a clinical setting [published online ahead of print 2019/01/23]. *Frontiers in Neuroscience*. 2018;12:1005.

39. Qi M, Li Y, Wu A, et al. Multi-sequence MR image-based synthetic CT generation using a generative adversarial network for head and neck MRI-only radiotherapy [published online ahead of print 2020/02/07]. *Medical Physics*. 2020;47:1880–1894.

40. Florkow MC, Zijlstra F, Willemsen K, et al. Deep learning-based MR-to-CT synthesis: The influence of varying gradient echo-based MR images as input channels [published online ahead of print 2019/10/09]. *Magnetic Resonance in Medicine*. 2020;83(4):1429–1441.

41. Tie X, Lam SK, Zhang Y, Lee KH, Au KH, Cai J. Pseudo-CT generation from multi-parametric MRI using a novel multi-channel multi-path conditional generative adversarial network for nasopharyngeal carcinoma patients [published online ahead of print 2020/02/06]. *Medical Physics*. 2020;47:1750–1762.

42. Leynes AP, Yang J, Wiesinger F, et al. Zero-echo-time and dixon deep pseudo-CT (ZeDD CT): Direct generation of pseudo-CT images for pelvic PET/MRI attenuation correction using deep convolutional neural networks with multiparametric MRI [published online ahead of print 2017/11/01]. *Journal of Nuclear Medicine: Official Publication, Society of Nuclear Medicine.* 2018;59(5):852–858.

43. Gong K, Yang J, Kim K, El Fakhri G, Seo Y, Li Q. Attenuation correction for brain PET imaging using deep neural network based on Dixon and ZTE MR images. *Physics in Medicine & Biology.* 2018;63(12):125011.

44. Kazemifar S, McGuire S, Timmerman R, et al. MRI-only brain radiotherapy: Assessing the dosimetric accuracy of synthetic CT images generated using a deep learning approach [published online ahead of print 2019/04/25]. *Radiotherapy and Oncology: Journal of the European Society for Therapeutic Radiology and Oncology.* 2019;136:56–63.

45. Dong X, Lei Y, Tian S, et al. Synthetic MRI-aided multi-organ segmentation on male pelvic CT using cycle consistent deep attention network [published online ahead of print 2019/10/22]. *Radiotherapy and Oncology: Journal of the European Society for Therapeutic Radiology and Oncology.* 2019;141:192–199.

46. Harms J, Lei Y, Wang T, et al. Paired cycle-GAN-based image correction for quantitative cone-beam computed tomography. *Medical Physics.* 2019;46(9):3998–4009.

47. Lei Y, Dong X, Wang T, et al. Whole-body PET estimation from low count statistics using cycle-consistent generative adversarial networks [published online ahead of print 2019/09/29]. *Physics in Medicine and Biology.* 2019;64(21):215017.

48. Lei Y, Harms J, Wang T, et al. MRI-only based synthetic CT generation using dense cycle consistent generative adversarial networks. *Medical Physics.* 2019;46(8):3565–3581.

49. Liu Y, Lei Y, Wang Y, et al. Evaluation of a deep learning-based pelvic synthetic CT generation technique for MRI-based prostate proton treatment planning [published online ahead of print 2019/09/06]. *Physics in Medicine and Biology.* 2019;64(20):205022.

50. Liu Y, Lei Y, Wang Y, et al. MRI-based treatment planning for proton radiotherapy: Dosimetric validation of a deep learning-based liver synthetic CT generation method [published online ahead of print 2019/05/31]. *Physics in Medicine and Biology.* 2019;64(14):145015.

51. Shafai-Erfani G, Lei Y, Liu Y, et al. MRI-based proton treatment planning for base of skull tumors [published online ahead of print 2020/01/31]. *International Journal of Particle Therapy.* 2019;6(2):12–25.

52. Li B, Lee HC, Duan X, et al. Comprehensive analysis of proton range uncertainties related to stopping-power-ratio estimation using dual-energy CT imaging. *Physics in Medicine & Biology.* 2017;62(17):7056–7074.

53. Hofmann M, Steinke F, Scheel V, et al. MRI-based attenuation correction for PET/MRI: A novel approach combining pattern recognition and atlas registration [published online ahead of print 2008/10/18]. *Journal of Nuclear Medicine: Official Publication, Society of Nuclear Medicine.* 2008;49(11):1875–1883.

54. Yang X, Fei B. Multiscale segmentation of the skull in MR images for MRI-based attenuation correction of combined MR/PET. *Journal of the American Medical Informatics Association.* 2013;20(6):1037–1045.

55. Arabi H, Zeng G, Zheng G, Zaidi H. Novel adversarial semantic structure deep learning for MRI-guided attenuation correction in brain PET/MRI [published online ahead of print 2019/07/03]. *European Journal of Nuclear Medicine and Molecular Imaging.* 2019;46(13):2746–2759.

56. Torrado-Carvajal A, Vera-Olmos J, Izquierdo-Garcia D, et al. Dixon-VIBE deep learning (DIVIDE) pseudo-CT synthesis for pelvis PET/MR attenuation correction [published online ahead of print 2018/09/01]. *Journal of Nuclear Medicine: Official Publication, Society of Nuclear Medicine.* 2019;60(3):429–435.

57. Jang H, Liu F, Zhao G, Bradshaw T, McMillan AB. Technical note: Deep learning based MRAC using rapid ultrashort echo time imaging [published online ahead of print 2018/05/16]. *Medical Physics.* 2018;45:3697–3704.

58. McKenzie EM, Santhanam A, Ruan D, O'Connor D, Cao M, Sheng K. Multimodality image registration in the head-and-neck using a deep learning-derived synthetic CT as a bridge [published online ahead of print 2019/12/20]. *Medical Physics.* 2019;47:1094–1104.

59. Wang T, Lei Y, Fu Y, et al. A review on medical imaging synthesis using deep learning and its clinical applications [published online ahead of print 2020/12/12]. *Journal of Applied Clinical Medical Physics.* 2021;22:11–36.

60. Xiang L, Wang Q, Nie D, et al. Deep embedding convolutional neural network for synthesizing CT image from T1-Weighted MR image. *Medical Image Analysis.* 2018;47:31–44.

61. Maspero M, Savenije MHF, Dinkla AM, et al. Dose evaluation of fast synthetic-CT generation using a generative adversarial network for general pelvis MR-only radiotherapy [published online ahead of print 2018/08/16]. *Physics in Medicine and Biology.* 2018;63(18):185001.

62. Lei Y, Harms J, Wang T, et al. MRI-only based synthetic CT generation using dense cycle consistent generative adversarial networks [published online ahead of print 2019/05/22]. *Medical Physics.* 2019;46(8):3565–3581.

63. Liu F, Yadav P, Baschnagel AM, McMillan AB. MR-based treatment planning in radiation therapy using a deep learning approach [published online ahead of print 2019/03/13]. *Journal of Applied Clinical Medical Physics.* 2019;20(3):105–114.

64. Liu Y, Lei Y, Wang T, et al. MRI-based treatment planning for liver stereotactic body radiotherapy: Validation of a deep learning-based synthetic CT generation method [published online ahead of print 2019/06/14]. *The British Journal of Radiology.* 2019;92(1100):20190067.

65. Neppl S, Landry G, Kurz C, et al. Evaluation of proton and photon dose distributions recalculated on 2D and 3D Unet-generated pseudoCTs from T1-weighted MR head scans [published online ahead of print 2019/07/05]. *Acta Oncologica (Stockholm, Sweden).* 2019;58(10):1429–1434.

66. Olberg S, Zhang H, Kennedy WR, et al. Synthetic CT reconstruction using a deep spatial pyramid convolutional framework for MR-only breast radiotherapy [published online ahead of print 2019/07/17]. *Medical Physics.* 2019;46(9):4135–4147.

67. Wang Y, Liu C, Zhang X, Deng W. Synthetic CT generation based on T2 weighted MRI of nasopharyngeal carcinoma (NPC) using a deep convolutional neural network (DCNN) [published online ahead of print 2019/12/19]. *Frontiers in Oncology.* 2019;9:1333.

68. Koike Y, Akino Y, Sumida I, et al. Feasibility of synthetic computed tomography generated with an adversarial network for multi-sequence magnetic resonance-based brain radiotherapy [published online ahead of print 2019/12/12]. *Journal of Radiation Research.* 2020;61(1):92–103.

69. Spuhler KD, Gardus J, 3rd, Gao Y, DeLorenzo C, Parsey R, Huang C. Synthesis of patient-specific transmission data for PET attenuation correction for PET/MRI neuroimaging using a convolutional neural network [published online ahead of print 2018/09/01]. *Journal of Nuclear Medicine: Official Publication, Society of Nuclear Medicine.* 2019;60(4):555–560.

70. Schilling KG, Blaber J, Huo Y, et al. Synthesized b0 for diffusion distortion correction (Synb0-DisCo) [published online ahead of print 2019/05/11]. *Magnetic Resonance Imaging.* 2019;64:62–70.

71. Brou Boni KND, Klein J, Vanquin L, et al. MR to CT synthesis with multicenter data in the pelvic area using a conditional generative adversarial network. *Physics in Medicine & Biology.* 2020;65(7):075002.

4

CBCT/CT-Based Image Synthesis

Hao Zhang

Memorial Sloan Kettering Cancer Center, New York, NY, USA

CONTENTS

4.1 Synthetic CT from CBCT Images

Cone-beam computed tomography (CBCT) plays an important role in image-guided radiation therapy. However, the image quality of CBCT can be degraded with noise and various artifacts including the streaks, ring, beam hardening, scattering, metal, and motion artifacts [1, 2]. Numerous correction approaches have been proposed to suppress these artifacts and improve CBCT image quality, including hardware-based and software-based methods. Nowadays, clinical CBCT images are substantially improved because of these efforts. Furthermore, iterative CBCT becomes available in the clinic [3, 4], which can reduce noise and artifacts and yield higher image quality than the standard CBCT. But generally, CBCT images are still not comparable to planning CT (pCT) images, especially on Hounsfield units (HU) accuracy and soft tissue differentiation. HU accuracy influences relative electron density conversion and dose calculation accuracy while soft tissue contrast affects organs-at-risk and target contouring. Therefore, although CBCT-based adaptive radiation therapy (ART) is desired in the clinic, the inferior image quality impedes its wide application to treatment planning. To improve the dose calculation accuracy, one conventional approach is to transform the pCT to onboard CBCT using rigid or deformable registration, and then calculate the dose using the deformed planning CT (dpCT). However, the registration between pCT and CBCT is rarely perfect and may introduce uncertainty to structure segmentation and dose calculation. Inspired by the success of deep learning, researchers explored synthesizing CT images from onboard CBCT images. The synthetic CT (sCT) has the same high quality as pCT while preserving anatomical structures as CBCT, which overcomes the limitations of current CBCT and enables online ART.

Synthesizing CT (target image B) from CBCT (source image A) can be formulated as a classical image-to-image translation problem. Deep learning approaches, especially

DOI: 10.1201/9781003243458-6

convolutional neural networks (CNNs), are state-of-the-art for such translation problems. Essentially, it is to train a mapping $M_{A \rightarrow B}$ that generates images which are indistinguishable from target domain images. CNNs use large datasets to learn the optimal parameters of this nonlinear mapping iteratively. Many CNN-based architectures have been investigated for image synthesis from CBCT to CT, while U-net and generative adversarial networks (GAN) are the most commonly used. The configuration of training data can be 2D (single slice or 2D patch), 2D+, multi-2D, 2.5D, and 3D (whole volume or 3D patches) [5]. While 3D training may capture more spatial relationships, the graphics processing unit (GPU) memory and computational load can be limitations in practice.

Generally, the training of CNNs for sCT generation from CBCT can be classified as supervised and unsupervised. Supervised training requires paired (exactly aligned) CBCT and pCT images, which are rarely available in the clinic because they are not acquired simultaneously. Thus, the pCT images are usually deformed to register with the CBCT. Preprocessing steps, such as body mask and image resampling, are usually performed to exclude non-anatomical structures and facilitate the registration. For instance, Kida et al. [6], Li et al. [7], Yuan et al. [8], and Chen et al. [9] used the dpCT and CBCT pairs for supervised training of U-net-based frameworks for sCT generation and yielded promising results. Meanwhile, there were also studies demonstrating that the U-net-generated sCT may be blurred or lose detailed information at the tissue boundaries. Also, conditional GAN (cGAN) and pix2pix model (one type of cGAN) were also exploited to generate sCT images from CBCT in a supervised way [10, 11]. It was demonstrated they outperform the U-net frameworks in terms of HU and dose calculation accuracy. But the registration accuracy between CBCT and pCT is not always guaranteed due to the low image quality of CBCT and potential anatomical changes. It was reported that supervised learning using inexactly matched CBCT and pCT may not preserve anatomical structures and generate unreliable sCT images, because of the pixel-to-pixel correspondence in the training loss functions. Unsupervised training allows the use of unpaired (unaligned) CBCT and pCT images, and therefore, is more widely explored for sCT generation. Especially, the cycle-GAN uses a cycle consistency loss and aims to solve the image-to-image translation problem without requiring paired images. Many studies have applied cycleGAN for CBCT-to-CT synthesis in an unsupervised way, which is illustrated in Figure 4.1. Liang et al. [12] demonstrated the efficacy of using cycleGAN with unpaired CBCT and pCT images for sCT generation, and showed it outperforms two other unsupervised approaches of deep

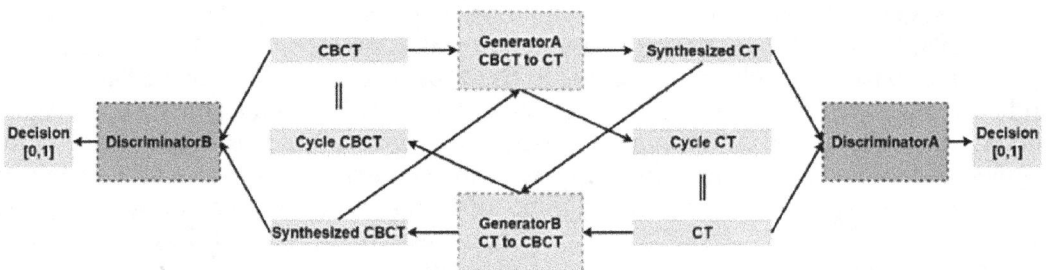

FIGURE 4.1
CycleGAN architecture is used to convert CBCT images to sCT images. (Figure reprinted from Liang et al. [12] with permission.)

convolutional GANs and progressive growing GANs. Similarly, Kurz et al. [13] and Kida et al. [14] demonstrated the effectiveness of using cycleGAN for unsupervised training. Harms et al. [15] integrated a residual block into the cycleGAN framework and employed a compound loss function to improve the network performance. Also, the authors suggested performing a rigid registration between CBCT and pCT images for better sCT generation, although the cycleGAN was initially designed for unpaired datasets. Liu et al. [16] added an attention gate to the generator network of cycleGAN, called deep-attention cycleGAN, to learn the structural variations. Deng et al. [17] integrated the respath concept into the cycleGAN, called respath-cycleGAN, to further improve performance. Tien et al. [18] combined the cycleGAN and deblur GAN to improve sCT generation from chest CBCT images. Besides cycleGAN, Zhang et al. [19] investigated an unsupervised 2.5D pix-2pix GAN model with feature mapping, Gao et al. [20] explored an unsupervised attention-guided GAN, Liu et al. [21] developed an unsupervised GAN with disentangled representation, while Chen et al. [22] proposed an unsupervised style-transfer-based approach for sCT generation. These studies claimed their models outperformed the cycle-GAN in terms of sCT image quality. Generally, unsupervised learning has been proved to be more reliable than supervised learning in improving sCT image quality and preserving the anatomical structures.

When comparing different deep learning models for sCT generation, it should be noted that their performances are dependent on multiple factors such as network architecture and depth, loss functions, training schedules, image quality, and configuration of training data. Therefore, we may also see contradictory conclusions on the relative performance of different models [10, 11, 23].

So far, deep-learning-based sCT generation from CBCT has been successfully applied to anatomical sites including head and neck (H&N), pelvis, breast, thorax, and abdomen. H&N images typically contain less anatomical motions and therefore the generated sCT are more likely to resemble the pCT images, as shown in Figure 4.2. In contrast, the abdomen region is more complex due to organ motions and variability in bowel filling. Therefore, the sCT images may still have residual artifacts which cannot be fully mitigated, as shown in Figure 4.3. Overall, the sCT image quality is substantially improved compared to the CBCT images and is slightly inferior to the dpCT images. Most studies compared mean absolute error (MAE) in the HU and dosimetric difference of sCT and CBCT, but few of them investigated the impact on organs-at-risk and target segmentation with the sCT images [24]. Especially, whether some unexpected fake structures and residual artifacts in sCT would significantly affect segmentation accuracy. Thus, more research is needed before we could implement sCT based ART in clinic.

For most studies, the pCT and CBCT images were acquired with fixed scanner settings and contain similar anatomical sites of the body. Several studies have also evaluated the generalizability of the synthetic models [9, 25]. For instance, it was found that site-specific models (e.g, H&N) may generate reasonable but suboptimal results if applied to a different anatomical site (e.g., pelvis). This makes sense because the pelvis CBCT images typically have more severe scattering artifacts than H&N CBCT images. A generic network trained from all the anatomical sites together may be preferred in such scenarios [26]. Also, patient images from different institutions or by different vendors may have different image contrast, noise level, and HU accuracy. Liang et al. [25] explored three practical solutions based on transfer learning to solve this generalization problem. They found that all three approaches (target, combined, and adapted) can achieve good performance when there are sufficient data in the target dataset.

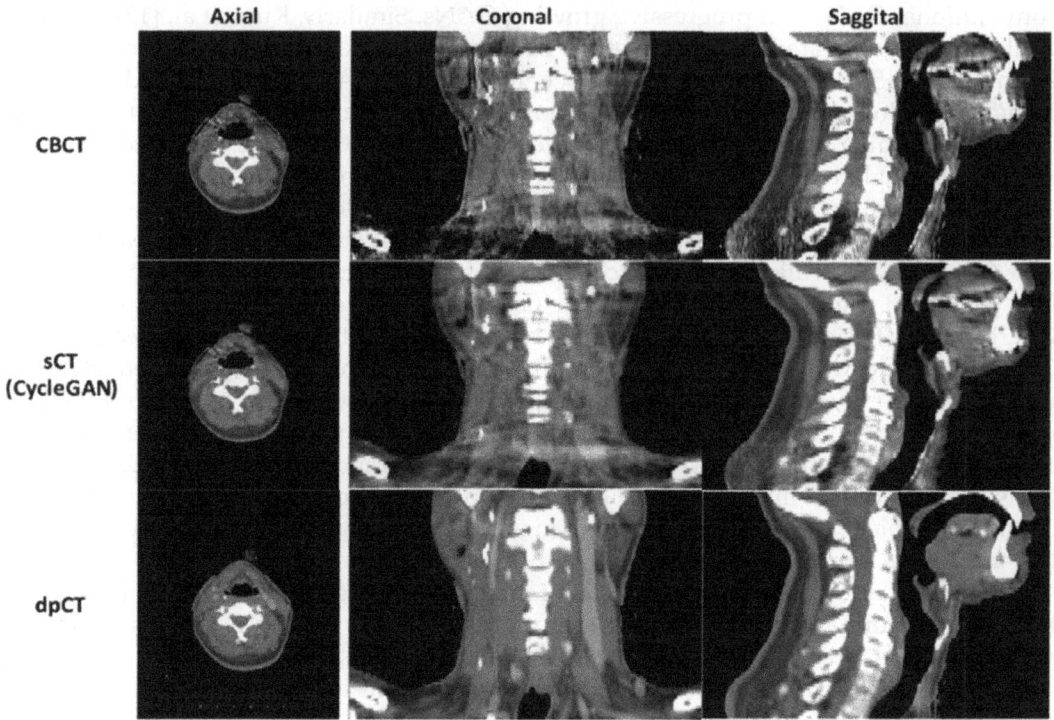

FIGURE 4.2
CBCT, sCT (CycleGAN), and dpCT images from axial, coronal and sagittal plane from one H&N test patient. Display window is (−300, 500) HU. (Figure reprinted from Liang et al. [12] with permission.)

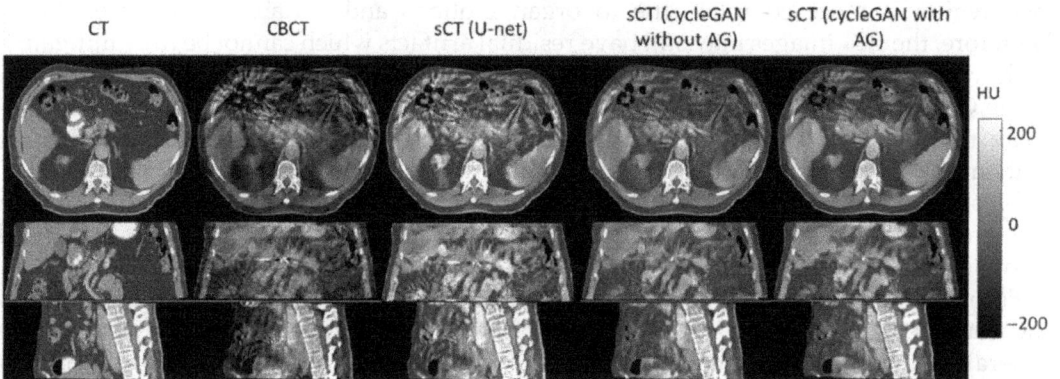

FIGURE 4.3
Summary of CBCT to sCT results in a representative pancreatic cancer patient. (Figure reprinted from Liu et al. [16] with permission.)

4.2 Synthetic MRI from CT/CBCT Images

Magnetic resonance imaging (MRI) provides superior soft tissue contrast than CT and CBCT, but is not prescribed to every patient due to availability and insurance issues. MRI-guided linear accelerators, which enable online ART, are still unavailable in most cancer centers due to the high cost and several technical challenges. To overcome this, researchers have also investigated to synthesize MRI images from the CT or CBCT images using deep learning approaches.

For instance, Dong et al. [27] employed a 3D cycleGAN architecture (using 3D patches) to learn the mapping from CT to MRI. The training CT and MRI (T2-weighted) images were deformably registered before being fed into the synthesis neural network. Also, since the CT and MRI images have drastically different physical properties, training a CT-to-MRI mapping with conventional deep learning models is challenging. To address this issue, a 3D cycleGAN architecture was employed because it contains both targeted transformation (CT-to-MRI) and inverse transformation (MRI-to-CT) to improve the image quality of synthetic MRI (sMRI). The sMRI images can be used to enhance the organ boundaries that might not be visible in the CT images. Because of the bony structure information provided by CT and superior soft tissue information provided by sMR, they demonstrated that sMRI-aided organs-at-risk automatic segmentation on male pelvis CT images can achieve improved accuracy than other deep-learning-based CT segmentation methods. Similarly, Liu et al. [28] used 3D cycleGAN to synthesize MRI (T1-weighted) images from CT. The sMRI images were then employed for 19 organs-at-risk segmentation on CT images of H&N patients. Figure 4.4 illustrates some representative slices of planning CT and sMRI images for one H&N patient.

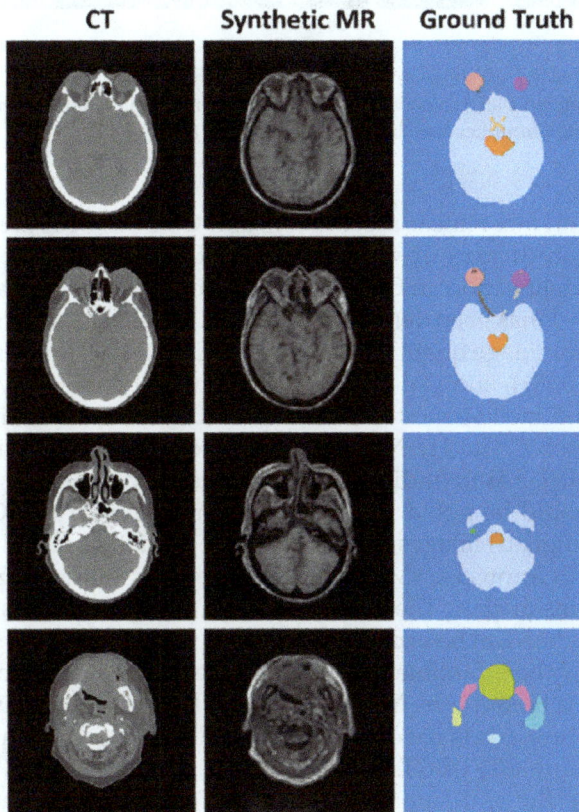

FIGURE 4.4
CT and sMRI images in a representative H&N cancer patient. (Figure reprinted from Liu et al. [28] with permission.)

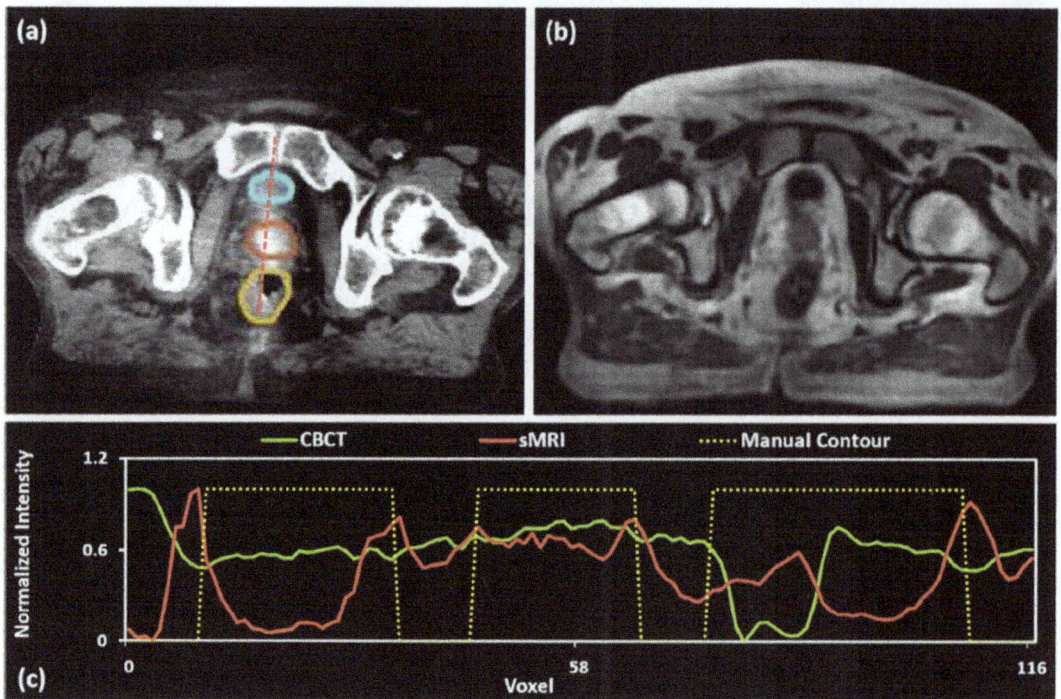

FIGURE 4.5
Visual results of generated sMRI. (a) shows the original CBCT image at the axial level. (b) shows the generated sMRI. (c) shows the normalized plot profile of CBCT, sMRI, and manual contour of the red dashed line in (a), respectively. (Figure reprinted from Lei et al. [31] with permission.)

CBCT is equipped on most linear accelerators and CBCT images are widely available in image-guided radiation therapy. If we can achieve fast and accurate organs-at-risk segmentation and target delineation on CBCT directly, online ART will be available for a larger population of patients. It has been demonstrated that it is feasible to synthesize MRI images from CBCT with decent image quality. Using paired CBCT-MRI images and cycleGAN, Fu et al. [29], Lei et al. [30, 31], and Dai et al. [32] demonstrated that they can generate sMRI images from CBCT and facilitate automated and accurate organs-at-risk segmentation on CBCT images of the pelvis and H&N patients. Figure 4.5 shows one image of CBCT and sMRI and how the sMRI aids in finding the boundaries of different organs.

Nevertheless, it should be noted that the sMRI images from CT/CBCT shown above used paired (exactly aligned) images and site-specific models (e.g., pelvis). Inaccurate registration could result in imprecise sMRI generation, which in turn degrade the segmentation accuracy of organs-at-risk. Therefore, if accurate registration is not achievable for some image datasets, they can be excluded to ensure the good performance of the trained model [29]. Also, it may be infeasible to apply a model trained from one anatomical site to another anatomical site because the MRI images of different organs have considerably different features. More research is needed to validate the networks' generality in the future for sMRI generation from CT/CBCT.

Furthermore, while the sMRI images from CBCT can improve the contrast of organ boundaries, they have limitations in enhancing the intra-organ soft tissue contrast. For instance, a liver tumor and normal liver tissue have similar intensity on the CBCT images,

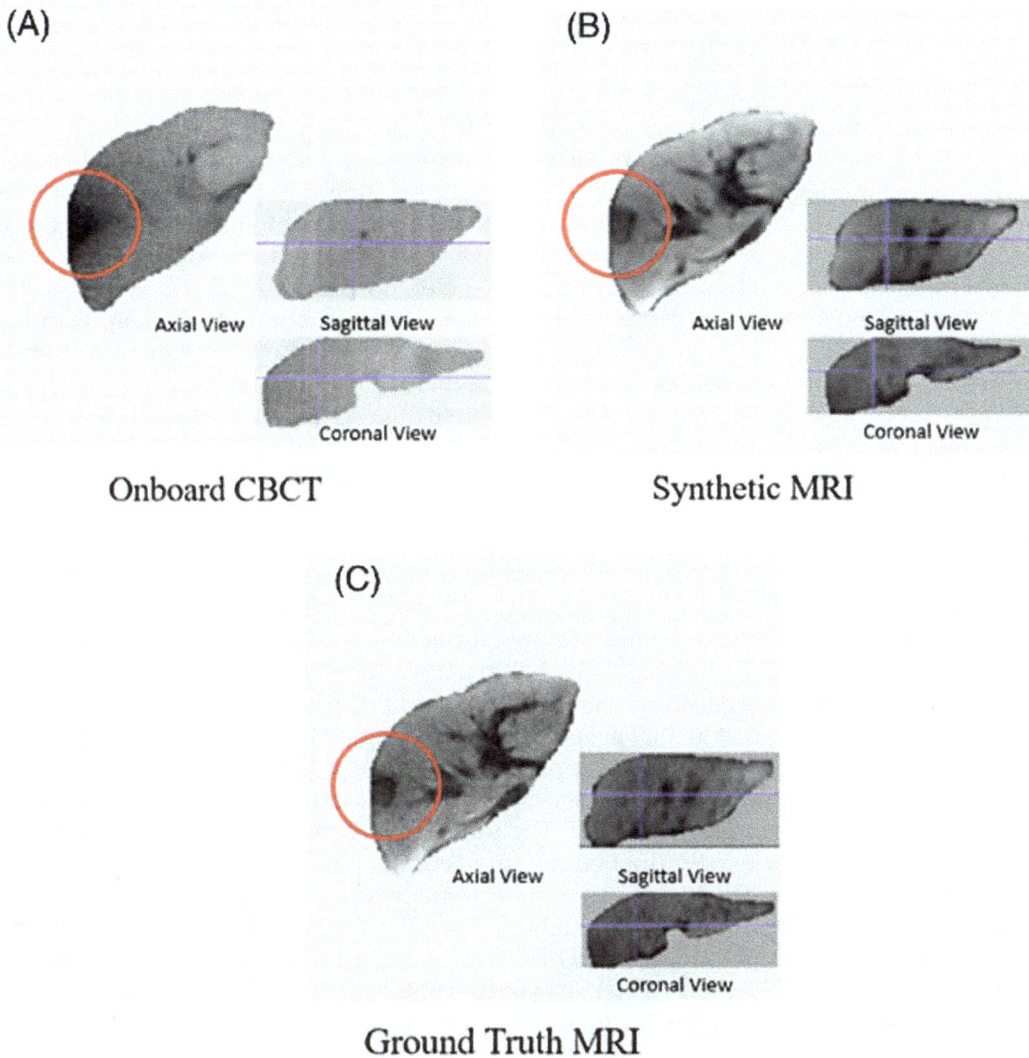

FIGURE 4.6
(a) onboard CBCT. (b) sMRI from onboard CBCT using patient-specific model. (c) ground truth MRI. The red circle highlights the region that contains the liver tumor. (Figure reprinted from Zhang et al. [33] with permission.)

so the sMRI images from the CBCT cannot artificially create contrast information for a liver tumor. To overcome this limitation, Zhang et al. [33] proposed to use patient-specific CBCT-MRI pairs to train a U-net model for sMRI generation from CBCT. Specifically, the CBCT-MRI pairs were registered and then augmented using translations and rotations to generate adequate patient-specific training data. They demonstrated the feasibility of using the patient-specific prior information to generate sMRI in only the liver region for accurate liver tumor localization. Figure 4.6 illustrates one example of the onboard CBCT, sMRI, and ground truth MRI for one liver patient. Apparently, the sMRI enhanced the contrast of both liver tissue and tumor while reducing scattering artifacts in CBCT images. This study also shows the potential to democratize MRI guidance on linear accelerators equipped with only CBCT.

4.3 Synthetic DECT from Single-Energy CT

Dual-energy CT (DECT) acquires two spectrally distinct datasets and therefore enables materials differentiation and quantification. Besides radiology, DECT is also very useful in radiation oncology such as: (1) tumor identification, characterization, and delineation; (2) functional sparing of organs-at-risk (e.g., functional lung); (3) improved dose calculation especially for proton planning due to accurate estimation of stopping-power ratios; (4) treatment response assessment. However, DECT scanners are also more expensive than standard CT simulators and thus are not widely available in clinics. A few studies have investigated to synthesize DECT images from single-energy CT via deep learning. In DECT, the low-energy image and high-energy image share the same anatomical structures. Liao et al. [34] proposed a cascade deep CNN to learn the nonlinear mapping from the measured low-energy CT images to the desired high-energy CT images. Zhao et al. [35] found their DECT images used for model training are noisy and it is beneficial to use denoising CNN to first improve the DECT images. With noise significantly reduced DECT images, a U-net model was then trained to synthesize high-energy CT images from measured low-energy CT data, as shown in Figure 4.7. They demonstrated that the synthetic DECT could provide clinically valuable high-quality virtual non-contrast (VNC) and iodine maps. Similarly, Liu et al. [36] used a U-net model to perform translation between low- and high-energy CT images and found that the low-to-high CT synthesis was better than the high-to-low CT synthesis. It was shown that the deep learning synthesized DECT images have improved signal-to-noise ration and can produce low-noise virtual monoenergetic images (VMI). Lyu et al. [37] proposed a fully sampled low-energy and single-view high-energy (FLESH) DECT imaging approach. Specifically, the denoised DECT images and the single-view dual-energy projections are used to train the projection domain CNN and material decomposition CNN. The trained networks use the input low-energy CT images and the single-view dual-energy projections to synthesize the corresponding high-energy CT images. The high-energy projection introduces a penalty to constrain the projection generated by the synthetic high-energy CT images to be consistent with the measurement. It was shown that the FLESH-DECT is superior to the direct low-energy to high-energy CT synthesis. Charyyev et al. [38] used a residual attention GAN to synthesize DECT images from single-energy CT images for proton radiotherapy. Residual blocks with

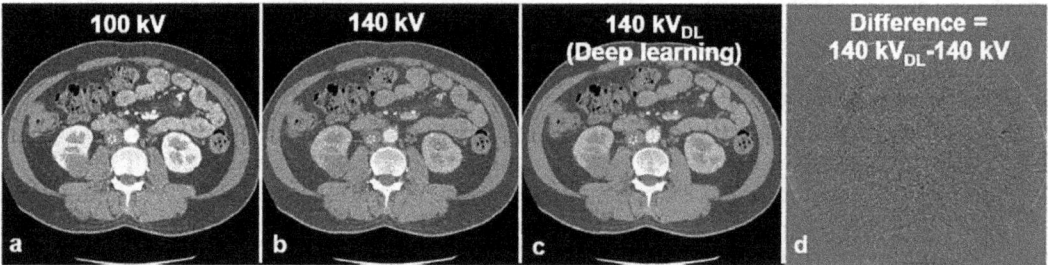

FIGURE 4.7
(a) Original low-energy CT image. (b) Original high-energy CT image. (c) Synthetic high-energy CT images from original low-energy CT images via deep learning. (d) Difference image between the synthetic and original 140 kV images. All images are displayed in (C = 0 HU and W = 500 HU). (Figure reprinted from Zhao et al. [35] with permission.)

attention gates were used to force the model to focus on the difference between DECT and single-energy CT images. They demonstrated that the synthesized DECT images could be used for proton stopping-power ratios map generation.

4.4 Discussion

This chapter briefly reviews CBCT and CT-based image synthesis using deep learning techniques. It highlights some of the studies on sCT, sMRI, and sDECT from CBCT/CT images. With new designs of neural network architectures and more training data, we are expecting to see improved results and clinical translation of these approaches in the future. In addition, there is also other research on CBCT/CT-based synthesis, such as synthetic megavoltage computed tomography (MVCT) from CT [39] and synthetic relative stopping-power (RSP) map from CBCT [40]. We could see more similar clinical applications as this field is evolving quickly and many publications are coming out.

References

[1] R. Schulze et al., "Artefacts in CBCT: A review," *Dentomaxillofacial Radiol.*, vol. 40, no. 5, pp. 265–273, 2011.

[2] A. Nagarajappa, N. Dwivedi, and R. Tiwari, "Artifacts: The downturn of CBCT image," *J. Int. Soc. Prev. Community Dent.*, vol. 5, no. 6, p. 440, 2015.

[3] H. Washio et al., "Accuracy of dose calculation on iterative CBCT for head and neck radiotherapy," *Phys. Medica*, vol. 86, no. June, pp. 106–112, 2021.

[4] Y. Hu, M. Arnesen, and T. Aland, "Characterization of an advanced cone beam CT (CBCT) reconstruction algorithm used for dose calculation on Varian Halcyon linear accelerators," *Biomed. Phys. Eng. Express*, vol. 8, no. 2, p. 025023, 2022.

[5] M. F. Spadea, M. Maspero, P. Zaffino, and J. Seco, "Deep learning based synthetic-CT generation in radiotherapy and PET: A review," *Med. Phys.*, vol. 48, no. 11, pp. 6537–6566, 2021.

[6] S. Kida et al., "Cone beam computed tomography image quality improvement using a deep convolutional neural network," *Cureus*, vol. 10, no. 4, p. e2548, 2018.

[7] Y. Li et al., "A preliminary study of using a deep convolution neural network to generate synthesized CT images based on CBCT for adaptive radiotherapy of nasopharyngeal carcinoma," *Phys. Med. Biol.*, vol. 64, no. 14, p. 145010, 2019.

[8] N. Yuan et al., "Convolutional neural network enhancement of fast-scan low- dose cone-beam CT images for head and neck radiotherapy," *Phys. Med. Biol.*, vol. 65, no. 3, p. 035003, 2021.

[9] L. Chen, X. Liang, C. Shen, S. Jiang, and J. Wang, "Synthetic CT generation from CBCT images via deep learning," *Med. Phys.*, vol. 47, no. 3, pp. 1115–1125, 2020.

[10] X. Xue et al., "Cone beam CT (CBCT) based synthetic CT generation using deep learning methods for dose calculation of nasopharyngeal carcinoma radiotherapy," *Technol. Cancer Res. Treat.*, vol. 20, pp. 1–10, 2021.

[11] Y. Zhang et al., "Generating synthesized computed tomography from CBCT using a conditional generative adversarial network for head and neck cancer patients," *Technol. Cancer Res. Treat.*, vol. 21, pp. 1–8, 2022.

[12] X. Liang et al., "Generating synthesized computed tomography (CT) from cone-beam computed tomography (CBCT) using CycleGAN for adaptive radiation therapy," *Phys. Med. Biol.*, vol. 64, no. 12, p. 125002, 2019.

[13] C. Kurz et al., "CBCT correction using a cycle-consistent generative adversarial network and unpaired training to enable photon and proton dose calculation," *Phys. Med. Biol.*, vol. 64, no. 22, p. 225004, 2019.

[14] S. Kida et al., "Visual enhancement of cone-beam CT by use of CycleGAN," *Med. Phys.*, vol. 47, no. 3, pp. 998–1010, 2020.

[15] J. Harms et al., "Paired cycle-GAN-based image correction for quantitative cone-beam computed tomography," *Med. Phys.*, vol. 46, no. 9, pp. 3998–4009, 2019.

[16] Y. Liu et al., "CBCT-based synthetic CT generation using deep-attention cycleGAN for pancreatic adaptive radiotherapy," *Med. Phys.*, vol. 47, no. 6, pp. 2472–2483, 2020.

[17] L. Deng, J. Hu, J. Wang, S. Huang, and X. Yang, "Synthetic CT generation based on CBCT using respath-cycleGAN," *Med. Phys.*, vol. 49, no. 8, pp. 5317–5329, 2022.

[18] H. J. Tien, H. C. Yang, P. W. Shueng, and J. C. Chen, "Cone-beam CT image quality improvement using Cycle-Deblur consistent adversarial networks (Cycle-Deblur GAN) for chest CT imaging in breast cancer patients," *Sci. Rep.*, vol. 11, no. 1, pp. 1–12, 2021.

[19] Y. Zhang et al., "Improving CBCT quality to CT level using deep learning with generative adversarial network," *Med. Phys.*, vol. 48, no. 6, pp. 2816–2826, 2021.

[20] L. Gao et al., "Generating synthetic CT from low-dose cone-beam CT by using generative adversarial networks for adaptive radiotherapy," *Radiat. Oncol.*, vol. 16, no. 1, pp. 1–16, 2021.

[21] J. Liu et al., "CBCT-based synthetic CT generation using generative adversarial networks with disentangled representation," *Quant. Imaging Med. Surg.*, vol. 11, no. 12, pp. 4820–4834, 2021.

[22] L. Chen, X. Liang, C. Shen, D. Nguyen, S. Jiang, and J. Wang, "Synthetic CT generation from CBCT images via unsupervised deep learning," *Phys. Med. Biol.*, vol. 66, no. 11, p. 115019, 2021.

[23] X. Wang et al., "Synthetic CT generation from cone-beam CT using deep-learning for breast adaptive radiotherapy," *J. Radiat. Res. Appl. Sci.*, vol. 15, no. 1, pp. 275–282, 2022.

[24] X. Liang, H. Morgan, D. Nguyen, and S. Jiang, "Deep learning–based CT-to-CBCT deformable image registration for autosegmentation in head and neck adaptive radiation therapy," *J. Artif. Intell. Med. Sci.*, vol. 2, no. 1–2, p. 62, 2021.

[25] X. Liang, D. Nguyen, and S. B. Jiang, "Generalizability issues with deep learning models in medicine and their potential solutions: illustrated with cone-beam computed tomography (CBCT) to computed tomography (CT) image conversion," *Mach. Learn. Sci. Technol.*, vol. 2, no. 1, p. 015007, 2021.

[26] M. Maspero et al., "CBCT-to-CT synthesis with a single neural network for head-and-neck, lung and breast radiotherapy," *Radiother. Oncol.*, vol. 152, p. S161, 2020.

[27] X. Dong et al., "Synthetic MRI-aided multi-organ segmentation on male pelvic CT using cycle consistent deep attention network," *Radiother. Oncol.*, vol. 141, pp. 192–199, 2019.

[28] Y. Liu et al., "Head and neck multi-organ auto-segmentation on CT images aided by synthetic MRI," *Med. Phys.*, vol. 47, no. 9, pp. 4294–4302, 2020.

[29] Y. Fu et al., "Pelvic multi-organ segmentation on cone-beam CT for prostate adaptive radiotherapy," *Med. Phys.*, vol. 47, no. 8, pp. 3415–3422, 2020.

[30] Y. Lei et al., "Male pelvic CT multi-organ segmentation using synthetic MRI-aided dual pyramid networks," *Phys. Med. Biol.*, vol. 66, no. 8, p. 085007, 2021.

[31] Y. Lei et al., "Male pelvic multi-organ segmentation aided by CBCT-based synthetic MRI," *Phys. Med. Biol.*, vol. 65, no. 3, p. 035013, 2020.

[32] X. Dai et al., "Head-and-neck organs-at-risk auto-delineation using dual pyramid networks for CBCT-guided adaptive radiotherapy," *Phys. Med. Biol.*, vol. 66, no. 4, p. 045021, 2021.

[33] Z. Zhang, Z. Jiang, H. Zhong, K. Lu, F. F. Yin, and L. Ren, "Patient-specific synthetic magnetic resonance imaging generation from cone beam computed tomography for image guidance in liver stereotactic body radiation therapy," *Precis. Radiat. Oncol.*, vol. 6, no. 2, pp. 110–118, 2022.

[34] Y. Liao et al., "Pseudo dual energy CT imaging using deep learning-based framework: basic material estimation," *Proc. SPIE Med. Imaging*, vol. 10573, p. 105734N, 2018.

[35] W. Zhao et al., "Dual-energy CT imaging using a single-energy CT data is feasible via deep learning," *arXiv*, arXiv:1906.04874v3, 2019, [Online]. Available: http://arxiv.org/abs/1906.04874

[36] C. K. Liu, C. C. Liu, C. H. Yang, and H. M. Huang, "Generation of brain dual-energy CT from single-energy CT using deep learning," *J. Digit. Imaging*, vol. 34, no. 1, pp. 149–161, 2021.

[37] T. Lyu et al., "Estimating dual-energy CT imaging from single-energy CT data with material decomposition convolutional neural network," *Med. Image Anal.*, vol. 70, p. 102001, 2021.

[38] S. Charyyev et al., "Learning-based synthetic dual energy CT imaging from single energy CT for stopping power ratio calculation in proton radiation therapy," *Br. J. Radiol.*, vol. 95, no. 1129, p. 20210644, 2022.

[39] J. Scholey et al., "Improved accuracy of relative electron density and proton stopping power ratio through CycleGAN machine learning," *Phys. Med. Biol.*, vol. 67, no. 10, p. 105001, 2022.

[40] J. Harms et al., "Cone-beam CT-derived relative stopping power map generation via deep learning for proton radiotherapy," *Med. Phys.*, vol. 47, no. 9, pp. 4416–4427, 2020.

5

CT-Based DVF/Ventilation/Perfusion Imaging

Ge Ren, Yu-Hua Huang, Jiarui Zhu, Wen Li, and Jing Cai
Hong Kong Polytechnic University, Kowloon, Hong Kong

CONTENTS

5.1 Introduction

As one of the most commonly used imaging modalities, computed tomography (CT) scan can provide anatomical information for medical diagnosis or treatment purposes. Compared to other imaging modalities, CT is advantageous in the following aspects: (i) CT scan has better

DOI: 10.1201/9781003243458-7

accessibility for medical institutions worldwide; (ii) CT data has more stable imaging regularity and is more recognizable with high-resolution anatomical information. For example, CT is regularly considered for attenuation and scatter correction of single-photon emission computed tomography (SPECT) [1]; (iii) CT has better interpretation on lung intensity alterations, which manifest multifarious pulmonary diseases, such as pulmonary inflammation and fibrosis [2].

However, CT does not account for any functional or motional information, which is also important for multiclinical applications. CT-based DVF/ventilation/perfusion imaging techniques aim to generate a given image with functional or motional information so that the desired features become perceivable for clinical application. This process allows the observer to see additional information that may not be observable in the raw CT images. This chapter covers the main three CT-based medical image synthesis techniques, including CT-based deformable vector field (DVF) imaging (CTDI), CT-based ventilation imaging (CTVI), and CT-based perfusion imaging (CTPI). We collected and categorized the correlated studies of each topic. A thorough review of each imaging method was presented, in which studies were discussed based on conventional methods and recent effects. Lastly, we summarized these methods on the future development trend for clinical application.

5.2 CT-based DVF Imaging (CTDI)

Deformable image registration (DIR) is a registration process that locally transforms an image set into a reference image according to matched features [14]. The result of DIR can be quantified as DVF, which contains the local changes information in the density of tissue as well as gross displacement of tissue due to anatomic change. DVF can identify the spatial correspondence between two or among multiple sets of images. Because of the large numbers of degrees of freedom (DOI) parameters, the DVF generated by DIR is able to indicate the free-form deformation of soft tissues. The vector of the DVF can describe the motion of an organ or the movement of the organ tissue.

The CTDI methods have been well explored in recent decades (see Table 5.1 for details). CT-based DVF also has been widely applied in multiple clinical issues, including real-time motion tracking [15], target localization [16], multimodality image fusion [17], gated treatment planning [18], and treatment response evaluations [19]. Another important application is to calculate the lung functional images from the CT-based DVF (Figure 5.1). Since four-dimensional (4D) CT is routinely utilized to observe the motion of thoracic and abdominal organs, CTDI is mainly based on different phases of 4DCT. Other typical CT modalities for CTDI are cone-beam CT (CBCT), and planning CT. The CTDI methods can be categorized into conventional CTDI methods and DL-based CTDI methods. In this section, we separately present the established DVF imaging methods based on lung CT images.

5.2.1 Conventional CTDI Methods

Conventional CTDI methods rely on the conventional DIR algorithms, which generate DVF with a pre-designed mathematical model and an iterative procedure [20]. A typical conventional DIR method is mainly composed of three parts: a deformation model, an objective function, and an optimization strategy [14]. The goal of DIR is to find the optimal

TABLE 5.1

CT-based DVF Imaging Methods

Authors	Publication Year	Region of Interest (ROI)	DIR Method	Application
Coselmon et al. [3]	2004	Lung	Thin-plate splines	Treatment planning
Zhang et al. [4]	2004	Lung	Finite element	Motion tracking
Zhen et al. [5]	2012	Lung (CT-CBCT)	Diffeomorphic Demons	Treatment planning
Min et al. [6]	2014	Lung	Optical flow & Thin-plate splines	Motion tracking
Yang et al. [7]	2010	Lung	Spatial smoothing	Treatment planning
Eppenhof et al. [8]	2018	Lung	CNN & Thin-plate Splines	Overall soft tissue alignment
Eppenhof et al. [9]	2019	Lung	U-Net	Overall soft tissue alignment
Sentker et al. [10]	2018	Thoracic	CNN	Treatment planning
Sokooti et al. [11]	2017	Lung	U-Net	Overall soft tissue alignment
Jiang et al. [12]	2020	Lung (CT-CBCT&CBCT-CBCT)	Multiscale-Joint-CNN	Overall soft tissue alignment
Fu et al. [13]	2020	Lung	LungRegNet	Overall soft tissue alignment

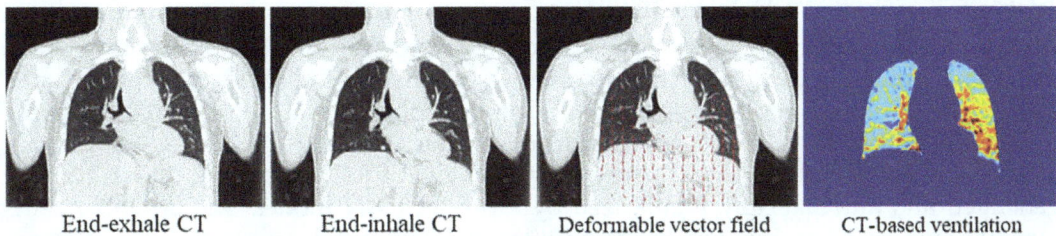

End-exhale CT   End-inhale CT   Deformable vector field   CT-based ventilation

FIGURE 5.1
Deformable vector field (DVF) generated by end-inhale/end-exhale CT and the CT-based ventilation images computed from the CT-based DVF.

DVF that deforms the moving image into a fixed image. When registering, the objective function guides the optimization strategy to iteratively optimize the parameters of the deformation model. Hence, the deformation model and the objective function determine the performance of a conventional CT DIR method, while the optimization strategy determines the computational efficiency. The deformation model of DIR introduces anatomical clinical prior knowledge by imposing task-specific constraints on the transformation. It also dictates the nature of the transformation, resulting in a parametrized and inherently smooth DVF. Representative state-of-art methods include spline-based method [21], finite element (FE)-based method [22], Diffeomorphic Demons method [23], optical flow method [24], and spatial smoothing method [7].

Combining the local and global motional features, the spline-based method uses spline-based flow fields with control vertices to define the flow within each spline patch. Coselmon et al. used the spline model to perform model deformation on lung CT between inhale and exhale breathing states [3]. Zhang et al. developed a 3D finite element model for lung deformation by matching the surface of the organ with structural material properties for DIR on different breathing phases of lung 4DCT [4]. Zhen et al. developed a non-parametric diffeomorphic transformation model with Diffeomorphic Demons as the optimization procedure on the entire space of displacement fields, which holds great promise to improve the CBCT artifacts and intensity inconsistency [5]. Min et al. developed an inversely consistent motion field for both registration directions of the target pair in a step-by-step process using modified optical flow algorithms in a multigrid and multiple-pass framework. The optical flow method performed well for 4DCT images [6]. Yang et al. also developed an out-of-plan image volumes extension transformation model to generate DVF between daily CT and planning CT of the lung with different sizes of the field of views and superior-inferior coverage [7].

CT-based DVF imaging is usually optimized based on intensity consistency between the target pair. The intensity consistency is formulated as intensity-based similarity metrics, such as sum-of-squares distance (SSD), mean square distance (MSD), cross-correlation (CC), and mutual information (MI). The parameters of the deformation model are fixed by iteratively updating the intensity-based similarity function. Additionally, conventional DIR methods in the CT domain are routinely utilized for lung motion tracking and achieve accurate registration based on an intensity similarity function. To deeply examine the clinical value of such methods, deformable physical phantoms are needed to simulate organ motion and evaluate the DIR performance [25].

5.2.2 Deep Learning-based CTDI Methods

We have witnessed the pervasive applications of deep learning techniques in medical image analysis. Deep learning technique uses several layers of nonlinear processing units for feature extraction and transformation. Convolutional neural network (CNN) is the most commonly utilized architecture for deep learning. For the CTDI tasks, deep learning has also demonstrated excellent feature extraction ability. In this process, DVF is directly generated and automatically regularized by a pre-set CNN model. The optimization is mainly based on a loss function based on intensity similarity between the output image and a desired warped image transformed by a ground truth DVF. However, the lack of training datasets with ground truth has been an important challenge for DL-based DIR. To address this challenge, supervised networks with pseudo ground truth and unsupervised networks have been proposed to estimate the DVF without prior knowledge.

5.2.2.1 Supervised DVF Synthesis Network

Supervised DVF synthesis network refers the CNN model that is trained the with ground truth DVF as reference. Eppenhof et al. developed a three-dimensional (3D) CNN to generate DVF between the inhale-exhale lung CT images [8]. The network directly learns the transformation between the input images, and outputs three maps for the x, y, and z components of the DVF, which is represented by a thin-plate spline transformation grid. Their neural network was further improved by using a U-Net architecture [9]. The new network was trained using synthetic random transformation. Affine pre-registration was used prior to CNN transformation prediction. On DIRLAB datasets, their method is able to decrease the target registration error (TRE) from 4.02 ± 3.08 mm to 2.17 ± 1.89 mm with the U-Net architecture. Although this TRE is lower than the conventional DIR approaches, their studies show that CNN can directly predict DVF. Sentker et al. used DVF generated from conventional DIR methods, including PlastiMatch [26], NiftyReg [27], and VarReg [28] as ground truth [10]. The loss function of the CNN model was the difference between the synthesized DVF and ground truth DVF for inhale-exhale lung CT registration. On DIRLAB, the TRE of their model was 2.50 ± 1.16 mm. In addition, Sokooti et al. generated artificial DVF using a respiratory motion model to simulate ground truth DVF for DIR [11]. Multiview patch-based training was employed, and the U-Net with TRE and Jacobian determinants were used as the evaluation metrics. In terms of TRE, the realistic model-based transformation performed better than random transformations. On the SPREAD and DIRLAB datasets, their model achieved TRE of 2.32 mm and 1.86 mm, respectively.

5.2.2.2 *Unsupervised DVF Synthesis Network*

A typical unsupervised DVF synthesis network uses fixed/moving images as inputs and warped images as output. The DVF is an intermediate product of the model and is applied to the moving image to generate warped images. During the training, the warped images are compared to fixed images to calculate loss function. DVF smoothness constraint is usually used to regularize the smoothness of the intermediate DVF. Jiang et al. developed an unsupervised multiscale framework for lung CT DIR [12]. Three CNN models were used in sequence for different scale concentrations. Without fine-tuning, the trained model also works well on CT-CBCT and CBCT-CBCT registration. On DIRLAB datasets, their model obtained an average TRE of 1.66 ± 1.44 mm. Fu et al. developed a two-step unsupervised CNN model for lung DIR [13]. A CoarseNet was developed to distort the moving image globally after whole-image registration on the down-sampled image. Using the patch-based FineNet model, image patches from the globally warped moving image were registered to image patches from the stationary image. Prior to DIR, vessel improvement was conducted to increase registration accuracy. Their model reached a TRE of 1.59 ± 1.58 mm on average, outperforming conventional DIR approaches.

5.3 CT-based Ventilation Imaging (CTVI)

The primary function of the lungs is gas exchange between the circulatory system and the external environment [29]. In respiration, oxygen from incoming air enters the blood, and carbon dioxide, waste gas from the metabolism, leaves the blood. Lung ventilation refers to airflow that reaches the alveoli. The CTVI is an experimental method to obtain regional pulmonary ventilation surrogates from the standard noncontrast CT images. This is feasible because respiratory-related CT scans, such as 4DCT, breath-hold CT, or respiratory-gated CT, are able to capture the breathing-induced CT characteristics changes in the lung area. From this perspective, CTVI can be regarded as an additional "bonus" information for lung cancer patients treated with radiotherapy, as thoracic 4DCT scans have been used for lung cancer radiotherapy. After nearly two decades of development, a large number of various implementations have been proposed (see Table 5.2 for details). In this part, we mainly introduce the existing DIR-based CTVI metrics and other CTVI methods based on their features.

TABLE 5.2

CT-based Ventilation Imaging Methods

Authors	Publication Year	Proposed CTVI Method	Input Image	Reference Modality	Subjects
Guerrero et al. [1]	2005	DIR-based HU model	BHCT	Measured tidal volume on CT	22 lung cancer patients
Reinhardt et al. [2]	2008	DIR-based Volume Model (Analytic Jacobian)	Respiratory-gated CT	Xenon CT	5 sheep
Zhang et al. [3]	2009	DIR-based Volume Model (Geometric Jacobian)	4DCT	N/A	12 lung cancer patients
Kipritidis et al. [9]	2015	Attenuation	4DCT	Galligas PET	25 lung cancer patients
Tian et al. [10]	2019	Attenuation (Simplified Implementation)	Time-average 4DCT	DTPA-SPECT	50 lung or esophageal cancer patients
Castillo et al. [5]	2019	Integrated Jacobian Formulation (IJF)	4DCT	N/A	10 public cases + 5 lung cancer patient
Castillo et al. [7]	2019	Mass-conserving Volume Change (MCVC)	4DCT	N/A	10 public cases
Szmul et al.[8]	2019	Multilayer Supervoxels	4DCT	Xe-MRI	3 cases
Zhong et al. [11]	2019	Deep Learning (FCN)	4DCT	DIR-based CTVI	82 lung cancer patients
Liu et al. [12]	2020	Deep Learning (2D U-Net)	4DCT	DTPA-SPECT	50 lung or esophageal cancer patients
Castillo et al. [6]	2020	Parametric IJF and MCVC	4DCT	DTPA-SPECT	15 lung cancer patients
Cazoulat et al. [4]	2021	Biomechanical Model	4DCT	DTPA-SPECT	6 lung cancer patients

5.3.1 Classical DIR-based CTVI Methods

5.3.1.1 HU-based Method

Simon [30] proposed that the CT value change (in Hounsfield Units, HU) between registered inhalation and exhalation breathing states corresponds to regional volume changes. The HU-based method was built on the assumption that the changes in local air content in the lungs caused by breathing behavior provide a biomarker that reflects the lung ventilation ability. Simon's method assumes that the air fraction of a particular lung region of CT image is given by Equation 5.1:

$$F_{\text{air}} = -\frac{HU}{1000} \tag{5.1}$$

The fractional change in air content within a specified volume is then processed by Equation 5.2:

$$\frac{\Delta V}{V} = \frac{F_2 - F_1}{F_1(1 - F_2)} \tag{5.2}$$

where ΔV is the regional volume change due to inhale the air, V is the volume of air within the lung area defined in exhale CT image, F_1 is the air fraction in the exhale lung and F_2 is the corresponding air fraction in the inhale lung [31]. With the DVFs given by performing the DIR algorithm between the end-inhale and the end-exhale CT image, the CTVI can be expressed by Equation 5.3:

$$\text{CTVI}_{\text{DIR}-\Delta HU} = 1000 \frac{\overline{HU}_{\text{inhale}} - HU_{\text{exhale}}}{HU_{\text{exhale}}\left(1000 + \overline{HU}_{\text{inhale}}\right)} \tag{5.3}$$

where HU_{exhale} is the HU value of voxel in the end-exhale phase CT, and, and $\overline{HU}_{\text{inhale}}$ is the average of all HU values corresponding to the set of end-inhale phase voxels that mapped into the end-exhale phase voxels under consideration.

5.3.1.2 Volume-based Method

The volume-based method is also named as Jacobian-based method [32], and is based on the assumption that the ventilation is proportional to the local volume change [33]. For analytic mathematical implementation, local lung volume change can be estimated directly using the determinant of the Jacobian of the transformation T (given by DIR algorithm) between two 4DCT phase images, according to the Jacobian of the DIR (Equation 5.4).

$$J(\vec{x}) = \det \begin{bmatrix} 1 + \dfrac{\partial T_1(\vec{x})}{\partial x_1} & \dfrac{\partial T_1(\vec{x})}{\partial x_2} & \dfrac{\partial T_1(\vec{x})}{\partial x_3} \\[2ex] \dfrac{\partial T_2(\vec{x})}{\partial x_1} & 1 + \dfrac{\partial T_2(\vec{x}_{\text{in}})}{\partial x_2} & \dfrac{\partial T_2(\vec{x})}{\partial x_3} \\[2ex] \dfrac{\partial T_3(\vec{x})}{\partial x_1} & \dfrac{\partial T_3(\vec{x})}{\partial x_2} & 1 + \dfrac{\partial T_3(\vec{x})}{\partial x_3} \end{bmatrix} \tag{5.4}$$

It assumes an initial voxel volume of unity. If the determinant of the Jacobian is also 1, then no expansion or contraction of the voxel occurs. Likewise, if the determinant is greater than 1, there exists regional volume expansion, while if the determinant is less than 1, there exists regional volume contraction. Furthermore, the specific volume change at each voxel position (i.e., ventilation) can be given simply by Equation 5.5:

$$\text{CTVI}_{\text{Jac}} = J(x) - 1 \tag{5.5}$$

Besides, there is another geometric implementation for estimating regional lung volume change from DVF [34]. It calculates the tetrahedra volume that results from displacing the eight voxel vertices from their initial cubic configuration according to the DVF. The main difference of these two methods is in numerical implementation. It can be mathematically proved that the geometric implementation for computing the volume of a deformed voxel is equivalent to a first-order finite difference implementation of the analytic Jacobian [35].

5.3.2 Improved DIR-based CTVI Methods

5.3.2.1 Hybrid Method

A hybrid method has been reported as a modification of the original HU-based model [36]. This method combines information from both HU value changes and regional volume changes simultaneously to perform a density correction for each voxel for tissue compression. The hybrid CTVI can be calculated using Equation 5.6:

$$\text{CTVI}_{\text{Hybrid}} = \frac{HU_{\text{ex}}^*(x) - HU_{\text{in}}^*(x+v)}{HU_{\text{in}}^*(x+v) + 1000} \times \frac{HU_{\text{ex}}(x) + 1000}{1000} \quad (5.6)$$

where the $HU_{\text{ex}}^*(x)$ is formulated by Equation 5.7:

$$HU_{\text{ex}}^*(x) = \frac{HU_{\text{ex}}(x)}{Jac(x,v)} \quad (5.7)$$

5.3.2.2 Biomechanical Model

Morfeus is a finite-element-based DIR algorithm [37]. For lung CT registration, Morfeus allows the definition of heterogeneous elastic properties inside the lung while controlling local deformation based on biomechanical features. This algorithm can provide more accurate estimation of DVFs, and also can obtain the strain and stress distribution in the lung. As an extension of Morfeus, Cazoulat et al. proposed an approach to use the mechanical stress distribution of the lung as the surrogate for assessing lung ventilation [38]. This method used the strain map given by DIR between the end-exhale to the end-inhale phase of 4DCT scans, combined with the mechanical elasticity of lung tissue to calculate the stress distribution, so as to assess the lung ventilation distribution. This method presented a significantly higher and also more consistent correlation with ground truth than the conventional DIR-based CTVI methods.

5.3.2.3 Integrated Jacobian Formulation (IJF) Method

Jacobian metric, as a mathematical implementation of volume-change model, uses the first derivative of the DVF to estimate the change in volume of the voxels. But the real CT images is discretized by the digital voxel grid into a matrix. Therefore, we cannot compute the derivative directly from this mathematical representation (because this is used for smooth continuous functions). The actual mathematical implementation is to use the forward difference with expression in Equation 5.8:

$$\frac{\partial \phi_i(\mathbf{x}_k)}{\partial x_j} \approx \phi_i(\mathbf{x}_k + \mathbf{e}_j) - \phi_i(\mathbf{x}_k) \quad (5.8)$$

This kind of approximation could occur an O (1) uncertainty, which could be as large as the voxel grid spacing. Such a large possible uncertainty makes the performance of the original Jacobian method very sensitive to DIR results and image quality.

   As an improved implementation of volume model-based ventilation estimation, the purpose of IJF method [39, 40] is to devise a more robust numerical method for estimating voxel volume changes, derived from DIR. In Ref [40], DIR methods are firstly used to obtain the deformation field between end-inhale and end-exhale phase. Then on the end-exhale phase, lung region is defined as several subregion controlled by randomly generated equally spaced point clouds. Through the hit-or-miss algorithm, the corresponding subregions on the end-inhale are accordingly defined and grown appropriately to meet the requirements of Gaussian statistics and standard error analysis. At the same time, the subregional-wise volume change ratio is also obtained. Finally, by solving a constrained linear least squares problem, the voxel-wise ventilation distribution is recovered. Since this version of the IJF method will consume huge computing resources due to the recovery of each unknown voxel in the lung, another simplified IJF implementation using the moving least squares was present in [39]. In this method, we only need to optimize the value of a certain number of sparse control points, and then calculate the voxel-wise ventilation through Shepard's class of moving least squares approximation. Compared with the classical Jacobian CTVI, the IJF method has higher repeatability and stability, and its robustness to the DIR solutions and image quality has been significantly improved.

5.3.2.4 Mass-Conserving Volume Change (MCVC) Method

Similarly, the "coarse-to-fine" concept from the IJF method can also be used to improve the robustness of the HU-based CTVI metrics. According to the change of subregional HU values, Castillo et al. [41] presented a mass-conserving volume change (MCVC) method for CTVI mapping, which is a robust implementation of the HU model. This method recovers voxel-wise volume changes from a series of subregional volume change measurements, which are defined in terms of mean density ratios. The resulting CTVIs generated by this method has also been demonstrated to have a good correlation with the IJF-derived CTVIs, showing that incorporating robustness into the ventilation calculation leads to more consistent cross-algorithm (i.e., HU-based v.s. volume-change-based) results. Additionally, they also proposed a simplified version that requires fewer computing resources in the literature [39], using moving least squares concept.

5.3.2.5 Multilayer Supervoxels Estimation Method

As an application of patch-based approaches, Szmul et al. proposed another robust CTVI method that tracks changes in the intensity of supervoxels [42]. With DIR-derived DVF, end-exhale and end-inhale 4DCT phase images are represented by multilayer supervoxel representations. The mean intensity difference for each supervoxel is calculated for supervoxel-wise ventilation mapping. Finally all difined supervoxel layers are averaged to obtain voxel-wise ventilation distribution. This patch-based CTVI method may be physiologically more consistent with lung anatomy than the classical HU-based model.

5.3.3 Other CTVI Methods

5.3.3.1 Attenuation Method

The attenuation metric [43] is the first proposed non-DIR CTVI calculation method, which is developed to estimated lung ventilation directly from 4DCT without performing DIR algorithms. Assuming that physiological ventilation (i.e., blood-gas exchange) should relate to the regional product of tissue and air densities, this method could be described using the following Equations 5.9 and 5.10:

$$\text{CTVI}_{\overline{\text{HU}}}(\mathbf{x}) = \sum_{\phi=1}^{N} V_\phi(\mathbf{x})/N \tag{5.9}$$

where

$$V_\phi(\mathbf{x}) = \begin{cases} f_\phi^{\text{Air}}(\mathbf{x}) \times f_\phi^{\text{Tissue}}(\mathbf{x}), & \text{if } \mathbf{x} \in L(\phi) \\ 0, & \text{if } \mathbf{x} \notin L(\phi) \end{cases}$$

$$= \begin{cases} \dfrac{\text{HU}_\phi(\mathbf{x})}{-1000} \times \dfrac{\text{HU}_\phi(\mathbf{x}) + 1000}{1000}, & \text{if } \mathbf{x} \in L(\phi) \\ 0, & \text{if } \mathbf{x} \notin L(\phi) \end{cases} \tag{5.10}$$

In these equations, $\text{HU}_\phi(\mathbf{x})$ is the HU value at $\mathbf{x}$ position of 4DCT phase bin ϕ. For HU values in the range of $(-1000, 0)$, the terms $f_\phi^{\text{Air}}(\mathbf{x})$ and $f_\phi^{\text{Tissue}}(\mathbf{x})$ represent the air and tissue fraction of each lung voxel, respectively. $L(\phi)$ refers to the phase-specific lung mask, which defines the lung region of each 4DCT phase image. The resulting $\text{CTVI}_{\overline{\text{HU}}}$ is the ventilation distribution defined on the time-averaged 4DCT geometry.

The attenuation metric could provide a straightforward and reproducible means to mapping CTVI via 4DCT. However, since this method does not account for the motion of lung tissue, it can be expected to exhibit generally poor spatial accuracy [36].

Inspired by this study, in 2019 Tian et al. [44] proposed a simplified implementation of the attenuation metric, which needs only a time-average 4DCT (AVG CT) image instead of the full series of 4DCT phase images, which can be formulated as Equation 5.11:

$$\text{CTVI}_{\text{AVG}}(\mathbf{x}) = \begin{cases} \dfrac{\text{HU}_{\text{AVG}}(\mathbf{x})}{-1000} \times \dfrac{\text{HU}_{\text{AVG}}(\mathbf{x}) + 1000}{1000}, & \text{if } \mathbf{x} \in L_{\text{AVG}} \\ 0, & \text{if } \mathbf{x} \notin L_{\text{AVG}} \end{cases} \tag{5.11}$$

here $\text{HU}_{\text{AVG}}(\mathbf{x})$ refers to the HU value at $\mathbf{x}$ position of AVG CT, and L_{AVG} is the lung mask defined on AVG CT. Compared to the original attenuation metric, this approach has less input requirement with improved computation efficiency.

5.3.3.2 Deep Learning Method

In recent years, the application of deep learning (DL) in the field of computer vision has achieved great success, related techniques have been gradually applied to the diagnosis, segmentation and generation of medical images. In 2019, Zhong et al. first reported their research of applying DL technic to the CTVI generation [45]. In this preliminary study, two phases of the 4DCT series (i.e., end-inhale phase and end-exhale phase) were used as input to feed into a nine-layer fully convolutional network, while the corresponding HU model generated CTVI was used as predicting label. Although there was a lack of "real" ground truth acquired via NM imaging, they firstly demonstrated the feasibility of using deep CNN to generate ventilation images from 4DCT. Later in the same year, Liu et al [46] presented an improved DL method for CTVI generation based on the well-tested popular 2D U-Net. Their framework was trained and evaluated with 50 patients who underwent both 4DCT and SPECT-V scans, and was demonstrated to have a significantly better performance than those conventional DIR-based CTVI algorithms.

5.4 CT-based Perfusion Imaging (CTPI)

Perfusion refers to blood that reaches tissue via the capillaries. For the diagnosis of lung diseases, together with ventilation, perfusion indicates the pulmonary gas exchange between airspaces and blood, which serves as a core function of the lung [47]. In clinical practice, perfusion image is routinely used as a clinical pulmonary function testing standard to measure the blood circulation within the lung [48]. Changes in lung perfusion images physiologically manifest changes in anatomical lung parenchyma [49]. In general, lung perfusion imaging can supplement anatomical images and present legible distribution of functional lung regions and enhance our understanding of the underlying pulmonary functional information [50]. In clinical practice, lung perfusion images can be acquired by nuclear medicine imaging methods with the aid of contrast agents, such as single-photon emission computed tomography (SPECT) with technetium-99m-labeled macro aggregated albumin (MAA) [51]. However, the nuclear imaging methods are limited available for the radiotherapy department. The complicated scanning procedure required for nuclear imaging makes these methods resource-intensive, inherently risky, inefficient, and technically challenging. With the development of new techniques, CTPI was proposed and achieved promising results as compared with the conventional SPECT perfusion (Figure 5.2). In this part, we mainly introduce emerging CTPI techniques and applications of perfusion-based lung cancer radiation therapy (see Table 5.3 for details).

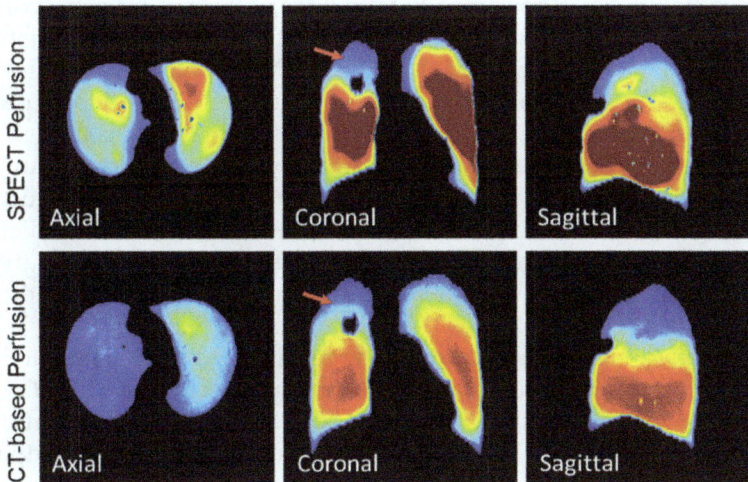

FIGURE 5.2
Illustration of the CT-based perfusion images in comparison with ground truth SPECT perfusion.

TABLE 5.3

CT-based Perfusion Imaging Methods

Authors	Publication Year	Region of Interest (ROI)	Method	Application
Edward et al. [52]	2021	Lung	DIR	Radiation therapy
Jang et al. [53]	2019	Lung	2D-U-Net	Radiation therapy
Ren et al. [54]	2019	Lung	3D-U-Net	Radiation therapy
Ren et al. [55]	2021	Lung	3D-U-Net	Radiation therapy
Porter et al. [56]	2021	Lung	3D-Resnet	Radiation therapy
Woo et al. [57]	2019	Lung	cGAN	Disease diagnosis
Santini et al. [58]	2018	Cardiac	2D-U-Net	Disease diagnosis
Kim et al. [59]	2021	Abdominal	cGAN & CNN	Disease diagnosis

5.4.1 DIR-based CTPI Methods

Edward et al. developed a DIR-based CTPI model [52] upon their previous HU-based CTVI approach [41]. The lung breathing can proportionately change blood/air volume variation within the lung. Based on this, this DIR-based CTPI method mathematically formulates pulmonary perfusion based on gross regional mass variations between inhale and exhale CT images at a voxel level. This regional mass variations are indicated by the intensity changes measured by HU values changes by the Jacobian factor of DIR transformation, with the assumption that mass within the lung volume remains constant throughout the breath cycle. The CT-based perfusion image generation is implied by numerical methods based on a linear system of equations integrating subregional magnitude mass changes for the voxel volume. The DIR-based CT-perfusion method showed promising results with a median Spearman correlation range of 0.49–0.57.

5.4.2 Deep Learning-based CTPI Methods

Deep learning-based CTPI methods mostly follow the pattern of directly synthesizing perfusion images from anatomic CT images using a CT-to-perfusion translation. These deep learning-based CTPI models are designed as 2D or 3D CNN architecture and trained with supervised learning. The CT images are used as input while SPECT perfusion images as ground truth. The multi convolutional layers of the deep learning model extract high-dimensional pulmonary perfusion features, and recognize complicated lung anatomical and vascular structures based on textural information from CT imaging modality without prior knowledge of the relationship [60]. Compared with the nuclear medicine perfusion imaging methods, deep learning-based CTPI methods significantly simplify the acquisition procedure of perfusion images, which is easier, faster, less costly, and more accessible for clinical application.

The existing studies have achieved promising results: Jang et al. generated SPECT lung perfusion from inhalation CT images acquired during collection of SPECT/CT with a 2D conditional generative adversarial network (GAN) model and achieved multiscale structural similarity (SSIM) of 0.87 [53]. Ren et.al developed a 3D attention residual neural network to generate discretized synthetic SPECT lung perfusion images from single free breathing CT images [54] and continued to modify the deep learning model design with expanded cohort, achieving a moderate-to-high voxel-wise correlation (Spearman's correlation coefficients, 0.6733 ± 0.1728) and high structural similarity (SSIM, 0.7635 ± 0.0697) [55]. Porter generated SPECT pulmonary lung perfusion images from clinical 4DCT images with a 3D residual neural network, achieving a Spearman correlation coefficient of 0.70 and a Pearson correlation coefficient of 0.66 [56].

The fundamental idea of the deep learning-based CTPI methods is to find sufficient underlying vascular functional information from the CT images. On one hand, more perceptive neural networks can be designed to extract multiscale lung structural features and reveal more deep anatomical patterns. On the other hand, more imaging modalities can be used as supervising ground truth to fully utilize the special information indication ability of different imaging modalities.

5.4.3 CTPI Techniques for Other Anatomies

Besides the profound results on synthesizing lung perfusion images, several deep learning-based CTPI techniques have been applied on different anatomies, such as head and neck, cardiac, and abdomen. For perfusion imaging of other anatomies, injection of iodine-based contrast medium is necessary, which enhances the visibility of vascular structures during radiographic procedures. However, the risk of the adverse reaction of iodine contrast medium is widely concerned [61]. To eliminate the use of contrast medium, deep learning algorithms were proposed. Woo et al. applied a conditional GAN model to predict the iodine contrast medium-enhanced perfusion images for lymph node cancer patients [57]. They used histogram equalization to adjust the intensities between input contrast-free CT and contrast-enhanced CT to improve the prediction accuracy. Their method achieved an SSIM of 0.908 ± 0.047 between synthesized perfusion images and ground truth perfusion images. To enhance the visibility of cardiac chambers, Santini et al. proposed a loss improved deconvolutional network to synthesize the virtual cardiac perfusion images [58]. Their study obtained a dice overlap of 0.88 ± 0.03 between the predicted and real perfusion CT images. Kim et al. investigated the feasibility of deep learning on abdominal CT-perfusion synthesis, they trained a deep learning algorithm to generate the virtual perfusion images in patients who visited the emergency department with acute abdominal pain [59]. Their study showed the potential to enhance the diagnosis accuracy and confidence in patients with acute abdominal pain. However, the above-mentioned studies only trained and evaluated their models using single hospital data. The performance of these models on data from other hospitals still need to be explored.

5.5 Clinical Applications of CT-based DVF/Ventilation/Perfusion Imaging

With the additional information provided by CT-based image synthesis, multiply clinical tasks can be completed easily, fast, and less costly. For the radiology department, the ventilation /perfusion images are often used to find the blood clots or diagnose other diseases in the lung [62]. If the ventilation/perfusion surrogate images can be directly synthesized from the corresponding CT images, it may strongly accelerate the diagnosis of lung disease, such as pulmonary embolism, chronic obstructive pulmonary disease (COPD), etc. However, there are uncertainties associated with current CT-based imaging methods. The quality of CT-based ventilation/perfusion has not been explored toward the diagnostic application.

As an improving technique, current CT-based imaging is limited to the radiotherapy application because of a relatively lower accuracy requirement. The current radiation therapy (RT) planning is based on anatomic imaging and assumes organs have a homogenous function. High energy X-rays for treatment need to pass through normal tissues to reach tumors so as to cause damage to surrounding organs. However, this does not account for the fact that the organ function can be heterogeneous. For example, for lung cancer patients, lung cancer is often accompanied by other lung diseases, such as COPD. These patients commonly have obvious regional lung function heterogeneity, low lung compensatory capacity, and poor tolerance to thoracic radiotherapy, which easily lead to radiation-reduced lung injury. In order to realize personalized plans and minimal lung toxicity in radiotherapy planning (e.g. functional lung avoidance radiotherapy, or FLART), we need information from pulmonary function imaging. Although the ventilation imaging techniques based on PET or SPECT scanners can directly map the patient's lung ventilation distribution through radioactive tracer, this kind of scanning can only be performed in the radiology department in most cases. For lung cancer radiotherapy, if PET and SPECT-based lung function images are still used to guide the implementation of FLART, additional money and time costs will be required, which will bring great limitations to this technology. As CT scan is the standard for dose calculation in treatment planning, the CT-based functional images may expedite the clinical application of functional avoidance radiation therapy.

Currently, there are three ongoing clinical trials in the United States (NCT02528942, NCT02308709, and NCT02843568) investigating the clinical efficacy of functionally guided RT for lung cancer. In addition, Matuszak et al integrated SPECT perfusion in dose optimization of 15 patients, and found that the mean dose to high functional region decreased from 12.6 ± 4.9 Gy to 9.9 ± 4.4 Gy [63]. Waxweiler et al observed an average decrease of the mean dose to the functional lung by 2.8 Gy in functionally guided planning [64]. Yamamoto et al reported a 5.0% decrease in the dose to the functional lung by functionally guided planning [65]. Functional avoidance radiotherapy was found to improve dose-volumetric outcomes for functional lung.

In addition to FLART, the CT-based imaging can also be used for the radiation outcome assessment. This application aims to assess dose dependence of radiotherapy-induce functional changes in the normal organs. Functional and standard dose-volume parameters can be used as predictors of post radiation pulmonary toxicity or risk of developing pulmonary complications in cancer patients undergoing chemoradiation or radiation therapy.

In clinical practice, lung functional images have been widely utilized for post-RT toxicity evaluation. After RT treatment, the CT-based images also can provide information for motional and functional evaluation of the treatment outcome. Multiparameter logistic/linear model, based on dose or time dependence of lung regions, are generally formed to measure the relationships between local dose and blood flow reduction with radiotherapy treatment, as validated toxicity model for hypofractionated schedules for stereotactic body radiotherapy (SBRT) [66], predictive model of population dose response curves (DRC) for thoracic radiotherapy [67], or toxicity scoring model for radiation-induced pulmonary morbidity [68].

Dose-volume histogram (DVH) in radiotherapy of lung cancer describes 3D dose distribution and is commonly used for prediction of radiation pneumonitis [69]. DVH parameters can be collected from treatment planning CT scans whereas pre-treatment perfusion images are commonly used for calculation of perfusion dose-volume parameters in a functional DVH (FDVH). Clinically, FDVH contains the vascular information and is capable of detecting pulmonary emboli or abnormal blood flow [62], thus indicating the presence of respiratory diseases, such as pulmonary embolism, pulmonary hypertension, or chronic obstructive disease [70]. Different mathematic models have been designed for more accurate risk estimation of radiation pneumonitis (RP). Studies have integrated DVH and FDVH as establishment of lung perfusion score measurements [71] or for predicting symptomatic radiation pneumonitis framework [72].

5.6 Concluding Remarks

This chapter focused on the introduction of CT-based DVF/ventilation/perfusion imaging techniques for medical image synthesis. These developing techniques that uses different CT data (4DCT, planning CT, diagnostic CT, etc.) and image processing to calculate a surrogate for additional lung information. These synthesis techniques are attractive in clinical practice because they are easier, faster, less costly, and more accessible for acquiring functional or motional images. Now most methods are still based on retrospective dataset and have a large uncertainty. We encourage scientists in the field to carry out a large-cohort study specifically on patients and incorporate more clinical information, whenever possible, to fully contextualize the results in the context of clinical application.

References

1. Damen, E.M.F., et al., Quantifying local lung perfusion and ventilation using correlated spect and Ct data. *J Nucl Med*, 1994. **35**(5): pp. 784–792.
2. Anthimopoulos, M., et al., Lung pattern classification for interstitial lung diseases using a deep convolutional neural network. *IEEE Trans Med Imaging*, 2016. **35**(5): pp. 1207–1216.
3. Coselmon, M.M., et al., Mutual information based CT registration of the lung at exhale and inhale breathing states using thin-plate splines. *Med Phys*, 2004. **31**(11): pp. 2942–2948.
4. Zhang, T., et al., Technical note: A novel boundary condition using contact elements for finite element based deformable image registration. *Med Phys*, 2004. **31**(9): pp. 2412–2415.
5. Zhen, X., et al., CT to cone-beam CT deformable registration with simultaneous intensity correction. *Phys Med Biol*, 2012. **57**(21): pp. 6807–6826.
6. Min, Y., et al., 4D-CT Lung registration using anatomy-based multi-level multi-resolution optical flow analysis and thin-plate splines. *Int J Comput Assist Radiol Surg*, 2014. **9**(5): pp. 875–889.
7. Yang, D., et al., Technical note: deformable image registration on partially matched images for radiotherapy applications. *Med Phys*, 2010. **37**(1): pp. 141–145.
8. Eppenhof, K.A., et al., Deformable Image Registration Using Convolutional Neural Networks. In *SPIE Medical Imaging*. Vol. 10574. 2018: SPIE. Bellingham: The International Society for Optics and Photonics (SPIE).
9. Eppenhof, K.A.J. and J.P.W. Pluim, Pulmonary CT registration through supervised learning with convolutional neural networks. *IEEE Trans Med Imaging*, 2019. **38**(5): pp. 1097–1105.
10. Sentker, T., F. Madesta, and R. Werner, *Deep Learning-Based Fast 4D CT Image Registration*. 2018. Cham: Springer International Publishing.
11. Sokooti, H., et al. *Nonrigid Image Registration Using Multi-scale 3D Convolutional Neural Networks*. 2017. Cham: Springer International Publishing.

12. Jiang, Z., et al., A multi-scale framework with unsupervised joint training of convolutional neural networks for pulmonary deformable image registration. *Phys Med Biol*, 2020. **65**(1): p. 015011.

13. Fu, Y., et al., LungRegNet: An unsupervised deformable image registration method for 4D-CT lung. *Med Phys*, 2020. **47**(4): pp. 1763–1774.

14. Oh, S. and S. Kim, Deformable image registration in radiation therapy. *Radiat Oncol J*, 2017. **35**(2): pp. 101–111.

15. Yang, D., et al., 4D-CT motion estimation using deformable image registration and 5D respiratory motion modeling. *Med Phys*, 2008. **35**(10): pp. 4577–4590.

16. Li, F., et al., Comparison of internal target volumes defined on 3-dimensional, 4-dimensonal, and cone-beam CT images of non-small-cell lung cancer. *Onco Targets Ther*, 2016. **9**: pp. 6945–6951.

17. Brock, K.K., et al., Use of image registration and fusion algorithms and techniques in radiotherapy: Report of the AAPM Radiation Therapy Committee Task Group No. 132. *Med Phys*, 2017. **44**(7): pp. e43–e76.

18. Lin, H., et al., Dosimetric study of a respiratory gating technique based on four-dimensional computed tomography in non-small-cell lung cancer. *J Radiat Res*, 2014. **55**(3): pp. 583–588.

19. MacManus, M., et al., Anatomic, functional and molecular imaging in lung cancer precision radiation therapy: treatment response assessment and radiation therapy personalization. *Transl Lung Cancer Res*, 2017. **6**(6): pp. 670–688.

20. Sotiras, A., C. Davatzikos, and N. Paragios, Deformable medical image registration: a survey. *IEEE Trans Med Imaging*, 2013. **32**(7): pp. 1153–1190.

21. Richard Szelisk, J.C., Spline-based image registration. *Int J Comput Vision*, 1997. **22**(3): pp. 199–218.

22. Brock, K.K., et al., Accuracy of finite element model-based multi-organ deformable image registration. *Med Phys*, 2005. **32**(6): pp. 1647–1659.

23. Vercauteren, T., et al., Diffeomorphic demons: efficient non-parametric image registration. *Neuroimage*, 2009. **45**(1 Suppl): pp. S61–S72.

24. Yang, D., et al., A fast inverse consistent deformable image registration method based on symmetric optical flow computation. *Phys Med Biol*, 2008. **53**(21): pp. 6143–6165.

25. Jina Chang, T.-S.S. and Dong-Soo Lee, Development of a deformable lung phantom for the evaluation of deformable registration. *J Appl Clin Med Phys*, 2010. **11**: pp. 281–286.

26. Modat, M., et al., Fast free-form deformation using graphics processing units. *Comput Methods Programs Biomed*, 2010. **98**(3): pp. 278–284.

27. Shackleford, J.A., N. Kandasamy, and G.C. Sharp, On developing B-spline registration algorithms for multi-core processors. *Phys Med Biol*, 2010. **55**(21): pp. 6329–6351.

28. Werner, R., et al., Estimation of lung motion fields in 4D CT data by variational non-linear intensity-based registration: A comparison and evaluation study. *Phys Med Biol*, 2014. **59**(15): pp. 4247–4260.

29. Powers, K.A. and A.S. Dhamoon, Physiology, Pulmonary Ventilation and Perfusion. In *StatPearls*. 2020. Treasure Island (FL): StatPearls Publishing LLC.

30. Simon, B.A., Non-invasive imaging of regional lung function using x-ray computed tomography. *J Clin Monit Comput*, 2000. **16**(5–6): pp. 433–442.

31. Guerrero, T., et al., Quantification of regional ventilation from treatment planning CT. *Int J Radiat Oncol Biol Phys*, 2005. **62**(3): pp. 630–634.

32. Vinogradskiy, Y., CT-based ventilation imaging in radiation oncology. *BJR Open*, 2019. **1**(1): p. 20180035.

33. Reinhardt, J.M., et al., Registration-based estimates of local lung tissue expansion compared to xenon CT measures of specific ventilation. *Med Image Anal*, 2008. **12**(6): pp. 752–763.

34. Zhang, G.G., et al., Derivation of high-resolution pulmonary ventilation using local volume change in four-dimensional CT data. *World Congress on Medical Physics and Biomedical Engineering, Vol 25, Pt 4: Image Processing, Biosignal Processing, Modelling and Simulation, Biomechanics*, 2010. **25**: pp. 1834–1837.

35. Castillo, R., et al., Ventilation from four-dimensional computed tomography: density versus Jacobian methods. *Phys Med Biol*, 2010. **55**(16): pp. 4661–4685.

36. Kipritidis, J., et al., The VAMPIRE challenge: A multi-institutional validation study of CT ventilation imaging. *Med Phys*, 2019. **46**(3): pp. 1198–1217.

37. Cazoulat, G., et al., Biomechanical deformable image registration of longitudinal lung CT images using vessel information. *Phys Med Biol*, 2016. **61**(13): pp. 4826–4839.

38. Cazoulat, G., et al., Mapping lung ventilation through stress maps derived from biomechanical models of the lung. *Med Phys*, 2021. **48**(2): pp. 715–723.

39. Castillo, E., et al., Technical note: On the spatial correlation between robust CT-ventilation methods and SPECT ventilation. *Med Phys*, 2020. **47**(11): pp. 5731–5738.

40. Castillo, E., et al., Robust CT ventilation from the integral formulation of the Jacobian. *Med Phys*, 2019. **46**(5): pp. 2115–2125.

41. Castillo, E., Y. Vinogradskiy, and R. Castillo, Robust HU-based CT ventilation from an integrated mass conservation formulation. *Med Phys*, 2019. **46**(11): pp. 5036–5046.

42. Szmul, A., et al., Patch-based lung ventilation estimation using multi-layer supervoxels. *Comput Med Imaging Graph*, 2019. **74**: pp. 49–60.

43. Kipritidis, J., et al., Estimating lung ventilation directly from 4D CT Hounsfield unit values. *Med Phys*, 2016. **43**(1): p. 33.

44. Tian, Y., et al., Availability of a simplified lung ventilation imaging algorithm based on four-dimensional computed tomography. *Phys Med*, 2019. **65**: pp. 53–58.

45. Zhong, Y., et al., Technical note: Deriving ventilation imaging from 4DCT by deep convolutional neural network. *Med Phys*, 2019. **46**(5): pp. 2323–2329.

46. Liu, Z., et al., A deep learning method for producing ventilation images from 4DCT: First comparison with technegas SPECT ventilation. *Med Phys*, 2020. **47**(3): pp. 1249–1257.

47. Mistry, N.N., et al., Pulmonary perfusion imaging in the rodent lung using dynamic contrast-enhanced MRI. *Magn Reson Med*, 2008. **59**(2): pp. 289–297.

48. Frost, A., et al., Diagnosis of pulmonary hypertension. *Eur Respir J*, 2019. **53**(1): 1801904.

49. Guerrero, T., et al., Dynamic ventilation imaging from four-dimensional computed tomography. *Phys Med Biol*, 2006. **51**(4): pp. 777–791.

50. Ruaro, B., et al., Monitoring the microcirculation in the diagnosis and follow-up of systemic sclerosis patients: Focus on pulmonary and peripheral vascular manifestations. *Microcirculation*, 2020. **27**(8): p. e12647.

51. Eslick, E.M., M.J. Stevens, and D.L. Bailey, SPECT V/Q in lung cancer radiotherapy planning. *Semin Nucl Med*, 2019. **49**(1): pp. 31–36.

52. Castillo, E., et al., Quantifying pulmonary perfusion from noncontrast computed tomography. *Med Phys*, 2021. **48**(4): pp. 1804–1814.

53. Jang, B.S., et al., Generation of virtual lung single-photon emission computed tomography/CT fusion images for functional avoidance radiotherapy planning using machine learning algorithms. *J Med Imaging Radiat Oncol*, 2019. **63**(2): pp. 229–235.

54. Ge Ren, W.Y.H., Jing Qin, and Jing Cai, Deriving lung perfusion directly from CT image using deep convolutional neural network: A preliminary study. In *Artificial Intelligence in Radiation Therapy (Lecture Notes in Computer Science)* ed D. Nguyen et al., 2019. pp. 102–109. Cham: Springer International Publishing.

55. Ren, G., et al., Deep learning-based computed tomography perfusion mapping (DL-CTPM) for pulmonary CT-to-perfusion translation. *Int J Radiat Oncol Biol Phys*, 2021. **110**(5): pp. 1508–1518.

56. Porter, E.M., et al., Synthetic pulmonary perfusion images from 4DCT for functional avoidance using deep learning. *Phys Med Biol*, 2021. **66**(17): p. 175005.

57. Woo, S.-K., Generation of contrast-enhanced computed tomography image using deep learning network. *J Korea Soc Comput Inf*, 2019. **24**(3): pp. 41–47.

58. Santini, G., et al., Synthetic contrast enhancement in cardiac CT with Deep Learning. *arXiv preprint*, arXiv:1807.01779, 2018.

59. Kim, S.W., et al., The feasibility of deep learning-based synthetic contrast-enhanced CT from nonenhanced CT in emergency department patients with acute abdominal pain. *Scientific Reports*, 2021. **11**(1): p. 20390.

60. LeCun, Y., Y. Bengio, and G. Hinton, Deep learning. *Nature*, 2015. **521**(7553): pp. 436–444.

61. Singh, J. and A. Daftary, Iodinated contrast media and their adverse reactions. *J Nucl Med Technol*, 2008. **36**(2): pp. 69–74; quiz 76–7.
62. Elojeimy, S., et al., Overview of the novel and improved pulmonary ventilation-perfusion imaging applications in the era of SPECT/CT. *Am J Roentgenol*, 2016. **207**(6): pp. 1307–1315.
63. Matuszak, M.M., et al., Priority-driven plan optimization in locally advanced lung patients based on perfusion SPECT imaging. *Adv Radiat Oncol*, 2016. **1**(4): pp. 281–289.
64. Waxweiler, T., et al., A complete 4 DCT-ventilation functional avoidance virtual trial: Developing strategies for prospective clinical trials. *J Appl Clin Med Phys*, 2017. **18**(3): pp. 144–152.
65. Yamamoto, T., et al., Changes in regional ventilation during treatment and dosimetric advantages of CT ventilation image guided radiation therapy for locally advanced lung cancer. *Int J Radiat Oncol* Biol* Phys*, 2018. **102**(4): pp. 1366–1373.
66. Scheenstra, A.E., et al., Local dose-effect relations for lung perfusion post stereotactic body radiotherapy. *Radiother Oncol*, 2013. **107**(3): pp. 398–402.
67. Zhang, J., et al., Radiation-induced reductions in regional lung perfusion: 0.1–12 year data from a prospective clinical study. *Int J Radiat Oncol Biol Phys*, 2010. **76**(2): pp. 425–432.
68. Farr, K.P., et al., Loss of lung function after chemo-radiotherapy for NSCLC measured by perfusion SPECT/CT: Correlation with radiation dose and clinical morbidity. *Acta Oncol*, 2015. **54**(9): pp. 1350–1354.
69. Bucknell, N.W., et al., Functional lung imaging in radiation therapy for lung cancer: A systematic review and meta-analysis. *Radiother Oncol*, 2018. **129**(2): pp. 196–208.
70. Dongqing Wang, J.S., Jingyu Zhu, Xiaohong Li, Yanbo Zhen, and Songtao Sui, Functional dosimetric metrics for predicting radiation-induced lung injury in non-small cell lung cancer patients treated with chemoradiotherapy. *Radiation Oncology*, 2012. **7**: p. 69.
71. Gayed, I.W., J. Chang, E.E. Kim, R. Nuñez, B. Chasen, et al., Lung perfusion imaging can risk stratify lung cancer patients for the development of pulmonary complications after chemoradiation. *J Thorac Oncol*, 2008. **3**: pp. 858–864.
72. Dhami, G., et al., Framework for radiation pneumonitis risk stratification based on anatomic and perfused lung dosimetry. *Strahlenther Onkol*, 2017. **193**(5): pp. 410–418.

6

Imaged-Based Dose Planning Prediction

Dan Nguyen

UT Southwestern Medical Center, Dallas, TX, USA

CONTENTS

6.1 Introduction

Radiation therapy treatment planning has greatly improved over the past few decades, particularly with the advent of intensity-modulated radiation therapy (IMRT)[1-7] and volume-modulated arc therapy (VMAT)[8-14]. This has entirely shifted the treatment paradigm and improved the overall patient outcome and quality of life. However, with these improvements comes additional treatment complexity, leading to increased treatment planning time and higher plan variability. Optimization has become an integral part of IMRT and VMAT, but the non-convex nature of the algorithms used in the commercial treatment planning systems does not guarantee the global optimum, and the trajectory in how the planner navigates tuning the hyperparameters affects the final plan.

The current clinical treatment planning workflow is still performed meticulously in a trial-and-error fashion. After the patient is imaged and the contours are drawn, the treatment planner uses the commercial treatment planning system to generate a treatment plan. This involves iteratively fine-tuning hyperparameters—such as structure weights, dose-volume points, etc.—and running the optimization engine several times until the plan seems acceptable to the planner. The physician then reviews the plan and provides further feedback on further improvements they want. The planner then continues to step through the optimization process and fine-tunes the hyperparameters even further. This feedback loop between the physician and planner continues until the physician is satisfied with the plan and approves it for treatment. This is largely because the physician's objective is dependent on the accumulation of their knowledge and experiences, and is largely simplified into mathematical objectives for the optimizer to run.

In the last decade, major developments in artificial intelligence (AI) technologies and algorithms, and in particular deep learning (DL), have completely reshaped almost every major field—anywhere from self-driving cars to home goods. Healthcare has begun to

DOI: 10.1201/9781003243458-8

also see major developments with the use of AI technologies. Radiation oncology has great potential for large improvement using these AI technologies, especially with the improvements in computer vision and imaging[15-18], as well as decision-making[19-24]. Research in radiation oncology has already begun to investigate the use of DL models in many areas including diagnosis, imaging, segmentation, treatment planning, quality assurance, treatment delivery, and follow-up. In particular, this chapter will focus on the use of the medical image and segmentation data for dose prediction using AI technologies and algorithms.

6.2 Status

Prior to the recent wave of DL progress, initial efforts went into knowledge-based planning (KBP)[25-42]. These KBP algorithms took historical plan data and handcrafted features—including spatial information of organs at risk (OAR) and planning target volumes (PTV), distance-to-target histograms (DTH), overlapping volume histograms (OVH), structure shapes, number of delivery fields, etc.[34-43]—for teaching a machine learning model to predict either the structures' dose-volume histograms (DVH) or specific dose-volume constraints. These earlier KBP frameworks were limited in their data complexity, and the outputs were typically limited to 1D or 2D data, leaving the remainder of the dose distribution up to the expertise of the physician and planner to generate a final deliverable plan. In addition, the handcrafted features largely needed to be determined by trial and error, potentially yielding a sub-optimal set of features. Furthermore, since these features were manually handcrafted, it may have resulted in a loss of subtle but crucial information.

Research-wise, the major DL algorithms for dose prediction typically have fallen into one of two categories: 1) clinical dose prediction, and 2) pareto optimal dose prediction. The overall goal of clinical dose prediction models is to try to predict a clinically acceptable dose distribution. Typically, these frameworks take on a KBP flavor since they are learning how to predict these clinical dose distributions by learning from a large pool of historical clinically treated patients. Compared to traditional KBP algorithms, which could only predict DVHs or specific dose-volume criteria, the modern DL models are capable of predicting 3D dose distributions, which provide a much more precise description of the plan. Clinical dose prediction models have been developed for several cancer sites, including prostate[44-48], lung[49,50], and head and neck[51-54]. Figure 6.1 shows an example set of dose distributions, which include the ground truth dose and the DL dose prediction for a head and neck cancer patient. Pareto optimal dose prediction allows for the navigation of tradeoffs between the various PTV and OARs. These frameworks are more similar to MCO[55,56] with respect to being able to provide realistic tradeoffs. The major advantage of the Pareto optimal dose prediction is in its real-time inference capability, typically sub-second for most models. This allows for them to be used immediately after segmentation by clinicians in deciding the best tradeoffs for the patient. Compared to the clinical dose predictors, these models require a much larger dataset to be trained from, needing multiple plans representing the various tradeoffs for the same patient. These models need another input as well to specify what kind of tradeoff the user is

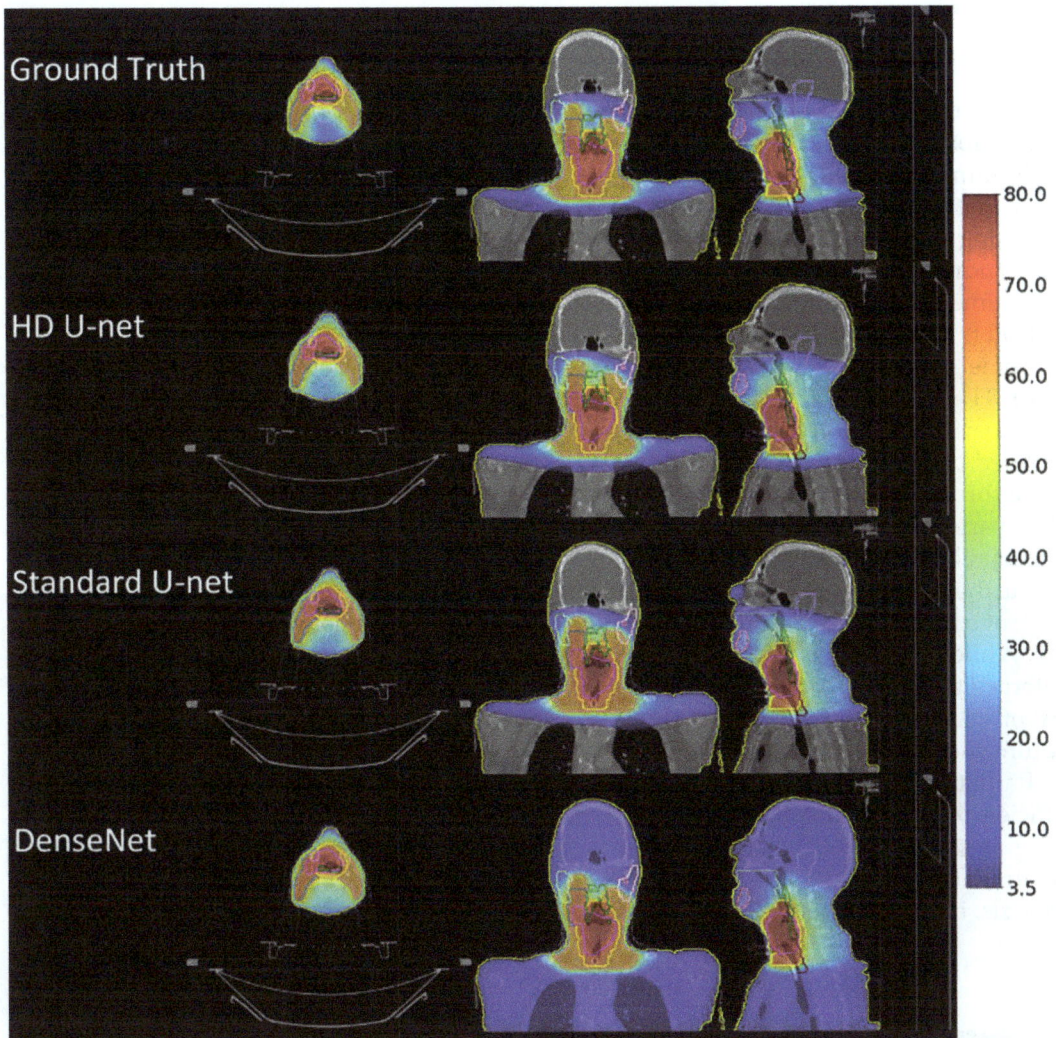

FIGURE 6.1

Dose washes of example patients from the test pool. The color bar is shown in units of Gy. The clinical ground truth dose is shown on the top row, followed by the dose predictions of the proposed model, HD U-net, and two other comparative models, Standard U-net and DenseNet. Low dose cutoff for viewing was chosen to be 5% of the highest prescription dose (3.5 Gy)[51].

looking for in regard to tradeoffs, which can be employed as structure weights[57,58], or by tuning the DVH directly[59], for example.

These dose prediction models can be used for various purposes such as for physician directive guidance and even for automatic treatment planning[53], which are still largely in development. In addition, by utilizing ensemble techniques or Monte Carlo Dropout[60], these DL-based dose prediction models can generate an uncertainty estimate—a way for the model to say "I don't know" when given data that it is not seen before[61]. The addition of uncertainty is crucial as the model goes from the research and development stage to the clinical deployment stage.

6.3 Current Challenges and Future Perspectives

AI technologies, and in particular DL algorithms, have made extensive progress in dose prediction, but there are still major hurdles to overcome. As these algorithms become clinically implemented, there are challenges of lateral and longitudinal variation in the plan data that is used to develop these models. Lateral variation refers to the differences in planning styles and trade-off preferences among different clinicians and institutions, while longitudinal variation refers to the evolution in planning styles and trade-off preferences over time. The dose prediction models will be biased toward the clinicians and the time period that the data is collected from.

Some major growing areas of AI are transfer learning (TL) and continuous learning (CL). TL is focused on leveraging the information from a source task/domain to achieve high performance on a target task/domain with potentially few target samples[62]. CL is focused on the ability of neural networks to continually obtain and adapt to new knowledge throughout their lifetime[63]. This can be used to address the lateral and longitudinal data variation problem. TL can also be used to quickly create models for every cancer site with a relatively lower data requirement.

Furthermore, as these models push toward implementation, there is a larger push to improve the model's interpretability and explainability, which are still large challenges for AI. While it's possible to add in uncertainty estimation[60,61] and attention modules[64,65], the interpretation of DL models is still largely left to the user. Interpretability and explainability of DL models is still on largely ongoing field of research that will be crucial for the future success of their implementation into the clinic.

Radiation therapy treatment planning will continue to strive for better plans at higher efficiency. With the explosion of AI, DL has begun to integrate itself into treatment planning at several levels. As these AI frameworks become more advanced and integrated, treatment planning will begin to shift toward more reliable, interpretable, and automated models, improving patient outcomes and quality of life.

References

1 Brahme, A. Optimization of stationary and moving beam radiation therapy techniques. *Radiotherapy and Oncology* **12**, 129–140 (1988).

2 Bortfeld, T., Bürkelbach, J., Boesecke, R. & Schlegel, W. Methods of image reconstruction from projections applied to conformation radiotherapy. *Physics in Medicine and Biology* **35**, 1423 (1990).

3 Bortfeld, T. R., Kahler, D. L., Waldron, T. J. & Boyer, A. L. X-ray field compensation with multileaf collimators. *International Journal of Radiation Oncology* Biology* Physics* **28**, 723–730 (1994).

4 Webb, S. Optimisation of conformal radiotherapy dose distribution by simulated annealing. *Physics in Medicine and Biology* **34**, 1349 (1989).

5 Convery, D. & Rosenbloom, M. The generation of intensity-modulated fields for conformal radiotherapy by dynamic collimation. *Physics in Medicine and Biology* **37**, 1359 (1992).

6 Xia, P. & Verhey, L. J. Multileaf collimator leaf sequencing algorithm for intensity modulated beams with multiple static segments. *Medical Physics* **25**, 1424–1434 (1998), doi:10.1118/1.598315.

7 Keller-Reichenbecher, M.-A. et al. Intensity modulation with the "step and shoot" technique using a commercial MLC: A planning study. *International Journal of Radiation Oncology* Biology* Physics* **45**, 1315–1324 (1999).

8 Yu, C. X. Intensity-modulated arc therapy with dynamic multileaf collimation: an alternative to tomotherapy. *Physics in Medicine and Biology* **40**, 1435 (1995).

9 Otto, K. Volumetric modulated arc therapy: IMRT in a single gantry arc. *Medical Physics* **35**, 310–317 (2008).

10 Xing, S. M. C., Wu, X., Takita, C., Watzich, M. & Xing, L. Aperture modulated arc therapy. *Physics in Medicine & Biology* **48**, 1333 (2003).

11 Earl, M., Shepard, D., Naqvi, S., Li, X. & Yu, C. Inverse planning for intensity-modulated arc therapy using direct aperture optimization. *Physics in Medicine and Biology* **48**, 1075 (2003).

12 Cao, D., Afghan, M. K., Ye, J., Chen, F. & Shepard, D. M. A generalized inverse planning tool for volumetric-modulated arc therapy. *Physics in Medicine & Biology* **54**, 6725 (2009).

13 Shaffer, R. et al. Volumetric modulated arc therapy and conventional intensity-modulated radiotherapy for simultaneous maximal intraprostatic boost: A planning comparison study. *Clinical Oncology* **21**, 401–407 (2009), doi:10.1016/j.clon.2009.01.014.

14 Palma, D. et al. Volumetric modulated arc therapy for delivery of prostate radiotherapy: Comparison with intensity-modulated radiotherapy and three-dimensional conformal radiotherapy. *International Journal of Radiation Oncology*Biology*Physics* **72**, 996–1001 (2008), doi:10.1016/j.ijrobp.2008.02.047.

15 LeCun, Y. et al. Backpropagation applied to handwritten zip code recognition. *Neural Computation* **1**, 541–551 (1989).

16 Krizhevsky, A., Sutskever, I. & Hinton, G. E. Imagenet classification with deep convolutional neural networks. *Advances in Neural Information Processing Systems* **25**, 1097–1105 (2012).

17 Girshick, R., Donahue, J., Darrell, T. & Malik, J. Rich feature hierarchies for accurate object detection and semantic segmentation. *Proceedings of the IEEE Conference on Computer Vision and Pattern Recognition*, 580–587 (2014).

18 Simonyan, K. & Zisserman, A. Very deep convolutional networks for large-scale image recognition. *arXiv preprint*, arXiv:1409.1556 (2014).

19 François-Lavet, V., Henderson, P., Islam, R., Bellemare, M. G. & Pineau, J. An introduction to deep reinforcement learning. *Foundations and Trends® in Machine Learning* **11**, 219–354 (2018).

20 Mnih, V. et al. Playing atari with deep reinforcement learning. *arXiv preprint*, arXiv:1312.5602 (2013).

21 Mnih, V. et al. Human-level control through deep reinforcement learning. *Nature* **518**, 529–533 (2015), doi: 10.1038/nature14236.

22 Watkins, C. J. & Dayan, P. Q-learning. *Machine Learning* **8**, 279–292 (1992).

23 Silver, D. et al. Mastering the game of go with deep neural networks and tree search. *Nature* **529**, 484 (2016), doi: 10.1038/nature16961.

24 Silver, D. et al. Mastering the game of go without human knowledge. *Nature* **550**, 354 (2017), doi: 10.1038/nature24270.

25 Zhu, X. et al. A planning quality evaluation tool for prostate adaptive IMRT based on machine learning. *Medical Physics* **38**, 719–726 (2011).

26 Appenzoller, L. M., Michalski, J. M., Thorstad, W. L., Mutic, S. & Moore, K. L. Predicting dose-volume histograms for organs-at-risk in IMRT planning. *Medical Physics* **39**, 7446–7461 (2012).

27 Wu, B. et al. Improved robotic stereotactic body radiation therapy plan quality and planning efficacy for organ-confined prostate cancer utilizing overlap-volume histogram-driven planning methodology. *Radiotherapy and Oncology* **112**, 221–226 (2014).

28 Shiraishi, S., Tan, J., Olsen, L. A. & Moore, K. L. Knowledge-based prediction of plan quality metrics in intracranial stereotactic radiosurgery. *Medical Physics* **42**, 908–917 (2015).

29 Li, N. et al. Automatic treatment plan re-optimization for adaptive radiotherapy guided with the initial plan DVHs. *Physics in Medicine and Biology* **58**, 8725 (2013).

30 Chanyavanich, V., Das, S. K., Lee, W. R. & Lo, J. Y. Knowledge-based IMRT treatment planning for prostate cancer. *Medical Physics* **38**, 2515–2522 (2011).

31 Good, D. et al. A knowledge-based approach to improving and homogenizing intensity modulated radiation therapy planning quality among treatment centers: an example application to prostate cancer planning. *International Journal of Radiation Oncology* Biology* Physics* **87**, 176–181 (2013).

32 Fogliata, A. et al. Assessment of a model based optimization engine for volumetric modulated arc therapy for patients with advanced hepatocellular cancer. *Radiation Oncology* **9**, 236 (2014).

33 Munter, J. S. & Sjölund, J. Dose-volume histogram prediction using density estimation. *Physics in Medicine & Biology* **60**, 6923 (2015).

34 Shiraishi, S. & Moore, K. L. Knowledge-based prediction of three-dimensional dose distributions for external beam radiotherapy. *Medical Physics* **43**, 378–387 (2016).

35 Wu, B. et al. Patient geometry-driven information retrieval for IMRT treatment plan quality control. *Medical Physics* **36**, 5497–5505 (2009), doi:10.1118/1.3253464.

36 Kazhdan, M. et al. A shape relationship descriptor for radiation therapy planning. *Medical Image Computing and Computer-Assisted Intervention* **12**, 100–108 (2009).

37 Wu, B. et al. Using overlap volume histogram and IMRT plan data to guide and automate VMAT planning: A head-and-neck case study. *Medical Physics* **40**, 021714 (2013), doi: 10.1118/1.4788671.

38 Wu, B. et al. Data-driven approach to generating achievable dose–volume histogram objectives in intensity-modulated radiotherapy planning. *International Journal of Radiation Oncology*Biology*Physics* **79**, 1241–1247 (2011), doi:10.1016/j.ijrobp.2010.05.026.

39 Tran, A. et al. Predicting liver SBRT eligibility and plan quality for VMAT and 4π plans. *Radiation Oncology* **12**, 70 (2017), doi:10.1186/s13014-017-0806-z.

40 Yuan, L. et al. Quantitative analysis of the factors which affect the interpatient organ-at-risk dose sparing variation in IMRT plans. *Medical Physics* **39**, 6868–6878 (2012), doi:10.1118/1.4757927.

41 Lian, J. et al. Modeling the dosimetry of organ-at-risk in head and neck IMRT planning: An inter-technique and interinstitutional study. *Medical Physics* **40**, 121704 (2013), doi:10.1118/1.4828788.

42 Folkerts, M. M., Gu, X., Lu, W., Radke, R. J. & Jiang, S. B. SU-G-TeP1-09: Modality-specific dose gradient modeling for prostate imrt using spherical distance maps of PTV and isodose contours. *Medical Physics* **43**, 3653–3654 (2016), doi:10.1118/1.4956999.

43 Folkerts, M. M. et al. Knowledge-based automatic treatment planning for prostate IMRT using 3-dimensional dose prediction and threshold-based optimization. *American Association of Physicists in Medicine* (2017).

44 Kandalan, R. N. et al. Dose prediction with deep learning for prostate cancer radiation therapy: Model adaptation to different treatment planning practices. *arXiv preprint*, arXiv:2006.16481 (2020).

45 Nguyen, D. et al. A feasibility study for predicting optimal radiation therapy dose distributions of prostate cancer patients from patient anatomy using deep learning. *Scientific Reports* **9**, 1076 (2019), doi:10.1038/s41598-018-37741-x.

46 Murakami, Y. et al. Fully automated dose prediction using generative adversarial networks in prostate cancer patients. *PloS One* **15**, e0232697 (2020).

47 Kajikawa, T. et al. A convolutional neural network approach for IMRT dose distribution prediction in prostate cancer patients. *Journal of Radiation Research* **60**, 685–693 (2019).

48 Sumida, I. et al. A convolution neural network for higher resolution dose prediction in prostate volumetric modulated arc therapy. *Physica Medica* **72**, 88–95 (2020).

49 Barragán-Montero, A. M. et al. Three-dimensional dose prediction for lung IMRT patients with deep neural networks: robust learning from heterogeneous beam configurations. *Medical Physics* **46**, 3679–3691 (2019), doi: 10.1002/mp.13597.

50 Shao, Y. et al. Prediction of three-dimensional radiotherapy optimal dose distributions for lung cancer patients with asymmetric network. *IEEE Journal of Biomedical and Health Informatics* **25**, 1120–1127 (2020).

51 Nguyen, D. et al. 3D radiotherapy dose prediction on head and neck cancer patients with a hierarchically densely connected U-net deep learning architecture. *Physics in Medicine & Biology* **64**, 065020 (2019), doi: 10.1088/1361-6560/ab039b.

52 Gronberg, M. P. et al. Dose prediction for head and neck radiotherapy using a three-dimensional dense dilated U-net architecture. *Medical Physics* **48**, 5567–5573 (2021).

53 Fan, J. et al. Automatic treatment planning based on three-dimensional dose distribution predicted from deep learning technique. *Medical Physics* **46**, 370–381 (2019).

54 Chen, X., Men, K., Li, Y., Yi, J. & Dai, J. A feasibility study on an automated method to generate patient-specific dose distributions for radiotherapy using deep learning. *Medical Physics* **46**, 56–64 (2019).

55 Craft, D. L., Hong, T. S., Shih, H. A. & Bortfeld, T. R. Improved planning time and plan quality through multicriteria optimization for intensity-modulated radiotherapy. *International Journal of Radiation Oncology* Biology* Physics* **82**, e83–e90 (2012).

56 Zarepisheh, M., Uribe-Sanchez, A. F., Li, N., Jia, X. & Jiang, S. B. A multicriteria framework with voxel-dependent parameters for radiotherapy treatment plan optimization. *Medical Physics* **41**, 041705 (2014), doi:10.1118/1.4866886.

57 Bohara, G., Barkousaraie, A. S., Jiang, S. & Nguyen, D. Using deep learning to predict beam-tunable pareto optimal dose distribution for intensity modulated radiation therapy. *arXiv preprint*, arXiv:2006.11236 (2020).

58 Nguyen, D. et al. Incorporating human and learned domain knowledge into training deep neural networks: A differentiable dose volume histogram and adversarial inspired framework for generating Pareto optimal dose distributions in radiation therapy. *Medical Physics* **47**, 837–849 (2019).

59 Ma, J. et al. Individualized 3D dose distribution prediction using deep learning. *Lecture Notes in Computer Science* **11850**, 110–118 (2019).

60 Gal, Y. & Ghahramani, Z. Dropout as a bayesian approximation: Representing model uncertainty in deep learning. *International Conference on Machine Learning*, **48**, 1050–1059 (2016) (PMLR).

61 Nguyen, D. et al. A comparison of Monte Carlo dropout and bootstrap aggregation on the performance and uncertainty estimation in radiation therapy dose prediction with deep learning neural networks. *Physics in Medicine & Biology* **66**, 054002 (2021).

62 Torrey, L. & Shavlik, J. in *Handbook of research on machine learning applications and trends: algorithms, methods, and techniques* 242–264 (IGI Global, Hershey, PA, 2010).

63 Parisi, G. I., Kemker, R., Part, J. L., Kanan, C. & Wermter, S. Continual lifelong learning with neural networks: A review. *Neural Networks* **113**, 54–71 (2019).

64 Xu, H. & Saenko, K. Ask, attend and answer: Exploring question-guided spatial attention for visual question answering. *European Conference on Computer Vision*, Proceedings, Part VII **14**, 451–466 (2016) (Springer International Publishing).

65 Vaswani, A. et al. Attention is all you need. *Advances in Neural Information Processing Systems* **30**, 5998–6008 (2017).

Section III

Applications of Intra-Modality Image Synthesis

7

Medical Imaging Denoising

Yao Xiao
The University of Texas MD Anderson Cancer Center, Houston, TX, USA

Kai Huang
The University of Texas Health Science Center at Houston, Houston, TX, USA

Hely Lin and Ruogu Fang
University of Florida, Gainesville, FL, USA

CONTENTS

DOI: 10.1201/9781003243458-10

7.1 Introduction

Noise disturbance is a fundamental characteristic of all medical imaging modalities. Noise disturbance may become a critical factor in determining the quality of the medical image, which is the basis for diagnosis, since noise could reduce the visibility of subtle but important structures, pathology, and changes, especially in low-contrast regions. It may change the result of the imaging-based diagnosis. Because of its importance, various kinds of noise removal (i.e., denoising) methods have been studied and applied to clinical imaging modalities. Now, the denoising process has become a mandatory and essential preprocessing step for further image analysis and clinical usage.

In this chapter, we first provide an overview of the denoising methods that are applied in imaging modalities which include computed tomography (CT), magnetic resonance (MR), ultrasound (US), and positron emission tomography (PET). Following the overview, two advanced deep-learning-based denoising techniques will be presented in detail.

7.2 Review of Medical Image Denoising Applications

7.2.1 Image Denoising Problem Statement

Mathematically, the problem of image denoising can be modeled as follows:

$$y = x + n$$

where y is the observed noisy image, x is the unknown clean image, and n is the additive white Gaussian noise (AWGN). The major challenges for image denoising are flat areas should be smooth, edges should be protected without blurring, textures should be preserved, and new artifacts should not be generated.

7.2.2 Denoising Methods

Image noise is a kind of random variation of brightness or color.

Image denoising methods represent a designed process of image transformation from the original image to a new image. In such a process, the role of noise pixels in the original image will be reduced or eliminated in the new image. There are many different image denoising methods and each method has a specific goal to achieve. Based on the ways of transforming from an original image to a new image, we can divide image denoising methods into two categories: (1) spatial domain methods and (2) transform domain methods. The category of spatial domain methods can be further divided into spatial domain filtering and variational denoising sub-categories.

7.2.2.1 Classical (Spatial Domain) Denoising Method

Spatial domain methods can be divided into two categories: spatial domain filtering and variational denoising methods.

7.2.2.1.1 Spatial Domain Filtering

Spatial filters make use of low pass filtering on pixel groups with the statement that the noise occupies a higher region of the frequency spectrum. For example, in linear filters, the convolution process is used for implementing the neighboring kernels as a neighborhood function. However, this may lead to the blurring of edges. Nonlinear filters help to preserve edges; the median filter is an example of a nonlinear filter. In the median filter, the ranking of the neighboring pixels is done according to the intensity or brightness level, and the value of the pixel under evaluation is replaced by the median value of the surrounding pixels. Spatial filters make use of low pass filtering on pixel groups with the statement that the noise occupies a higher region of the frequency spectrum. Normally, spatial filters eliminate noise to a reasonable extent but at the cost of image blurring, which in turn loses sharp edges. Example applications of spatial domain filtering for CT image denoising include linear filter, mean filter, wiener filter, nonlinear filter, and median filter[1], combining robust bilateral filtering with anisotropic diffusion filtering as anisotropic diffusion[2], and Bilateral Filter and Shearlet transform[3].

7.2.2.1.2 Variational Denoising Methods

Variational denoising methods use prior images and minimize an energy function to calculate the denoised image. The motivation for variational denoising methods is to maximize a posterior probability estimate. For the variational denoising methods, the key is to find a suitable image prior. Successful prior models include gradient priors, nonlocal self-similarity priors, sparse priors, and low-rank priors. The following are some variational denoising methods.

7.2.2.1.2.1 Total Variation (TV) Regularization
Total variation regularization is based on the statistical fact that natural images are locally smooth, and that pixel intensity gradually varies in most regions. The basic principle is that the signals with excessive and possibly spurious details have high total variation, which means the integral of the absolute image gradient is high. Based on this, reducing the total variation of the signal may remove unwanted details such as noise and preserve important details such as edges.

This method can effectively remove noise and retain sharp edges. However, textures tend to be over-smoothed; flat areas are approximated by a piecewise constant surface resulting in a stair-casing effect, which may cause the image to lose contrast. Example applications include brain tumor magnetic resonance imaging (MRI) image denoising[4,5].

7.2.2.1.2.2 Nonlocal Regularization
This method exploits the similarity between rectangle patches in the image to reduce noise. Unlike "local mean" filtering, which takes the mean value of a group of pixels surrounding a target pixel to smooth the image, nonlocal filtering takes a mean of all pixels in the image, weighted by how similar these pixels are to the target pixel. The resulting denoised image is recovered as a weighted linear combination of all the pixels in the noisy image. This results in much greater post-filtering clarity, and less loss of detail in the image compared to local mean algorithms. Example applications include Gamma, Poisson, and Gaussian noise models[6], dynamic PET images, and "deep learning prior" combined with regularization[7].

7.2.2.1.2.3 Sparse Representation   Sparse representation is a way to represent data that can be formed by a linear combination of some basic elements. A high-dimensional signal can be recovered with only a few linear measurements provided that the signal is sparse or nearly sparse. Image structures such as edges, corners, and textures can be represented as sparse vectors or patches. These elements in the library do not necessarily need to be orthogonal and can be an over-complete spanning set. This problem setting also allows the represented signal to be of higher dimensionality than one of the signals being observed. The above two properties allow multiple representations of the same signal but provide improvements in representational sparsity and flexibility. The sparse representation model can be learned from a dataset and from the image itself. Applications include brain CT and MRI[8], brain MRI[9], and multimodality image fusion and denoising[10].

7.2.2.1.2.4 Low-rank Minimization   Low-rank minimization tries to approximate high-dimensional data with low-dimensional data in a way that uses less efforts to represent. In a low-rank minimization problem, a cost function is used to measure the fit between the given data and the approximation data, with the constraint that the approximation data has reduced rank. The rank constraint is related to a constraint on the complexity of a model that fits the data. Low-rank minimization problems are similar to parameter optimization problems. However, many low-rank models have no explicit solution. Example applications include brain MRI[11,12], and multimodality image fusion and denoising[10].

7.2.2.2 Transform Techniques

Initially, transform domain methods were developed from the Fourier transform. However, since then, a variety of transform domain methods gradually emerged, such as cosine transform, wavelet domain methods, block-matching, and 3D filtering. Transform domain methods use the characteristics of image information and the difference in noise in the transform domain.

7.2.2.2.1 Transform Domain Filtering

Different from classical spatial domain filtering, transform domain filtering first transforms the given noisy image into another domain. Then, it removes the noisy information on the transformed image based on distinct characteristics of the image and its noise. For example, higher frequencies correspond to larger coefficients denoting the edge or details of the image, while the noise has lower frequencies and smaller coefficients. Based on different transform functions, transform domain filtering methods can be divided into data-adaptive and non-data-adaptive methods.

7.2.2.2.2 Data-adaptive Transform

The main drawback of this method is the high computational cost because they use sliding windows and require a sample of noise-free data or at least two image frames from the same scene. However, in some applications, it might be difficult to obtain noise-free training data.

7.2.2.2.2.1 Independent Component Analysis (ICA)   ICA is a method for solving blind source separation problems, which can separate data into informational components or source signals. The assumption is that different signals are from different physical processes such that they are statistically independent. Considering this, extracting these statistically independent signals from signal mixtures can be done when these extracted signals were from different physical processes. ICA methods have been applied in brain MRI image denoising for neuroscience research[13,14].

7.2.2.2.2.2 Principal Component Analysis (PCA)   PCA is a method for separating potentially correlated observations into linear uncorrelated components. The data can be decomposed into a low-rank component that captures patterns and a sparse component that indicates noise. Applications using PCA in medical image denoising include spatio-temporal arterial spin labeling perfusion MRI signal denoising[15], diffusion-weighted MRI[16], and multi-exponential MRI relaxometry[17].

7.2.2.2.3 Non-data-adaptive Transform

7.2.2.2.3.1 Spatial-frequency Domain   Spatial-frequency domain filtering methods use low pass filtering by designing a frequency domain filter that passes all frequencies lower than a cut-off frequency and attenuates all frequencies higher than that cut-off frequency. For example, in the Fourier transform, image information mainly spreads in the low-frequency domain, while noise spreads in the high-frequency domain. Thus, we can remove noise by selecting specific transform domain features and transforming them back to the image domain. Example applications include PET[18] and MRI[19].

7.2.2.2.3.2 Wavelet Domain   Wavelet transform decomposes the input data into a scale-space representation. It has been proven that wavelets can successfully remove noise while preserving the image characteristics, regardless of its frequency content. Similar to spatial domain filtering, filtering operations in the wavelet domain can also be subdivided into linear and nonlinear methods. However, the wavelet transform relies heavily on the selection of wavelet bases. If the selection is inappropriate, images shown in the wavelet domain cannot be well represented, causing a poor denoising effect. Example applications include multimodalities US, CT, MRI[20], and MRI CT[21].

7.2.2.2.4 Block-Matching and 3D (BM3D) Filtering

BM3D is a two-stage nonlocal collaborative filtering method in the transform domain. In this method, similar patches are stacked into 3D groups by block-matching, and the 3D groups are transformed into the wavelet domain. Then, hard thresholding or Wiener filtering with coefficients is employed in the wavelet domain. Finally, after an inverse transformation of coefficients, all estimated patches are aggregated to reconstruct the whole image. However, when the noise gradually increases, the denoising performance of BM3D greatly decreases, and artifacts are introduced, especially in flat areas.

Block-matching and 4D filtering (BM4D) methods are an extension of BM3D for volumetric data. The 4-D transformation applied to the group simultaneously exploits the local and nonlocal correlation of voxels. Example applications for BM3D include US[22] and CT MRI[23]. Example applications for BM4D include MRI[24] and CT[25].

7.2.2.3 Deep Learning (DL)-based Image Denoising Methods

Many state-of-the-art medical image-denoising methods are based on deep learning techniques. In deep learning, when training data sets that contain pairs of noisy and associated clean images are available, the deep-learning-based denoising methods will train the model by optimizing a loss function on the training set. This type of deep learning method belongs to supervised learning. Sometimes, when there are not sufficient pairs of noisy-clean images, semi-supervised (with limited paired images) and unsupervised learning (with noisy images only) methods will be considered. There are many different types of DL architectures such as deep neural networks, deep belief networks, deep reinforcement networks, recurrent neural networks, and convolutional neural networks (CNN). Deep learning has been used in medical image denoising in a variety of ways. Example applications are PET image denoising using unsupervised deep learning[26], dynamic PET Image Denoising[27], unpaired Image Denoising via Wasserstein Generative Adversarial Network (GAN) in Low-Dose CT Image[30], supervised learning with cycleGAN for low-dose FDG PET image denoising[28], low-dose CT Image Denoising Using Unpaired Deep Learning Methods[29], MR image denoising, and reconstruction using unsupervised deep learning[31].

7.3 Convolutional Neuron Network (CNN) for Medical Image Denoising and Resolution Restoration

Convolutional neural network (CNN) is specific type of deep learning. In CNN, each neuron in the hidden layer has connections with only a very small portion of neurons in the previous or next layer. This characteristic allows CNN to significantly reduce computation time and the need for it to be able to store and process very large amounts of data. It has been used in image data-related studies such as image recognition, image processing, and image classification. In this section, we will present how CNN be used for image denoising and high-resolution restoration.

First, we may analyze the noisy and low-resolution image data to determine how many kernels are needed and the sizes of the kernels. The ways for treating relatively larger areas of noise and tiny areas of noise are different. For a relatively larger area of noise, we need to use a relatively larger kernel size to capture it. For a tiny area of noise, since a relatively larger kernel is not sensitive to tiny areas of noise, we need to use a relatively smaller kernel to capture it. Such an analysis may lead us to include multiple kernels with different sizes in our CNN model design.

Next, we will design and develop our CNN-based denoising and high-resolution restoration system. The first layer in the CNN system is input layer. It is used to read and store medical images of data. After the input layer are our hidden layers. These hidden layers are organized into blocks of kernels. Each block of kernels includes one image padding layer and a follow up kernel layer. This is because for any sized image, after it passes through the kernel, also known as a convolutional matrix with a size larger than 1x1, the filtered image will be a size shrunk image. To make sure our input image keeps the same image size as the original image size, we need to add a padding layer in front of our kernel layer. To make sure the padding does not affect the filtered image, we set the kernel for the padding layer to be size 1x1 with a matrix value of 1. The values of padding pixels are 0s. For the padding, we need to add the same number of pixels in every direction of the image. The number of pixels to add in each direction is dependent on the size selected for the kernel. For example, if the size of the kernel is 7x7 then we will add 3 pixels of padding with the value of 0 in every direction of the image.

A kernel, also called kernel filter, is a matrix. It is used to transform the input image to a filtered image. It will capture the values of a patch in the input image and map it to one pixel value on the filtered image through the dot product between the values of the patch and the kernel matrix. We can slide the kernel over the image one pixel by one pixel and do kernel filtering. After the kernel slides through the whole image, we get a filtered image. This filtered image will be further mapped through rectified linear activation function also called Rectified Linear Unit (ReLU).

ReLU is defined as such a mapping: if the value is negative, it will be 0 and if the value is positive, it will be the value itself. ReLU plays a very important role in noise reduction or removal. After kernel values are updated to contains both negative and positive values, the pixels in the noisy area are very likely to be filtered as negative values. Then, ReLU does further mapping to make it 0. Pixels with values of 0 have no role in mapping the next filtered image in the next layer.

In CNN, the values of kernels are unknown parameters. When setting up the kernel, these unknown parameters are initialized with some values. The CNN model training process is also a process to keep updating the unknown parameters of the kernels. To update kernel parameters, all the noisy images from our training data set will pass through all the kernel layers and come to the pooling layer. In the pooling layer, all noisy images have been transformed into filtered images. These filtered images together will participate in the comparison with all the associated clean and high-resolution images. The difference is defined as the loss function. If the comparison passes through the criteria for denoising and resolution restoration, then the parameters of all kernels will be set down, and the CNN model system is trained. If the comparison does not pass through the criteria for denoising and resolution restoration, then the parameters of all kernels will be updated by minimizing the loss function and all filtered images in the pooling layer will be used as input images to pass through all the kernels again until the comparison passes the criteria.

To speed up the updating of kernel parameters, batch normalization may be used inside the kernel layers. It adds a batch normalization (BN) process. The kernel layer with kernel filter, ReLU activation function, and batch normalization is called CONV + ReLU +BN.

In the later section of this chapter, a CNN application for Radiation Reduction in CT Perfusion will be presented.

7.4 Different Medical Image Modalities

7.4.1 Computed Tomography (CT)

CT images are essentially a collection of X-ray images at various angles. They are produced by transmitting X-rays through the object at many angles. The attenuated X-rays, once detected, contain information about the attenuation coefficients of the object through the path the photons have traveled. Thus, in CT images, each pixel encodes the attenuation coefficients of the corresponding tissue's voxel. Noise is the variation of the attenuation coefficients detected between voxels; even if a homogeneous water phantom of a defined CT number is imaged, each imaged voxel would have a slightly different value due to noise. High noise reduces low-contrast detectability. Such information can be critically important in disease diagnosis and prevention. Noise is statistical; we can decrease it by increasing the number of photons received by a voxel, that is, increasing the radiation dose delivered. However, reducing noise in half would require a four-fold increase in dose. The radiation used in CT or X-rays is ionizing. Thus, it is important to limit the amount of dose delivered as an increased dose means increased risks for patients. A significant amount of research has been devoted to decreasing noise while keeping radiation dose at a safe level.

There are various sources for random noises in CT, including quantum noise caused by a limited number of photons detected, and electronic noise stemming from electrons in the electrical component of the device. Quantum noise can be modeled by Poisson distribution, and electronic noise can be modeled by Normal distribution.

Despite changing the acquisition parameters or physical components of CT devices, noise can be influenced by reconstruction algorithms and post-processing techniques. In the following section, we will explain reconstruction algorithms including filtered back-projected, iterative reconstruction, and image post-processing techniques including filter-based, modal-based, and deep-learning-based.

The noise of CT can be measured by computing the standard deviation of a homogeneous region in the CT image of a phantom.

7.4.1.1 Reconstruction

7.4.1.1.1 Filtered Back Projection

Filtered back projection is a common reconstruction technique used for CT. Back projection is a mathematical procedure where each projection profile is traced back to each unknown voxel of an imaged object to form simultaneous equations. A computer can solve those simultaneous equations quickly to obtain the values of each voxel.

However, this would result in blurred images due to the geometry of the back projection simply summing projections around 360 degrees results in characteristic $1/r$ blurring. Such blurring can be corrected using mathematical convolution with filters before back projection.

A ramp filter is the most common and basic filter to use. The filter amplifies the signal linearly by increasing frequencies. This is effective because the $1/r$ blurring is $1/f$ blurring in the frequency domain after Fourier transform. However, when high frequencies are amplified, noises are amplified as well. To overcome this limitation, in practice, modified ramp filters with decreased amplification in higher spatial-frequency regions where quantum noise dominates are used[32]. Optimal ramp filters may be different for various applications such as soft tissue or bone. Different CT manufacturers would give their filters different names based on the application.

Filtered back projection is highly computationally efficient[33] but may cause blurring and lower contrast[33]. Quantum noises can be amplified and create streak artifacts.

7.4.1.1.2 Iterative Reconstruction

Iterative reconstruction is an iterative approach to reconstructing images. The process starts with a guess of the image, then the guess is compared with the measured value. Based on the differences between the guess and measured values, corrections are made to the guess to bring them into an agreement. This process repeats until the reconstructed images and measured images become the same or within certain limits of each other. Iterative reconstruction is computationally intensive. However, with increased computational capability, most vendors provide reconstructive solutions in their CT scanners. Vendors provide different iterative algorithms that correct images in different manners. For example,

Siemens Healthineers offers Sinogram Affirmed Iterative Reconstruction (SAFIRE), and GE Healthcare offers Adaptive Statistical Iterative Reconstruction (ASiR)[34]. Iterative reconstruction is superior to filtered back projection in its ability to produce images with a higher signal-to-noise ratio. This means that noises can be reduced while maintaining good image quality, thus allowing patient radiation dosage to be reduced.

7.4.2 Magnetic Resonance Imaging (MRI)

MRI uses a magnetic field and radio waves to generate images of the body's tissues and organs. The human body contains billions of hydrogen atoms. When the human body is placed in a magnetic field, the atoms line up in the direction of the field and the positively charged hydrogen atoms become almost uniformly aligned. To deflect the aligned protons, a radio frequency wave is then pulsed into the body. When the protons return to their original position, energy is released. The intensity of the received signal is then plotted in grayscale, and the cross-sectional images are built up. The raw MR images are complex values that represent the Fourier transform of magnetization distribution. Noises in MRI images are primarily Rician. In low-intensity regions of magnitude images, Rician noise becomes Rayleigh noise. In high-intensity regions of magnitude images, Rician noise follows Gaussian distribution[35]. Noise in MRIs causes random fluctuations, which reduces the image contrast due to signal-dependent data bias. This disturbs the precise qualitative and quantitative evaluation as well as feature detection in MR images. The signal-to-noise ratio in MRI is high in most clinical applications, resulting in the manifestation of Gaussian noise. The primary sources of noise in MRI images are electronic interferences in the receiver circuits, radiofrequency emissions due to the thermal motion of the ions in the patient's body, and the measurement chain of MRI scanners, i.e., coils, electronic circuits etc[36].

The magnitude of the noise distribution of the MR signal is the square root of the sum of the squares of two independent Gaussian variables. The noise in MR images follows a stationary Rician distribution with constant noise power at each voxel[36].

7.4.2.1 MRI Denoising Approaches

Noises in MRI images can be reduced by signal averaging. However, signal averaging is not desirable as it requires increased speed during clinical MR imaging. Post-processing techniques have been used extensively in MRI denoising. Post-processing using linear filters including spatial and temporal filters can be used to remove Gaussian noises. However linear filters can introduce unwanted aliasing artifacts and cause edge blurring without increasing the signal-to-noise ratio. The blurring reduces the diagnostic value of images as small lesions may become less visible. Therefore, nonlinear filters were introduced to preserve sharp edges. Examples include Anisotropic Diffusion Filter (ADF)[37], Nonlocal Means (NLM)[38] filters, and combinations of domain and range filters. Nonlinear filters work to assign weights to pixels based on their statistical characteristics. This reduces blurring by separating pixels with non-similar patterns. ADF uses a partial differential equation and edges preserved by only averaging pixels orthogonal to the gradient of images[39]. NLM takes a weighted average of all the pixels and their neighboring pixels by considering the similarity between them. NLM operates under the assumption that there is a high degree of redundancy in natural images[38]. A combination of domain and range filters can also be used to denoise MRI images. Bilateral filtering, proposed by Tomasi and Manduchi[40], is a class of nonlinear filters that is a non-iterative alternative to ADF. Images are filtered using a weighted average by the geometric and photometric similarities of neighboring pixels[41]. Based on the concept of bilateral filtering, Wong proposed a trilateral filtering method that considered local structural similarity as well[41].

Another denoising technique commonly used in MRI is the transform domain. The approach includes wavelet transform, curvelet transform, and contourlet transform. As mentioned in the previous sections, the wavelet-based denoising technique transforms images to the wavelet domain to separate noise from features. After suppressing noise in the wavelet domain, signals are inversely transformed back to form denoised images. Weaver et al.[42] were among the first to apply basic wavelet transformation to MRI images. The limitation was that image features with smaller wavelet transforms than that of noises would be removed. Most of the wavelet transform work was performed based on additive Gaussian noise models. Nowak[35] proposed a method that explicitly accounts for Rician noise and achieved superior results than the standard wavelet method in situations of low signal-to-noise ratio. Wood and Johnson[43] used wavelet packet analysis to denoise images with a very low signal-to-noise (SNR) ratio contaminated by Rician noise and provided a more compact signal representation than wavelet transform. In addition to wavelet transform, curvelet transform was proposed by Starck et al[44] to overcome the inherent tradeoff between preserving fine edges and difficulties in estimating a large number of wavelet coefficients in wavelet transform. Contourlet transformation introduced by Do et al[45] can capture intrinsic geometrical structures in images and perform superior to the wavelet transformation in preserving smooth curves.

Another method in denoising MRI images is based on statistical estimation of noise variance in images. There are different approaches for estimating noises including maximum likelihood (ML), linear minimum mean square error estimate (LMMSE), phase error estimation, nonparametric estimation, and singularity function analysis. Sijbers et al [46] used ML to estimate Rician noise in MRI images. Aja-Fernandez et al used LMMSE for describing Rician noise[47]. Rajan et al expanded the ML solution to estimate non-central Chi distribution which is the noise characteristics displayed by magnitude images from multi-coil systems[48]. Tisdal and Atkins proposed a phase error estimation scheme that is specific to complex-valued MRI signals and offers the potential for image denoising without over-smoothing[49]. Awate and Whitaker[50] used nonparametric density estimation for character-izing neighborhood structures. The higher-order statistics of image neighborhoods are used as image priors within a Bayesian denoising framework. Singularity function analy-sis (SFA) was proposed by Luo et al[51,52] Luo et al used a 2D SFA model to reconstruct images based on partial spectrums and then achieved denoising through averaging of reconstructed images[52]. One-dimensional SFA was also experimented with to replace spec-tral data with low SNR while preserving spectral data with high SNR[51]. These methods allow efficient denoising with minimal distortion.

7.4.3 Positron Emission Tomography

PET is a nuclear medicine-based imaging technology that enables the visualization of dif-ferent functional and metabolism-related processes in the body. In PET, radioactive nuclide particles like 15O, 11C, 18F, and 13N are injected into the subject and their tracer concen-tration is recorded in the form of 3-D images with the help of computer analysis[36]. The diseased areas often have higher levels of chemical activity and the radioactive substances in these areas get collected. PET scans often spot these areas before other imaging scans can predict them. PET provides a uniform and intensity-independent noise distribution, which is Gaussian in nature, over the whole image. The prevalence of noise in these images limits the comprehension of the changes induced in the pathological lesions. It impairs the visualization potential and affects the precision of the quantitative procedure performed with these images.

The major sources of noise in PET are inherent random variations in the indefinite photon count, electronic recorders, and detector systems. The total number of events recorded by a PET system includes true coincidences, random coincidences, and scatter coincidences. The random and scattered coincidences are sources of noise. Random coincidences happen when detectors register photons from different nuclear transformations as from the same annihilation. The scatter coincidences occur when scattered photons were registered to come from the same annihilation.

Since random coincidences are proportional to the square of activity in the patient, they can be measured and subtracted from measured coincidences. However, scatter coincidences are difficult to correct since they are the results of true coincidences. Additionally, like CT, reconstruction algorithms and filters can reduce noise.

7.4.3.1 PET Denoising Approaches

The clinical use of iterative reconstruction has been widely accepted as computational speed advanced and more efficient algorithms were developed[53]. Like CT, compared to analytical reconstruction techniques such as filtered back projection, iterative reconstruction produces superior images with reduced noises. Further improvements on the SNR of PET in the image domain use post-processing filters such as Gaussian moving average filters[54]. However, such methods not only improve SNR but also smooth images, increase partial volume effects, and reduce quantification accuracy, which is critical for PET to be clinically useful. Edge-preserving filters such as bilateral filtering and total variation minimization were examined to reduce noises while maintaining accurate SUV values[54,55]. PET images can also be denoised in the frequency domain through wavelet or curvelet transforms[56,57].

7.4.4 Ultrasound

Medical ultrasound is a modality that uses sound waves with frequencies exceeding the range of human hearing to produce real-time images of tissues in the body. A typical medical ultrasound system consists of a transmitter probe pulsing signals ranging from 20KHZ to several gigahertz and a receiver display. The sound waves disperse through the body and reach between soft tissue or fluid tissue or bone. These waves are reflected by the coplanar surfaces in the body and can be displayed on the computer[36]. The speed at which sounds propagate depends on the stiffness and density of the medium. The differences in speed at tissue boundaries create contrast in ultrasound images, thus allowing different materials to be identified[32].

US images are inherent of lower visual quality due to varying temporal and acoustic variations. The visual quality of these images is further lowered by the presence of amplifier noise, impulsive noise, or speckle noise. Gaussian noise is the typical representative of amplifier noise occurring due to numerous thermal and electronic fluctuations in the circuits. Speckle noise is a kind of granular noise, resulting from constructive and destructive interference of the temporal coherent sound waves in the system. Maintaining optimal low-contrast resolution is critical to the detection of small objects in images. This requires low noise with a high contrast-to-noise ratio, especially when objects to an image are deep as the ultrasound beam attenuates, exponentially causing noise to increase, and contrast, to decrease with depth. Sources of noise can include electronic amplifiers of the system which generate electronic noises. Speckle noise is random and can therefore be reduced by averaging multiple imaging frames[32]. However, frame averaging would cause a lower frame rate and reduced spatial resolution.

7.4.4.1 US Denoising Approaches

Much of the denoising efforts for ultrasound images have been focusing on despeckling, that is, reducing the speckle noises. The presence of speckle noise significantly reduces the diagnostic values of ultrasound images. Speckle noises are multiplicative and locally correlated. Commonly used filters in ultrasound images include Lee[58,59], Frost[60], Kuan[61], and Gamma maximum a posteriori (MAP)[62,63], anisotropic diffusion filters[64], and bilateral filters[65]. Lee and Kuan's filters both are minimum mean square error filters that use local statistics as a basis for spatially enhancing image features. The filtered pixel is replaced by a value calculated from the neighboring pixels. Kuan's filter has a different weighting function than Lee's filter. Frost filter preserves edges in images by applying an exponentially damped filter that uses local statistics. The filtered pixels depend on the damping factor, distances from the filter centers, and the coefficient of variation, which is the ratio of the local variance to the mean of images. MAP filter considers a Gamma distributed intensity rather than a Gaussian distributed intensity optimal for Lee and Kuan filters. Speckle-reducing anisotropic diffusion (SRAD) was introduced to overcome the limitations that most filters are dependent on the size and shape of filter windows[64]. Bilateral filtering, another edge-preserving smoothing filter, has also been applied to ultrasound images[65].

Wavelet transforms have also been applied to the post-processing of ultrasound images. Wavelet methods can be divided into three methods[66]: thresholding[67,68], Bayesian estimation[69,70], and coefficients correlation methods[14]. Using thresholding methods, wavelets with coefficients smaller than the pre-determined thresholds would be removed. The challenge with this method is the difficulty in determining the appropriate threshold for applications. Using the Bayesian estimator, Xie et al estimated noise-free wavelet coefficients by shrinkage functions based on locally weighted averages[69]. Pizurica et al proposed to use prior knowledge about correlations of image features to perform classifications on coefficients. Such classifications are then used to separate image features and noises by estimating the statistical distribution of coefficients that represent significant image features[71].

7.5 Deep Learning Approaches for CT Denoising

7.5.1 Convolutional Neuron Network Approached Design Optimization for Medical Image Denoising

Convolutional Neuron Network (CNN) has been used in many aspects of image analysis and processing. Most of the designs come from designers' knowledge and experience with CNN models and medical image data sets. Is there a way to automatically create a high-quality CNN model design without relying on designers' knowledge and experience? We will explore the answer to the above question step by step. First, we need to identify what adjustable parameters can affect the CNN model's performance. We find that the number of kernels, sizes of kernels, order of kernel blocks, values of kernels, and type of activation function are notable parameters that affect CNN model performance. Can we adjust the above list of parameters automatically to find the best or close to the best CNN model design?

Using the traditional way, the answer is no. There are too many different sets of parameters in the CNN model design to be trained and tested. It will consume a very large amount of CPU workload and running time. A very powerful search technique, the Genetic Algorithm, came to researchers' vision.

The Genetic Algorithm (GA) is a technique used for searching optimization. This algorithm adopts the principle of natural evolution. In this algorithm, best-fit individuals are selected as seeds to produce the next generation, and poor-fit individuals are eliminated. The next generation is produced either through exchanging genes between the best-fit individuals or updating the genes of the selected best-fit individuals. Such a process will be repeated one generation at a time until the best criteria are reached.

How to use the GA technique in our CNN model design optimization for medical image denoising and resolution restoration? Theoretically, we can use the GA technique on our CNN model design optimization. However, it may still take a very long time and consume a large amount of CPU. It may still not be practical. To overcome this challenge, we need to use GA with deep customization to make it serves our needs.

"Deep Evolutionary Networks with Expedited Genetic Algorithms for Medical Image Denoising technique"[72] is an article that provides an example of how to use GA with deep customization in the CNN model design optimization process for the medical image denoising task.

The first customization shown in the article is that instead of using all the images from the training dataset, we use a randomly sampled set of images from the training dataset. This will reduce the CPU workload and consumption time significantly. Although the best CNN model design from this customization is a statistics sense of best, it still gives us a very meaningful foundation to find the CNN model design we need.

The second customization shown in the article is a strategy called the experience-based greedy exploration strategy. The core idea for this strategy is that all the good genes found from published articles or that have been proved through the GA process itself will be preserved and participate in producing the next generation.

The following picture comes from the above article "Deep Evolutionary Networks with Expedited Genetic Algorithms for Medical Image Denoising technique". It shows the customized GA process in the CNN design optimization process (Figure 7.1).

7.5.2 STIR-Net: Spatial-Temporal Image Restoration Net for CT Perfusion Radiation Reduction

Computed Tomography Perfusion (CTP) imaging is a cost-effective and fast approach to provide diagnostic images for acute stroke treatment. Its cine scanning mode allows the visualization of anatomic brain structures and blood flow; however, it requires contrast agent injection and continuous CT scanning over an extended time. In fact, the accumulative radiation dose to patients will increase health risks such as skin irritation, hair loss, cataract formation, and even cancer. Solutions for reducing radiation exposure include reducing the tube current and/or shortening the X-ray radiation exposure time. However, images scanned at lower tube currents are usually accompanied by higher levels of noise and artifacts. On the other hand, shorter X-ray radiation exposure time with longer scanning intervals will lead to image information that is insufficient to capture the blood flow dynamics between frames. Thus, it is critical for us to seek a solution that can preserve the image quality when the tube current and the temporal frequency are both low. We propose STIR-Net in

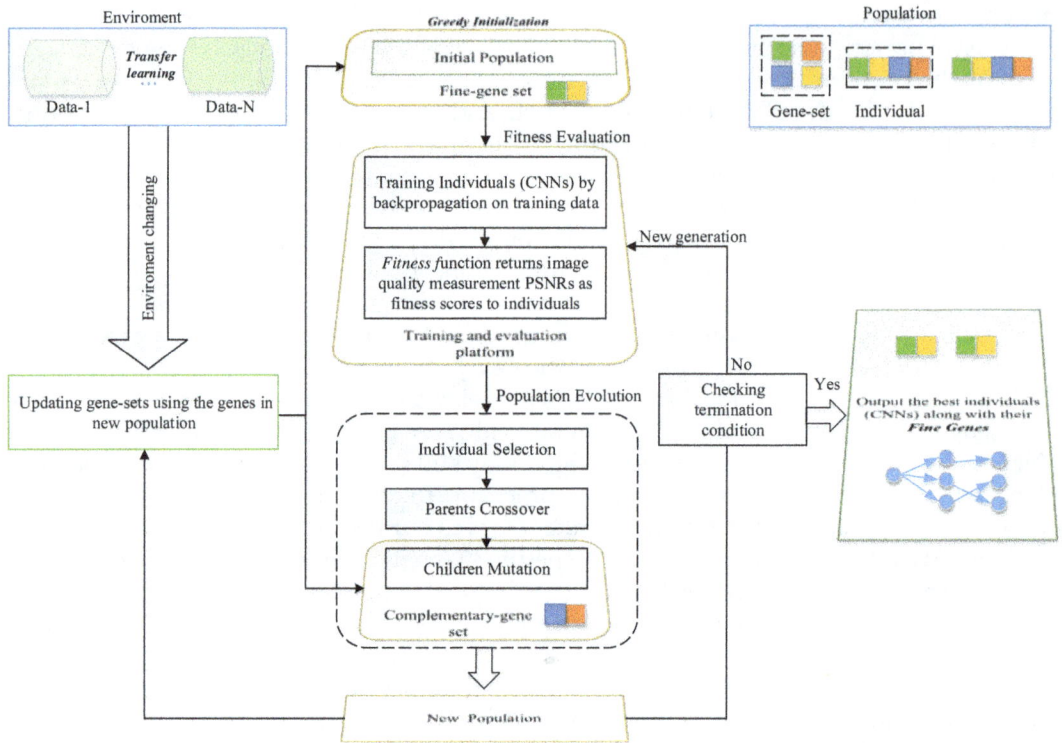

FIGURE 7.1

Overview of the proposed method composed of fitness evaluation and population evolution. The CNN architecture is trained on medical images using a fitness score. The individual networks labeled with the fitness scores are sent to individual selection. The surviving individuals are presented as parents for crossover and mutation. (Reprinted from Med. Image Anal., Liu P, El Basha MD, Li Y, Xiao Y, Sanelli PC, Fang R, Deep Evolutionary Networks with Expedited Genetic Algorithms for Medical Image Denoising, 2019, with permission from Elsevier.)

this section, an end-to-end spatial-temporal convolutional neural network structure, which exploits multi-directional automatic feature extraction and image reconstruction schema to recover high-quality CT slices effectively. With the inputs of low-dose and low-resolution patches at different cross-sections of the spatio-temporal data, STIR-Net blends the features from both spatial and temporal domains to reconstruct high-quality CT volumes. In this study, we finalize extensive experiments to appraise the image restoration performance at different levels of tube current and spatial and temporal resolution scales. The results demonstrate the capability of our STIR-Net to restore high-quality scans at as low as 11% of the absorbed radiation dose of the current imaging protocol, yielding an average of 10% improvement for perfusion maps compared to the patch-based log-likelihood method.

In this section, we propose an end-to-end Spatial-Temporal Image Restoration Net (STIR-Net) for CTP image restoration[73]. This structure consists of two main components: Super- Resolution Denoising Nets (SRDNs) and a multi-directional conjunction layer that addresses image super-resolution (SR) and denoising in both spatial and temporal cross-sections. The contributions of this work are five-fold:

1) SRDN's patch representation layer extracts features from both the spatial and temporal dimensions of the CTP volume as cross-sections, which allows our model to present spatial-temporal details at the same time.

2) SRDN has the ability to perform image SR and denoising individually and simultaneously. It also can handle multi-level noise and multi-scale resolution and sampling.

3) We integrate multiple SRDNs based on different cross-sections into a multidirectional network, which can boost the performance further than individual cross-sections.

4) The results of the experiments demonstrate the effectiveness of STIR in the recovery of low-radiation-dose CTP images. STIR-Net can provide practical solutions for radiation dose reduction from three aspects (low tube current, decreased temporal sampling rate, and poor spatial resolution) with comparable image quality to the standard dose protocol.

5) We also provide the comparisons of Cerebral Blood Flow (CBF) and Cerebral Blood Volume (CBV), these maps attest that our proposed method can provide comparable results to the existing methods.

SRDN is an end-to-end structure that learns from pair-wise LR/noisy patches with their original clean images and outputs high-quality CT images based on low-quality input images while testing. The structure of SRDN is shown in Figure 7.2. The main functional part of SRDN is built by stacking four modularized kernel regulation blocks (KR block). KR blocks are inspired by GoogLeNet[74] which has a combination of kernels of varying sizes. Specifically, each block is made up of two convolutional layers of size 1×1, one convolutional layer of size 7×7, and one convolutional layer of size 3×3 for regulating the features extracted by the 7×7 convolutional layer[73]. The combination of large and small filters is to balance the extraction of subtle and edge features. Moreover, each block is embedded with a skip connection, which allows reference to feature mapping from previous layers and boosts the network performance.

- **Serial connections**. Image classification needs to summarize diverse information into a linear classifier. On the contrary, image denoising needs to find the most prominent features for a progressive transformation. Therefore, we adopt three kernel sizes (e.g., 1×1, 3×3, and 7×7) in the KR-block module. Kernels of each size are placed in series to allow the small kernels to regulate the features extracted by the large kernel.

- **Small behind large**. Large kernels (e.g., 7×7) can extract certain features by observing a local region with more statistical pixel information. The small kernels (e.g., 3×3) are primarily used for exploiting deeper prior information from the underlying feature maps obtained by large preceding kernels. The subtle textures are especially highlighted during this regularization procedure. Large kernels excel in noise removal but may also smooth the whole image irrespective of its edges or details. Small kernels can preserve subtle textures, but noise pixels may detract from the information attained. Therefore, placing a small kernel behind the large one is a straightforward strategy to enhance denoiser regularization.

- **Feature blending**. The features extracted by large kernels contain both actual pixel values and noise, whereas the small kernel can capture real pixels while simultaneously ignoring much of the noise. At the end of a KR block, features captured by small kernels are blended with the features extracted from large kernels. To allow the locally highlighted features to be shared across neighboring KR blocks, feature blending is processed by pixel-wise summation (see Figure 7.2 top) rather than concatenation (e.g., in GoogLeNet). This helps with finding the most prominent features for a forward transformation. Eventually, the output of a KR block contains more accurate pixel information with less noise.

- **1 × 1 convolution.** The special usage of 1 × 1 convolution in a KR block is for two purposes: first, it reduces the dimensions inside KR-block modules, such as the first 1 × 1 convolution layer; second, it adds more non-linearity by having Parametric Rectified Linear Unit (PReLU) immediately after every 1 × 1 convolution and suffers less from overfitting due to smaller kernel size.

Convolutional networks learn a mapping function between corrupted image input and a corresponding noise-free image. The network contains L convolution layers (Conv), each of which implements a feature-extraction procedure. To ensure our network has rich feature representations, we use a considerable amount of large filters in the first two convolutional layers[73] to extract diverse and representative features for feature mapping and spatial transformation. We define densely convolutional features extracted from the *l*th layer as

$$x_l = \text{Conv}\left(y_l, f_l, n_l, c_l\right)_{f \geq 7 \times 7, n \geq 128} \tag{7.1}$$

where $l = 1...L$ indexes the layer, y_l, f_l, n_l, and c_l represent the l's input, the filter size, filter number, and channel number, respectively. x_l is the feature map extracted from y_l by Conv(·), which denotes convolution. As the top and bottom layers have different functional attention[73], the network can be decomposed into three parts (the bottom part is shown in Figure 7.2): feature extraction, feature regulation, feature mapping, and image reconstruction. In the proposed SRDN, the first two layers have the same volume: $(f_l, n_l, c_l) = (7, 128, 1)$.

Several KR blocks are cascaded to perform feature regulation, mapping, and transformation. Also, residual learning is performed here by skip connection, which connects the outputs of two adjacent KR blocks. The use of skip connections between KR blocks leads to faster and more stable training. The purpose of using a shortcut between the input and the end of the network is to incorporate more information from the original input into image reconstruction. This strategy helps relax the network interference difficulty because input data contains a lot of real pixel information that can be taken as a prior. To make SRDN more compact, we introduce two 1 × 1 composite units, referred to as "Shrinking" and "Expanding," shown in Figure 7.2. After dense convolutional feature-extraction layers, we reduce the number of feature maps by "Shrinking." After feature regulation and mapping, we expand feature maps such that there are sufficient various features that can be provided for image reconstruction. The convolutional layer before the last layer has the volume: $(f_l, n_l, c_l) = (3, 128, 1)$. We utilize a deconvolutional layer with the volume: $(f_l, n_l, c_l) = (3, 1, 1)$ as our last layer.

The combination of the various features extracted from multi-directional data enhances the network's capability for inference and generality. Since multi-directional inputs provide different perspectives of the 3D volume data, they cannot merely be regarded as more training data being fed into multi-networks. Instead, they complement each other nicely to encode the sparse features through the network.

Dense convolutions and kernel regulation strategy ensure diverse features from multi-directional brain CT images, which can be encoded as network representations. In this paper, we adopted three SRDNs to cope with three directional extracted data, respectively,

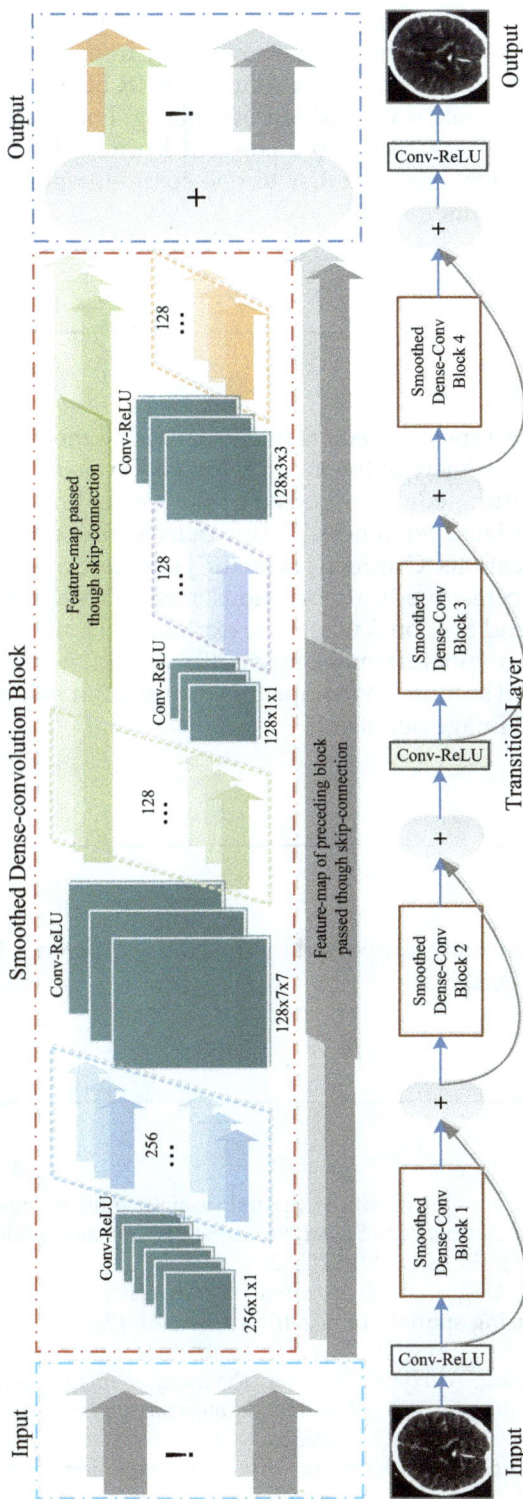

FIGURE 7.2

(Top) A kernel regulation block (KR block) with a massive of convolution computations (128 × 7 × 7) comprises two 1 × 1 convolution components for computation reduction and one 3 × 3 convolution module for regularizing the features extracted by the preceding large size kernels. The number of dark-gray blocks indicates the number of kernels in the current convolutional layers, and the size of dark-gray blocks represents the size of kernels and the density of convolution. The color arrows represent the number of feature-map outputs. **(Bottom)** SRDN is consisted of feature extraction, shrinking, regulation and mapping, expanding, and image reconstruction. Four KR blocks are embedded in the proposed SRDN. Reprint from Frontiers in Neurology by Xiao Y, Liu P, Liang Y, Stolte S, Sanelli P, Gupta A, Ivanidze J, Fang R. "STIR-net: deep spatial-temporal image restoration net for radiation reduction in CT perfusion" 2019 Jun 26; 10:647 under the terms of the Creative Commons Attribution License (CC-BY), which permits the re-use, distribution, and reproduction of material from published articles, provided the original authors and source are credited.

to form our STIR-Net: $Y \times T$, $X \times T$, and $X \times Y$. The structure of STIR-Net is shown in Figure 7.2. During training, the input and output layers are matched with pair-wise noise and label patches. The label here refers to the patches extracted from the original high radiation dose CTP volume ($X \times Y \times T$). Each SRDN contains 4 KR blocks that can fully encode the features from each directional data without overfitting. For the testing stage, the outputs of the three SRDN nets assemble into a conjoint learning layer. This layer blends various features from all SRDN nets together to one spatio-temporal volume by calculating the mean of the three outputs.

7.6 Summary and Discussion

In this chapter, we reviewed various types of medical image denoising methods for different medical imaging modalities. The review of the medical image denoising methods covers both conventional and latest cutting-edge methods. For the conventional method, we provided an introduction and associated references. This is reflected in Section 2: Review of Medical Image Denoising Applications. Our focus is on the latest cutting-edge methods and their important applications in real medical operations such as computed tomography. This can be seen in Section 3 and Section 5 where we explained the method in detail. We also explained important types of medical image modalities and where medical image noise comes from. This is reflected in section 4. We hope this material can help readers better understand the topic of medical image denoising.

Acknowledgment

This material is based upon work partially supported by the National Science Foundation under Grant No. IIS-1564892, IIS- 1908299.

References

1. Narasimha, C. & Rao, A. N. A comparative study: Spatial domain filter for medical image enhancement. in *2015 International Conference on Signal Processing and Communication Engineering Systems*, 291–295 (2015). doi:10.1109/SPACES.2015.7058268.
2. AnchalBudhiraja, S., Goyal, B., Dogra, A. & Agrawal, S. An efficient image denoising scheme for higher noise levels using spatial domain filters. *Biomed. Pharmacol. J.* **11**(2), 625–634 (2018).
3. Thakur, K., Damodare, O. & Sapkal, A. Hybrid method for medical image denoising using Shearlet transform and bilateral filter. in *2015 International Conference on Information Processing (ICIP)*, 220–224 (2015). doi:10.1109/INFOP.2015.7489382.
4. Kollem, S., Reddy, K. R. & Rao, D. S. Improved partial differential equation-based total variation approach to non-subsampled contourlet transform for medical image denoising. *Multimed. Tools Appl.* **80**, 2663–2689 (2021).

5. Bhadauria, H. S. & Dewal, M. L. Medical image denoising using adaptive fusion of curvelet transform and total variation. *Comput. Electr. Eng.* **39**, 1451–1460 (2013).

6. Jidesh, P. & Febin, I. P. Estimation of noise using non-local regularization frameworks for image denoising and analysis. *Arab. J. Sci. Eng.* **44**, 3425–3437 (2019).

7. Sun, H. et al. Dynamic PET image denoising using deep image prior combined with regularization by denoising. *IEEE Access* **9**, 52378–52392 (2021).

8. Yuan, Q. et al. Medical image denoising algorithm based on sparse nonlocal regularized weighted coding and low rank constraint. *Sci. Program.* **2021**, e7008406 (2021).

9. Bai, J., Song, S., Fan, T. & Jiao, L. Medical image denoising based on sparse dictionary learning and cluster ensemble. *Soft Comput.* **22**, 1467–1473 (2018).

10. Li, H., He, X., Tao, D., Tang, Y. & Wang, R. Joint medical image fusion, denoising and enhancement via discriminative low-rank sparse dictionaries learning. *Pattern Recognit.* **79**, 130–146 (2018).

11. Ji, L., Guo, Q. & Zhang, M. Medical image denoising based on biquadratic polynomial with minimum error constraints and low-rank approximation. *IEEE Access* **8**, 84950–84960 (2020).

12. Chen, Z., Zhou, Z. & Adnan, S. Joint low-rank prior and difference of Gaussian filter for magnetic resonance image denoising. *Med. Biol. Eng. Comput.* **59**, 607–620 (2021).

13. Rai, H. M. & Chatterjee, K. Hybrid adaptive algorithm based on wavelet transform and independent component analysis for denoising of MRI images. *Measurement* **144**, 72–82 (2019).

14. Li, H. et al. Denoising scanner effects from multimodal MRI data using linked independent component analysis. *NeuroImage* **208**, 116388 (2020).

15. Zhu, H., Zhang, J. & Wang, Z. Arterial spin labeling perfusion MRI signal denoising using robust principal component analysis. *J. Neurosci. Methods* **295**, 10–19 (2018).

16. Gurney-Champion, O. J. et al. Principal component analysis for fast and model-free denoising of multi b-value diffusion-weighted MR images. *Phys. Med. Biol.* **64**, 105015 (2019).

17. Does, M. D. et al. Evaluation of principal component analysis image denoising on multi-exponential MRI relaxometry. *Magn. Reson. Med.* **81**, 3503–3514 (2019).

18. Arabi, H. & Zaidi, H. Improvement of image quality in PET using post-reconstruction hybrid spatial-frequency domain filtering. *Phys. Med. Biol.* **63**, 215010 (2018).

19. Mustafi, A. & Ghorai, S. K. A novel blind source separation technique using fractional Fourier transform for denoising medical images. *Optik* **124**, 265–271 (2013).

20. Sidhu, K. S., Khaira, B. S. & Virk, I. S. Medical image denoising in the wavelet domain using Haar and DB3 filtering. *Int. Refereed J. Eng. Sci.* **1**, 1–8 (2012).

21. Juneja, M. et al. Denoising of magnetic resonance imaging using Bayes shrinkage based fused wavelet transform and autoencoder based deep learning approach. *Biomed. Signal Process. Control* **69**, 102844 (2021).

22. Gan, Y., Angelini, E., Laine, A. & Hendon, C. BM3D-based ultrasound image denoising via brushlet thresholding. in *2015 IEEE 12th International Symposium on Biomedical Imaging (ISBI)*, 667–670 (2015). doi:10.1109/ISBI.2015.7163961.

23. Bai, J., Sun, Y., Fan, T., Song, S. & Zhang, X. Medical image denoising based on improving K-SVD and block-matching 3D filtering. in *2016 IEEE Region 10 Conference (TENCON)*, 1624–1627 (2016). doi:10.1109/TENCON.2016.7848292.

24. Zhang, S. et al. An image denoising method based on BM4D and GAN in 3D shearlet domain. *Math. Probl. Eng.* **2020**, e1730321 (2020).

25. Huang, K., Tian, X., Zhang, D. & Zhang, H. Performance evaluation and optimization of BM4D-AV denoising algorithm for cone-beam CT images. in *Seventh International Conference on Graphic and Image Processing (ICGIP 2015)* vol. 9817 240–245 (SPIE, 2015).

26. Cui, J. et al. PET image denoising using unsupervised deep learning. *Eur. J. Nucl. Med. Mol. Imaging* **46**, 2780–2789 (2019).

27. Hashimoto, F., Ohba, H., Ote, K., Teramoto, A. & Tsukada, H. Dynamic PET image denoising using deep convolutional neural networks without prior training datasets. *IEEE Access* **7**, 96594–96603 (2019).

28. Zhou, L., Schaefferkoetter, J. D., Tham, I. W. K., Huang, G. & Yan, J. Supervised learning with cyclegan for low-dose FDG PET image denoising. *Med. Image Anal.* **65**, 101770 (2020).

29. Li, Z., Zhou, S., Huang, J., Yu, L. & Jin, M. Investigation of low-dose CT image denoising using unpaired deep learning methods. *IEEE Trans. Radiat. Plasma Med. Sci.* **5**, 224–234 (2021).

30. Yin, Z. et al. Unpaired image denoising via wasserstein GAN in low-dose ct image with multi-perceptual loss and fidelity loss. *Symmetry* **13**, 126 (2021).

31. Gong, K., Han, P., El Fakhri, G., Ma, C. & Li, Q. Arterial spin labeling MR image denoising and reconstruction using unsupervised deep learning. *NMR Biomed.* **35**, e4224 (2022).

32. Bushberg, J. T., Seibert, J. A., Leidholdt, E. M. & Boone, J. M. *The Essential Physics of Medical Imaging* (Lippincott Williams & Wilkins, Philadelphia, PA, 2012).

33. Kulathilake, K. A. S. H., Abdullah, N. A., Sabri, A. Q. M. & Lai, K. W. A review on deep learning approaches for low-dose computed tomography restoration. *Complex Intell. Syst.* (2021). doi:10.1007/s40747-021-00405-x.

34. Jensen, K., Martinsen, A. C. T., Tingberg, A., Aaløkken, T. M. & Fosse, E. Comparing five different iterative reconstruction algorithms for computed tomography in an ROC study. *Eur. Radiol.* **24**, 2989–3002 (2014).

35. Nowak, R. D. Wavelet-based rician noise removal for magnetic resonance imaging. *IEEE Trans. Image Process. Publ. IEEE Signal Process. Soc.* **8**, 1408–1419 (1999).

36. Goyal, B., Dogra, A., Agrawal, S. & Sohi, B. S. Noise issues prevailing in various types of medical images. *Biomed. Pharmacol. J.* **11**, 1227–1237 (2018).

37. Perona, P. & Malik, J. Scale-space and edge detection using anisotropic diffusion. *IEEE Trans. Pattern Anal. Mach. Intell.* **12**, 629–639 (1990).

38. Buades, A., Coll, B. & Morel, J.-M. A non-local algorithm for image denoising. in *2005 IEEE Computer Society Conference on Computer Vision and Pattern Recognition (CVPR'05)*, vol. 2, 60–65 (2005).

39. Mohan, J., Krishnaveni, V. & Guo, Y. A survey on the magnetic resonance image denoising methods. *Biomed. Signal Process. Control* **9**, 56–69 (2014).

40. Tomasi, C. & Manduchi, R. Bilateral filtering for gray and color images. in *Sixth International Conference on Computer Vision (IEEE Cat. No.98CH36271)*, 839–846 (1998). doi:10.1109/ICCV.1998.710815.

41. Wong, W. C. K. & Chung, A. C. S. A nonlinear and non-iterative noise reduction technique for medical images: concept and methods comparison. *Int. Congr. Ser.* **1268**, 171–176 (2004).

42. Weaver, J. B., Xu, Y. S., Healy, D. M. & Cromwell, L. D. Filtering noise from images with wavelet transforms. *Magn. Reson. Med.* **21**, 288–295 (1991).

43. Wood, J. C. & Johnson, K. M. Wavelet-packet denoising of magnetic resonance images: importance of Rician statistics at low SNR. *Magn. Reson. Med.* **41**, 631–635 (1999).

44. Starck, J.-L., Candès, E. J. & Donoho, D. L. The curvelet transform for image denoising. *IEEE Trans. Image Process. Publ. IEEE Signal Process. Soc.* **11**, 670–684 (2002).

45. Do, M. N. & Vetterli, M. The contourlet transform: an efficient directional multiresolution image representation. *IEEE Trans. Image Process.* **14**, 2091–2106 (2005).

46. Sijbers, J., den Dekker, A. J., Scheunders, P. & Van Dyck, D. Maximum-likelihood estimation of Rician distribution parameters. *IEEE Trans. Med. Imaging* **17**, 357–361 (1998).

47. Aja-Fernandez, S., Alberola-Lopez, C. & Westin, C.-F. Noise and signal estimation in magnitude MRI and Rician distributed images: a LMMSE approach. *IEEE Trans. Image Process. Publ. IEEE Signal Process. Soc.* **17**, 1383–1398 (2008).

48. Rajan, J., Veraart, J., Van Audekerke, J., Verhoye, M. & Sijbers, J. Nonlocal maximum likelihood estimation method for denoising multiple-coil magnetic resonance images. *Magn. Reson. Imaging* **30**, 1512–1518 (2012).

49. Tisdall, D. & Atkins, M. S. *MRI Denoising via Phase Error Estimation*. in (eds. Fitzpatrick, J. M. & Reinhardt, J. M.) 646 (2005). doi:10.1117/12.595677.

50. Awate, S. P. & Whitaker, R. T. Nonparametric Neighborhood Statistics for MRI Denoising. in *Information Processing in Medical Imaging* (eds. Christensen, G. E. & Sonka, M.) 677–688 (Springer, 2005). doi:10.1007/11505730_56.

51. Luo, J., Zhu, Y. & Hiba, B. Medical image denoising using one-dimensional singularity function model. *Comput. Med. Imaging Graph.* **34**, 167–176 (2010).

52. Luo, J., Zhu, Y. & Magnin, I. E. Denoising by averaging reconstructed images: Application to magnetic resonance images. *IEEE Trans. Biomed. Eng.* **56**, 666–674 (2009).
53. Tong, S., Alessio, A. M. & Kinahan, P. E. Noise and signal properties in PSF-based fully 3D PET image reconstruction: an experimental evaluation. *Phys. Med. Biol.* **55**, 1453–1473 (2010).
54. Hofheinz, F. et al. Suitability of bilateral filtering for edge-preserving noise reduction in PET. *EJNMMI Res.* **1**, 23 (2011).
55. Lee, J. A., Geets, X., Grégoire, V. & Bol, A. Edge-preserving filtering of images with low photon counts. *IEEE Trans. Pattern Anal. Mach. Intell.* **30**, 1014–1027 (2008).
56. Turkheimer, F. E., Aston, J. A. D., Banati, R. B., Riddell, C. & Cunningham, V. J. A linear wavelet filter for parametric imaging with dynamic PET. *IEEE Trans. Med. Imaging* **22**, 289–301 (2003).
57. Shih, Y.-Y., Chen, J.-C. & Liu, R.-S. Development of wavelet de-noising technique for PET images. *Comput. Med. Imaging Graph.* **29**, 297–304 (2005).
58. Lee, J.-S. Digital image enhancement and noise filtering by use of local statistics. *IEEE Trans. Pattern Anal. Mach. Intell.* **PAMI-2**, 165–168 (1980).
59. Lee, J.-S. Refined filtering of image noise using local statistics. *Comput. Graph. Image Process.* **15**, 380–389 (1981).
60. Frost, V. S., Stiles, J. A., Shanmugan, K. S. & Holtzman, J. C. A model for radar images and its application to adaptive digital filtering of multiplicative noise. *IEEE Trans. Pattern Anal. Mach. Intell.* **PAMI-4**, 157–166 (1982).
61. Kuan, D., Sawchuk, A., Strand, T. & Chavel, P. Adaptive restoration of images with speckle. *IEEE Trans. Acoust. Speech Signal Process.* **35**, 373–383 (1987).
62. Lopes, A., Touzi, R. & Nezry, E. Adaptive speckle filters and scene heterogeneity. *IEEE Trans. Geosci. Remote Sens.* **28**, 992–1000 (1990).
63. Lopes, A., Nezry, E., Touzi, R. & Laur, H. Structure detection and statistical adaptive speckle filtering in SAR images. *Int. J. Remote Sens.* **14**, 1735–1758 (1993).
64. Yu, Y. & Acton, S. T. Speckle reducing anisotropic diffusion. *IEEE Trans. Image Process.* **11**, 1260–1270 (2002).
65. Balocco, S., Gatta, C., Pujol, O., Mauri, J. & Radeva, P. SRBF: Speckle reducing bilateral filtering. *Ultrasound Med. Biol.* **36**, 1353–1363 (2010).
66. Narayanan, S. K. & Wahidabanu, R. S. D. A view on despeckling in ultrasound imaging. *Int. J. Signal Process.* **2**(3), 85–98 (2009).
67. Yue, Y., Croitoru, M. M., Bidani, A., Zwischenberger, J. B. & Clark, J. W. Nonlinear multiscale wavelet diffusion for speckle suppression and edge enhancement in ultrasound images. *IEEE Trans. Med. Imaging* **25**, 297–311 (2006).
68. Gupta, N., Swamy, M. N. S. & Plotkin, E. Despeckling of medical ultrasound images using data and rate adaptive lossy compression. *IEEE Trans. Med. Imaging* **24**, 743–754 (2005).
69. Xie, H., Pierce, L. E. & Ulaby, F. T. SAR speckle reduction using wavelet denoising and Markov random field modeling. *IEEE Trans. Geosci. Remote Sens.* **40**, 2196–2212 (2002).
70. Achim, A., Bezerianos, A. & Tsakalides, P. Novel Bayesian multiscale method for speckle removal in medical ultrasound images. *IEEE Trans. Med. Imaging* **20**, 772–783 (2001).
71. Pizurica, A., Philips, W., Lemahieu, I. & Acheroy, M. A versatile wavelet domain noise filtration technique for medical imaging. *IEEE Trans. Med. Imaging* **22**, 323–331 (2003).
72. Liu, P. et al. Deep evolutionary networks with expedited genetic algorithms for medical image denoising. *Med. Image Anal.* **54**, 306–315 (2019).
73. Xiao, Y. et al. STIR-net: Deep spatial-temporal image restoration net for radiation reduction in CT perfusion. *Front. Neurol.* **10**, (2019).
74. Szegedy, C. et al. Going deeper with convolutions. in *2015 IEEE Conference on Computer Vision and Pattern Recognition (CVPR)*, 1–9 (2015). doi:10.1109/CVPR.2015.7298594.

8

Attenuation Correction for Quantitative PET/MR Imaging

Se-In Jang and Kuang Gong

Harvard Medical School, Boston, MA, USA

CONTENTS

8.1 Introduction

Positron emission tomography (PET) is widely used in clinical studies related to oncology, cardiology, and neurology. Because of its high sensitivity, PET can reveal a molecular level of activities inside the body through the injection of specific radioactive tracers. Another advantage of PET imaging is that it can quantitively measure physiology *in vivo* based on further image analysis and kinetic modeling, such as blood flow, binding potential, glucose metabolism rate, and membrane potential.

To full reveal the quantitative merits of PET, correction for the object attenuation of annihilation photons is essential. In the early days, this was achieved through transmission-source scanning. Nowadays computed tomography (CT) images can be utilized to derive the attenuation map through a bilinear scaling transform to match the energy level with the wide adoption of PET/CT scanners [1]. The conversion formula from CT Hounsfield-unit (HU) to attenuation coefficient at voxel j is [2]

$$\mu_j = \begin{cases} 9.6e^{-5}\left(HU_j + 1000\right) & \text{if } HU_j < \text{Threshold,} \\ a\left(HU_j + b\right) & \text{if } HU_j > \text{Threshold.} \end{cases} \tag{8.1}$$

DOI: 10.1201/9781003243458-11

where HU_j is the HU of voxel j in the CT image. a, b and *Threshold* are values depending on the energy of the x-ray, given in [2]. Afterwards, the attenuation factors can then be calculated from the attenuation map μ by

$$a_{ii} = e^{-\sum_j l_{ij}\mu_j},$$

(8.2)

where l_{ij} denotes the interaction length of line of response (LOR) i with voxel j.

Since the early 2010s, simultaneously combined PET/MR scanners have been introduced in clinics because of magnetic resonance (MR)' functional imaging ability, high soft tissue contrast, as well as potentials for PET motion correction [3] and partial volume correction [4]. One challenge for PET/MR imaging is that the MR image cannot be utilized to generate the attenuation map μ directly or through simple transformations as the MR signal is not related to photon attenuation. To address this challenge, many methods have been developed in the last decades to generate attenuation maps from MR images through different approaches, such as image segmentation, image registration, joint estimation of the emission and attenuation images, machine learning methods, and recent deep learning methods. In this chapter, we will further discuss these different categories of methods. There are also many excellent review papers about MR-based PET attenuation correction [5–12] for further reading.

8.2 Methods for PET Attenuation Correction

8.2.1 Atlas-based Methods

An atlas-based approach enables template-based attenuation correction using the patient's MR image and the corresponding CT image. Specifically, this is related to image registration problems [13]. Image warping, also known as a linear image transformation (e.g., rotation, scaling, and shearing), addresses alignment issues between the MR and CT images obtained from the same patient. The linear transformation can be written as:

$$T(x) = Ax,$$

(8.3)

where $T(\cdot)$ is a linear transformation function, x is a column vector that includes x and y coordinates in the image space. According to different purposes for registration, A can be designed as follows:

$$A_{\text{rotation}} = \begin{bmatrix} \cos\theta & \sin\theta \\ -\sin\theta & \cos\theta \end{bmatrix}, A_{\text{scaling}} = \begin{bmatrix} s & 0 \\ 0 & s \end{bmatrix}, A_{\text{shearing}} = \begin{bmatrix} 1 & k_x \\ k_y & 1 \end{bmatrix},$$

(8.4)

where A_{rotation}, A_{scaling}, and A_{shearing} are matrices for rotation, scaling, and shearing. θ is an angle for a clockwise transformation, s is a scaling factor, and k_x and k_y are factors for the x and y axis-wise transformations, respectively. In [14], a single pair of the MR and CT images for each patient is registered based on the linear transformation to generate an averaged attenuation map of five male and five female patients. However, reconstruction errors are still generated from these different gender structures. To alleviate the sensitivity caused by the gender variation, two mean templates for males and females are separately prepared [15]. In [16], a subject-specific nonlinear transformation is developed using a cubic B-spline free-form deformation algorithm [17] given by:

$$\upsilon_x(x,y,z) = \sum_{l=0}^{3}\sum_{m=0}^{3}\sum_{n=0}^{3}\beta_l(u)\beta_m(v)\beta_n(w)P_x(i+l,j+m,k+n),$$ (8.5)

where υ_x is the deformation function at the voxel location (x, y, z), P_x is a coefficient related to each control point for the x dimension, (i, j, k) denote the coordinates in the voxel image, and (u, v, w) indicates the local coordinates of the voxel (x, y, z). $\beta_l(u)$ for the x direction is the B-spline basis function which can be written as:

$$\beta_0(u) = (1-u)^3/6,$$ (8.6)

$$\beta_1(u) = (3u^3 - 6u^2 + 4)/6,$$ (8.7)

$$\beta_2(u) = (-3u^3 + 3u^2 + 3u + 1)/6,$$ (8.8)

$$\beta_3(u) = u^3/6.$$ (8.9)

$\beta_m(v)$ and $\beta_n(w)$ are also defined similarly for the y and z directions, respectively. Note that (u, v, w) are normalized between 0 and 1. This B-spline deformation offers a better registration than the linear transformation to achieve more accurate attenuation correction. Instead of the subject-specific transformation, an inter-subject coordinate mapping with multiple MR and CT images pairs is presented using the cubic B-spline in [18]. The pseudo-CT image is then generated based on an intensity fusion using an exponential decay function given by:

$$w_i = e^{-\alpha R_i},$$ (8.10)

where α is a decay constant and R_i is a rank value to find the best matched image from the registered atlas images. The weights w_i for each image are adopted to conduct a weighted averaging for the intensity fusion.

Since the atlas sets are aligned based on the whole image, they cannot sufficiently recover local information due to the structural difference. This local information can be addressed by a similarity measure between local voxel patches instead of the whole image. In [19], a B-spline registration with an optical flow deformation [20] is proposed which observes alignment differences between the MR and CT images as:

$$I(x,y,z) = I(x+\Delta x, y+\Delta y, z+\Delta z),$$ (8.11)

where $I(x, y, z)$ is an intensity at location x, y, and z in a voxel image, Δx, Δy, and Δz are the alignment difference. The B-spline and the optical flow methods are utilized to reduce the atlas sets' alignment errors. The optical flow estimation is solved by the Lucas–Kanade method [21]. A voxel-based attenuation correction is then conducted to address the local information. These methods help to improve several cases, such as artifacts and truncations.

However, it still has difficulties to address inter-subject variations (e.g., different age, gender, and race [22]) that is not shown in the established atlas sets. Due to its matching sensitivity between unseen patients, the atlas-based approach has not been frequently applied to whole-body applications. Moreover, the atlas-based attenuation map calculation is very time-consuming due to the large number of datasets in the atlas sets.

8.2.2 Segmentation-based Methods

A segmentation-based approach is to efficiently synthesize attenuation maps based on segmentation that can classify MR images into several tissue classes (e.g., fat, water, air and bone). One of the simplest segmentation methods is to use a threshold method given by:

$$I(x,y) = \begin{cases} I(x,y) & if\ I(x,y) > \tau \\ 0 & if\ I(x,y) \le \tau \end{cases}' \tag{8.12}$$

where $I(x, y)$ is an intensity at location x and y in an image, and τ is a fixed threshold to segment the target class. In [23], for neurological applications, the thresholding technique is not only performed for bone and soft tissue segmentation using different thresholds, but also for noise and background removal. Morphological image processing (e.g., dilation and erosion) is then adopted to refine the segmentation results. The morphological dilation is given by:

$$(f \oplus b)(x,y) = \max_{(i,j)\in b}\left[f(x+i,y+j)+b(i,j)\right], \tag{8.13}$$

where f is an image and b is a structure function. indicates the dilation operation in mathematical morphology. The pixel value is the maximum value of all pixels in its neighborhood. For example, a binary image's pixel value is set to 1 if any neighboring pixels have the value 1. Therefore, the dilation operation makes the target segmentation more visible and fills in small holes. Similarly, the morphological erosion is given by:

$$(f \ominus b)(x,y) = \min_{(i,j)\in b}\left[f(x+i,y+j)-b(i,j)\right], \tag{8.14}$$

where $\ominus$ indicates the erosion operation. The pixel value is the minimum value of all pixels in its neighborhood. A binary image's pixel value is set to 0 if any of the neighboring pixels have the value 0. The erosion operation removes floating dots in the segmentation results. In [24], two different attenuation maps for a whole-body application are combined: a segmented attenuation map with the thresholding technique for four segmentation classes (e.g., background, lung, fat, and soft tissue) and an attenuation map without bone segmentation. The cortical bone segmentation is avoided because the manual settings of the threshold values are not performed well due to a very low-intensity and nonspecific signal on the MR image. Instead of the manual settings in the thresholding techniques, an automatic threshold technique is developed with a T1-weighted MR image to ease the segmentation of air, lung, and soft tissue and to reduce the misclassification of cortical bone and air [25]. In [26], air, skull, brain tissue, and nasal sinuses segmentation are conducted by a fuzzy C-means clustering method [27] which minimizes the following objective function:

$$\arg\min_{C} \sum_{i=1}^{n}\sum_{j=1}^{c} w_{i,j}^{m} \parallel x_i - c_j \parallel, \tag{8.15}$$

where $w_{i,j}^{m} = 1/\sum_{j=1}^{c}\left(\dfrac{\parallel x_i - c_j \parallel}{\parallel x_i - c_k \parallel}\right)^{\frac{2}{m-1}} \in [0,1]$ gives the degree to which element x_i belongs to a cluster c_j. m is a parameter, which is commonly set to 2. As an unsupervised learning method, this fuzzy C-means clustering method offers more reliable segmentation results than thresholding techniques. In [28], as a supervised learning method, a three-layered feed-forward neural network-based segmentation is performed using the T1-weighted MR images with prior anatomical information (e.g., region position and its rough shape).

Since the above methods obtains the segmentation tasks using the anatomical intensity values on conventional MRI sequences, it may suffer from the low signal intensity of cortical bones, which is difficult in distinguishing bone from the air. In [29], Ultra-short echo time (UTE) sequences are used for bone, soft tissue, and air segmentations without any anatomical information. Multiple UTE echo sequences are offered for more accurate performances than the single UTE sequences [30, 31].

8.2.3 Joint Estimation-based Methods

Attempts to jointly calculate the attenuation map and PET emission map has been investigated for decades [32]. One framework is to iteratively find the solutions of the attenuation and emission maps jointly through the Maximum Likelihood Estimation of Activity and Attenuation (MLAA) framework [33, 34]. The joint-estimation framework is non-conex and the solution is non-unique. Cross-talks between the attenuation and emission maps can occur. With the availability of time-of-flight (TOF) information, Defrise *et al.* have shown that the TOF information can help solve the crosstalk issue and can determine the activity distribution up to a scaling constant [35]. This has inspired new interests in the MLAA framework. For an extensive overview of the MLAA framework, the review paper by Berker and Li [12] is a good resource. Below the basics of MLAA and its further extensions are introduced.

In the MLAA framework, the PET attenuation map μ and activity image λ are jointly estimated by maximizing the log-likelihood of μ and, based on the measured TOF PET data y as

$$\hat{\mu}, \hat{\lambda} = \mathrm{argmax}_{\mu \geq 0, \lambda \geq 0} \sum_{i=1}^{M}\sum_{t=1}^{T} \log p(y_{i,t} \mid \mu, \lambda), \tag{8.16}$$

where M is the number of LORs and T is the number of TOF bins. When solving the objective function shown in (16), the strategy is to alternatively update $\hat{\mu}$ and $\hat{\lambda}$ [36]. After running multiple iterations, the final results can be obtained based on a pre-defined stopping criterion. Apart from estimating the attenuation map μ, we can also directly estimate the attenuation coefficient a defined in (2), as was proposed in the framework of MLACF [37]. In practice, the timing offsets and timing resolutions have shifts, which can compromise the joint-estimation accuracy. Rezaei *et al.* proposed to re-estimate these calibration factors to further improve the performance of MLAA [38].

With the TOF information, the joint estimation of attenuation and emission maps still cannot solve the scaling issue. Various methods have been proposed to further improve the accuracy of the MLAA framework based on additional information from the scan. Panin *et al.* proposed to use an external rotating line source to further constrain the attenuation image [39]. Rothfuss *et al.* showed that background radiation originating from the isotope lutetium-176 in the LSO scintillators could be used as a transmission source for attenuation correction [40]. The background radiation from lutetium-176 has also been further proposed to improve the quantitative accuracy of the MLAA framework [41, 42]. Additionally, Feng *et al.* proposed to utilize the single events, which contain depth-dependent attenuation factors compared to the coincidence events, to obtain a unique solution of joint attenuation and activity estimation [43].

Prior information from the MR images simultaneously acquired in the PET/MR scanner can also be explored to improve the MLAA framework. Salomon *et al.* (2011) proposed to utilize geometrical information obtained from the MR segmentation results to parameterize the attenuation map using the uniform regional basis functions [44]. Mehranian and Zaidi proposed to utilize the MR information based on the Gaussian mixture model to improve the estimation of the attenuation map [45]. The framework was further extended to include MR priors when estimating both emission and attenuation maps [46]. Benoit *et al.* proposed to utilize UTE and T1-weighted MR images to further tune parameters involved in the attenuation-map-estimation step to optimize the MLAA framework [47]. Ahn *et al.* proposed to use 4-class (water, fat, lung, and air) segmentation of MR images to derive the attenuation map first, through which to further constrain the attenuation map in the MLAA framework. Based on the attenuation-intensity prior derived from MR images, Rezaei *et al.* performed a comprehensive quantitative analysis using 34 brain [18F]-FDG patient scans, and showed that with an accurate TOF-based calibration, regional errors of MLAA framework has similar results as the state-of-the-art method in PET/MR [48].

Apart from the attenuation map estimation, the MLAA framework has been applied/ extended to estimate dual-energy CT images [49], remove the misalignment artifacts between CT and PET in cardiac PET studies [50], and respiratory-gated PET/CT studies [51]. It has also been utilized to determine the position, shape, and linear attenuation coefficient of metallic implants [52], and estimate flexible hardware components in PET/MR imaging [53].

8.2.4 Machine Learning-based Methods

A machine learning-based approach is a potential candidate for learning the relationship between the MR images and attenuation maps. In [54], a voxel patch-based pseudo-CT method is developed based on a combination of a fixed coordinate system as prior knowledge and a Gaussian kernel function to find the location of the closest CT patch to the MR patch in the atlas sets. For the pseudo-CT prediction, a kernel regularized least-squares method [55] for nonlinear regression is applied as follows:

$$w = k^T R^{-1} y, \tag{8.17}$$

where k is obtained from a kernel function, $k(x_i, x_j)$, which is a symmetric and positive semi-definite function that takes two arguments and produces a measure of the similarity between x_i and x_j. R is a covariance matrix of k. y is a reference vector, and w is a parameter vector. In their kernel function design, the voxel patch and its location are considered as follows:

$$k\left(x_i, x_j\right) = exp\left(\frac{-\parallel p_i - p_j \parallel^2}{2\sigma_{patch}^2}\right) \times exp\left(\frac{-\parallel l_i - l_j \parallel^2}{2\sigma_{loc}^2}\right), \tag{8.18}$$

where $x_i = (p_i, l_i)$, p is a voxel patch, l is the patch's location, and σ is a variance. In [56], their pseudo-CT prediction is improved by consideration of a five-class segmentation (e.g., air, lungs, fat tissue, a fat-nonfat tissue mixture, and nonfat tissue) in the Gaussian kernel function as follows:

$$k\left(x_i, x_j\right) = \exp\left(\frac{-\parallel p_{MR,i} - p_{MR,j} \parallel^2}{2\sigma_{MR\ patch}^2}\right) \times \exp\left(\frac{-\parallel p_{seg,i} - p_{seg,j} \parallel^2}{2\sigma_{seg\ patch}^2}\right) \times \exp\left(\frac{-\parallel l_i - l_j \parallel^2}{2\sigma_{loc}^2}\right), \tag{8.19}$$

where $x_i = (p_{MR,\ i},\ p_{seg,\ i},\ l_i)$, $p_{MR,\ i}$ is a voxel patch from the MR image, $p_{seg,\ i}$ is a voxel patch from the segmented image. In [57], Dixon-volume and UTE images are utilized with a support vector machine that can learn the mapping between MR and CT HU as follows:

$$w = \sum_{i=1}^{n} a_i y_i \phi\left(x_i\right), \tag{8.20}$$

where $a_i \geq 0$ is a Lagrange multiplier [58], y_i is a reference point, and $\phi(x_i)$ is a radial basis kernel function which indicates $k(x_i, x_j) = \phi(x_i) \cdot \phi(x_j)$. x_i includes mean, median, variance, minimum and maximum values of a voxel patch. Since the only use of the local feature extraction from each patch may lead to sub-optimal solutions, the use of both local and global feature extraction can be an effective way to achieve better performance. In [59], voxel-level (e.g., spatial voxel coordinates and pairwise voxel differences), sub-region level (e.g., Haar-like features [60]), and whole patch level (e.g., discrete cosine transform coefficients) features are extracted in different resolutions to obtain multi-scale local and global information. Based on these local and global features, a structured random forest method with an auto-context model [61] is then learned to generate patch-wise pseudo-CT prediction from the MR image patches. Although this machine learning-based approach effectively builds the relationship between the input and the output, careful design of feature extraction (e.g., kernel and different level features) is still necessary to accomplish promising performances.

8.2.5 Deep Learning-based Methods

With the success of deep learning in computer vision tasks, it has been extensively investigated for PET attenuation correction tasks because this approach can offer promising performances without feature extraction designs [62–65]. Researchers have focused on convolutional neural networks (CNNs) to generate the attenuation map directly from MR images [66–75]. In PET attenuation correction tasks, the training loss function of the CNN-based methods can be generally written as follows:

$$L_{CNN}\left(CT_p,CT_g\right)=\beta_1\sum_i\left|CT_g^i-CT_p^i\right|+\beta_2\times\sum_i\left\{\left|\nabla_xCT_g^i-\nabla_xCT_p^i\right|\right.\tag{8.21}$$
$$\left.+\left|\nabla_yCT_g^i-\nabla_yCT_p^i\right|+\left|\nabla_zCT_g^i-\nabla_zCT_p^i\right|\right\},$$

where the first term on the right side calculates the ℓ_1 distance between the ground truth, CT_g, and the predicted pseudo-CT, CT_p. Compared to the ℓ_2 distance calculation, the ℓ_1 distance prevents over-smoothness of the prediction outputs. The last three terms provide the gradient difference between CT_g and CT_p along the x, y, and z directions, respectively. β_1 and β_2 are the parameters to adjust each term.

In deep learning-based PET attenuation correction frameworks, the MR images are always directly utilized as the network input. Due to the strong feature extraction and representation power of the neural network, the final pseudo-CT generation results show better performance than that of atlas and segmentation-based methods. In [73], different input types, such as magnetization-prepared rapid acquisition with gradient echo (MPRAGE), Dixon, and multi-echo UTE, are studied. This study shows that the use of the multi-echo UTE offers the best performance as there are more tissue information available from the specifically designed multi-echo UTE sequence. Apart from employing MR images as the network input, simultaneous reconstructed emission, and attenuation images from TOF PET data based on the MLAA framework have also been adopted as the network input [71, 76–78]. There are also efforts to directly train the CNN networks to generate PET images with non-attenuated-corrected PET images [79–82]. This approach suggests direct quantitative PET reconstruction without attenuation correction is possible, but at the potential cost of more training datasets.

To address the over-smoothness of the CNN methods, generative adversarial network (GAN) [83, 84] based methods have also been explored [85–88]. Generally, GAN consists of two networks, namely generative and discriminative networks. The generative network, $G(\cdot)$, learns to create new pseudo images with a similar distribution as the training set, while the discriminative network, $D(\cdot)$, differentiates the generated images from the actual data distribution. However, these GAN-based methods may still suffer from data generation issues when the input and the ground truth are misaligned. The Cycle-GAN framework [89] tries to solve this misalignment issue. It offers two mappings, namely a forward cycle consistency, $G_F(\cdot):X\to Y$, and a backward cycle consistency $G_B(\cdot):Y\to X$ to prevent the learned mappings, $G_F(\cdot)$ and $G_B(\cdot)$, from contradicting each other. This process helps to solve misaligned data problems. In [87, 88], Cycle-GAN is utilized for MR-based AC, as shown in Figure 8.1. The Cycle-GAN framework employs four independent networks as follows: (1) a generative network G_F for pseudo-CT generation using MR images; (2) a generative network G_B for pseudo-MR generation using CT images; (3) a discriminative network D_{CT} evaluating whether the generated pseudo-CT image is similar to the true CT; (4) a discriminative network D_{MR} evaluating whether the generated pseudo-MR image is close to the true MR. The following objective function is generally used in the Cycle-GAN framework for PET attenuation correction to address the misalignment issue:

$$L=\sum_{i=1}^{M}\beta_dD_{CT}\left(G_F\left(MR_i\right)\right)+\beta_dD_{MR}\left(G_B\left(CT_i\right)\right)+\beta_c\left|MR_i-G_B\left(G_F\left(MR_i\right)\right)\right|+\beta_c\left|CT_i-G_F\left(G_B\left(CT_i\right)\right)\right|,$$

$$\tag{8.22}$$

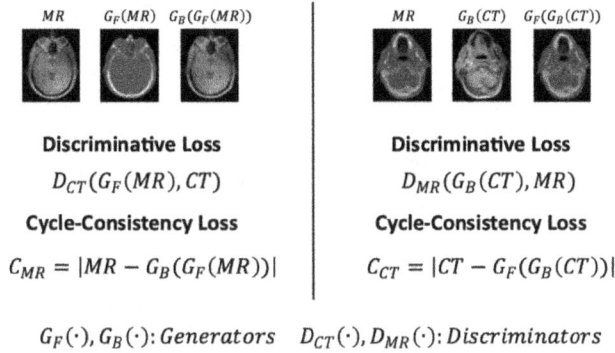

FIGURE 8.1
The schematic plot of the Cycle-GAN framework.

where the first two terms are the discriminative loss functions, and the last two are the cycle-consistency loss functions. $D_{CT}(G_F(MR_i))$ and $D_{MR}(G_B(CT_i))$ are utilized to check the quality of generated pseudo-CT and peudo-MR images, respectively. $|MR_i - G_B(G_F(MR_i))|$ forces $G_B(G_F(MR_i))$ to be close to the original MR image and $|CT_i - G_F(G_B(CT_i))|$ forces $G_F(G_B(CT_i))$ to be close to the original CT images. β_c and β_d are regularization parameters to leverage between the discriminative and cycle-consistency losses.

These deep learning-based methods eliminate various preprocessing steps typically involved with classical machine learning, e.g., alignment, segmentation, and feature extraction issues. Researchers can focus on designing network structures or proposing new training strategies to solve general or more specific PET attenuation correction problems.

8.3 Conclusion

From the atlas-based approach to the deep learning-based approach, attenuation correction for PET/MR imaging has been actively investigated. The deep learning-based approach does not involve any feature extraction stages that need careful designs and has shown better performances in various studies than other methods. Further developments of novel deep learning frameworks that are suitable for challenging PET attenuation correction scenarios, e.g., scenarios where respiratory or cardiac motion exists, are needed. In addition, more thorough evaluation studies of the deep learning methods using a large number of clinical datasets with different tracers are also needed.

References

1. Kinahan, P.E., et al., Attenuation correction for a combined 3D PET/CT scanner. *Medical Physics*, 1998. **25**(10): pp. 2046–2053.
2. Carney, J.P.J., et al., Method for transforming CT images for attenuation correction in PET/CT imaging. *Medical Physics*, 2006. **33**(4): pp. 976–983.

3. Catana, C., et al., MRI-assisted PET motion correction for neurologic studies in an integrated MR-PET scanner. *Journal of Nuclear Medicine*, 2011. **52**(1): pp. 154–161.

4. Chen, K.T., et al., An efficient approach to perform MR-assisted PET data optimization in simultaneous PET/MR neuroimaging studies. *Journal of Nuclear Medicine*, 2019. **60**(2): pp. 272–278.

5. Wagenknecht, G., et al., MRI for attenuation correction in PET: methods and challenges. *Magnetic Resonance Materials in Physics, Biology and Medicine*, 2013. **26**(1): pp. 99–113.

6. Bezrukov, I., et al. MR-based PET attenuation correction for PET/MR imaging. *Seminars in Nuclear Medicine* 2013. **43**(1), 45–59.

7. Keereman, V., et al., Challenges and current methods for attenuation correction in PET/MR. *Magnetic Resonance Materials in Physics, Biology and Medicine*, 2013. **26**(1): pp. 81–98.

8. Mehranian, A., H. Arabi, and H. Zaidi, Vision 20/20: magnetic resonance imaging-guided attenuation correction in PET/MRI: challenges, solutions, and opportunities. *Medical Physics*, 2016. **43**(3): pp. 1130–1155.

9. Hofmann, M., et al., Towards quantitative PET/MRI: a review of MR-based attenuation correction techniques. *European Journal of Nuclear Medicine and Molecular Imaging*, 2009. **36**(1): pp. 93–104.

10. Lee, J.S., A review of deep learning-based approaches for attenuation correction in positron emission tomography. *IEEE Transactions on Radiation and Plasma Medical Sciences*, 2020. **5**(2): pp. 160–184.

11. Catana, C., Attenuation correction for human PET/MRI studies. *Physics in Medicine & Biology*, 2020. **65**(23): p. 23TR02.

12. Berker, Y. and Y. Li, Attenuation correction in emission tomography using the emission data—a review. *Medical Physics*, 2016. **43**(2): pp. 807–832.

13. Chen, Y. and H. An, Attenuation correction of PET/MR imaging. *Magnetic Resonance Imaging Clinics*, 2017. **25**(2): pp. 245–255.

14. Kops, E.R. and H. Herzog. Alternative methods for attenuation correction for PET images in MR-PET scanners. In *2007 IEEE Nuclear Science Symposium Conference Record*. 2007. IEEE.

15. Kops, E.R. and H. Herzog. Template based attenuation correction for PET in MR-PET scanners. In *2008 IEEE Nuclear Science Symposium Conference Record*. 2008. IEEE.

16. Malone, I.B., et al., Attenuation correction methods suitable for brain imaging with a PET/MRI scanner: a comparison of tissue atlas and template attenuation map approaches. *Journal of Nuclear Medicine*, 2011. **52**(7): pp. 1142–1149.

17. Rueckert, D., et al., Nonrigid registration using free-form deformations: application to breast MR images. *IEEE Transactions on Medical Imaging*, 1999. **18**(8): pp. 712–721.

18. Burgos, N., et al., Attenuation correction synthesis for hybrid PET-MR scanners: application to brain studies. *IEEE Transactions on Medical Imaging*, 2014. **33**(12): pp. 2332–2341.

19. Schreibmann, E., et al., MR-based attenuation correction for hybrid PET-MR brain imaging systems using deformable image registration. *Medical Physics*, 2010. **37**(5): pp. 2101–2109.

20. Nagel, H.-H. and W. Enkelmann, An investigation of smoothness constraints for the estimation of displacement vector fields from image sequences. *IEEE Transactions on Pattern Analysis and Machine Intelligence*, 1986. **5**: pp. 565–593.

21. Lucas, B.D. and T. Kanade. *An iterative image registration technique with an application to stereo vision. Proceedings of Imaging Understanding Workshop*, pp. 121–130. 1981. Vancouver.

22. Adeloye, A., K.R. Kattan, and F.N. Silverman, Thickness of the normal skull in the American blacks and whites. *American Journal of Physical Anthropology*, 1975. **43**(1): pp. 23–30.

23. Le Goff-Rougetet, R., et al. Segmented MR images for brain attenuation correction in PET. in *Medical imaging 1994: image processing*. 1994. SPIE, Newport Beach, CA, United States.

24. Martinez-Möller, A., et al., Tissue classification as a potential approach for attenuation correction in whole-body PET/MRI: evaluation with PET/CT data. *Journal of Nuclear Medicine*, 2009. **50**(4): pp. 520–526.

25. Schulz, V., et al., Automatic, three-segment, MR-based attenuation correction for whole-body PET/MR data. *European Journal of Nuclear Medicine and Molecular Imaging*, 2011. **38**(1): pp. 138–152.

26. Zaidi, H., M.L. Montandon, and D.O. Slosman, Magnetic resonance imaging-guided attenuation and scatter corrections in three-dimensional brain positron emission tomography. *Medical Physics*, 2003. **30**(5): pp. 937–948.

27. Bezdek, J.C., R. Ehrlich, and W. Full, FCM: the fuzzy c-means clustering algorithm. *Computers & Geosciences*, 1984. **10**(2–3): pp. 191–203.

28. Wagenknecht, G., et al., Knowledge-based segmentation of attenuation-relevant regions of the head in T1-weighted MR images for attenuation correction in MR/PET systems. In *2009 IEEE Nuclear Science Symposium Conference Record (NSS/MIC)*. 2009. IEEE.

29. Keereman, V., et al., MRI-based attenuation correction for PET/MRI using ultrashort echo time sequences. *Journal of Nuclear Medicine*, 2010. **51**(5): pp. 812–818.

30. Catana, C., et al., Toward implementing an MRI-based PET attenuation-correction method for neurologic studies on the MR-PET brain prototype. *Journal of Nuclear Medicine*, 2010. **51**(9): pp. 1431–1438.

31. Berker, Y., et al., MRI-based attenuation correction for hybrid PET/MRI systems: a 4-class tissue segmentation technique using a combined ultrashort-echo-time/Dixon MRI sequence. *Journal of Nuclear Medicine*, 2012. **53**(5): pp. 796–804.

32. Censor, Y., et al., A new approach to the emission computerized tomography problem: simultaneous calculation of attenuation and activity coefficients. *IEEE Transactions on Nuclear Science*, 1979. **26**(2): pp. 2775–2779.

33. Nuyts, J., et al., Simultaneous maximum a posteriori reconstruction of attenuation and activity distributions from emission sinograms. *IEEE Transactions on Medical Imaging*, 1999. **18**(5): pp. 393–403.

34. Krol, A., et al., An EM algorithm for estimating SPECT emission and transmission parameters from emission data only. *IEEE Transactions on Medical Imaging*, 2001. **20**(3): pp. 218–232.

35. Defrise, M., A. Rezaei, and J. Nuyts, Time-of-flight PET data determine the attenuation sinogram up to a constant. *Physics in Medicine & Biology*, 2012. **57**(4): pp. 885–885.

36. Rezaei, A., et al., Simultaneous reconstruction of activity and attenuation in time-of-flight PET. *IEEE Transactions on Medical Imaging*, 2012. **31**(12): pp. 2224–2233.

37. Rezaei, A., M. Defrise, and J. Nuyts, ML-reconstruction for TOF-PET with simultaneous estimation of the attenuation factors. *IEEE Transactions on Medical Imaging*, 2014. **33**(7): pp. 1563–1572.

38. Rezaei, A., et al., Estimation of crystal timing properties and efficiencies for the improvement of (joint) maximum-likelihood reconstructions in TOF-PET. *IEEE Transactions on Medical Imaging*, 2019. **39**(4): pp. 952–963.

39. Panin, V., M. Aykac, and M. Casey, Simultaneous reconstruction of emission activity and attenuation coefficient distribution from TOF data, acquired with external transmission source. *Physics in Medicine & Biology*, 2013. **58**(11): p. 3649.

40. Rothfuss, H., et al., LSO background radiation as a transmission source using time of flight. *Physics in Medicine & Biology*, 2014. **59**(18): p. 5483.

41. Cheng, L., et al., Maximum likelihood activity and attenuation estimation using both emission and transmission data with application to utilization of Lu-176 background radiation in TOF PET. *Medical Physics*, 2020. **47**(3): pp. 1067–1082.

42. Teimoorisichani, M., et al. Development of a TOF-MLAA algorithm using LSO background radiation. In *2020 IEEE Nuclear Science Symposium and Medical Imaging Conference (NSS/MIC)*. 2020. IEEE.

43. Feng, T., J. Wang, and H. Li, Joint activity and attenuation estimation for PET with TOF data and single events. *Physics in Medicine & Biology*, 2018. **63**(24): p. 245017.

44. Salomon, A., et al., Simultaneous reconstruction of activity and attenuation for PET/MR. *IEEE Transactions on Medical Imaging*, 2010. **30**(3): pp. 804–813.

45. Mehranian, A. and H. Zaidi, Joint estimation of activity and attenuation in whole-body TOF PET/MRI using constrained Gaussian mixture models. *IEEE Transactions on Medical Imaging*, 2015. **34**(9): pp. 1808–1821.

46. Mehranian, A., H. Zaidi, and A.J. Reader, MR-guided joint reconstruction of activity and attenuation in brain PET-MR. *NeuroImage*, 2017. **162**: pp. 276–288.

47. Benoit, D., et al., Optimized MLAA for quantitative non-TOF PET/MR of the brain. *Physics in Medicine & Biology*, 2016. **61**(24): p. 8854.

48. Rezaei, A., et al., A quantitative evaluation of joint activity and attenuation reconstruction in TOF PET/MR brain imaging. *Journal of Nuclear Medicine*, 2019. **60**(11): pp. 1649–1655.

49. Wang, G., PET-enabled dual-energy CT: image reconstruction and a proof-of-concept computer simulation study. *Physics in Medicine & Biology*, 2020. **65**(24): p. 245028.

50. Presotto, L., et al., Simultaneous reconstruction of attenuation and activity in cardiac PET can remove CT misalignment artifacts. *Journal of Nuclear Cardiology*, 2016. **23**(5): pp. 1086–1097.

51. Bousse, A., et al., Maximum-likelihood joint image reconstruction/motion estimation in attenuation-corrected respiratory gated PET/CT using a single attenuation map. *IEEE Transactions on Medical Imaging*, 2015. **35**(1): pp. 217–228.

52. Fuin, N., et al., PET/MRI in the presence of metal implants: completion of the attenuation map from PET emission data. *Journal of Nuclear Medicine*, 2017. **58**(5): pp. 840–845.

53. Heußer, T., et al., MLAA-based attenuation correction of flexible hardware components in hybrid PET/MR imaging. *EJNMMI Physics*, 2017. **4**(1): pp. 1–23.

54. Hofmann, M., et al., MRI-based attenuation correction for PET/MRI: a novel approach combining pattern recognition and atlas registration. *Journal of Nuclear Medicine*, 2008. **49**(11): pp. 1875–1883.

55. James, G., et al., *An introduction to statistical learning*. 2013. Springer, New York, US.

56. Hofmann, M., et al., MRI-based attenuation correction for whole-body PET/MRI: quantitative evaluation of segmentation-and atlas-based methods. *Journal of Nuclear Medicine*, 2011. **52**(9): pp. 1392–1399.

57. Navalpakkam, B.K., et al., Magnetic resonance–based attenuation correction for PET/MR hybrid imaging using continuous valued attenuation maps. *Investigative Radiology*, 2013. **48**(5): pp. 323–332.

58. Breusch, T.S. and A.R. Pagan, The Lagrange multiplier test and its applications to model specification in econometrics. *The Review of Economic Studies*, 1980. **47**(1): pp. 239–253.

59. Huynh, T., et al., Estimating CT image from MRI data using structured random forest and auto-context model. *IEEE Transactions on Medical Imaging*, 2015. **35**(1): pp. 174–183.

60. Viola, P. and M. Jones. Rapid object detection using a boosted cascade of simple features. In *Proceedings of the 2001 IEEE Computer Society Conference on Computer Vision and Pattern Recognition. CVPR 2001*. 2001. IEEE.

61. Tu, Z. and X. Bai, Auto-context and its application to high-level vision tasks and 3d brain image segmentation. *IEEE Transactions on Pattern Analysis and Machine Intelligence*, 2009. **32**(10): pp. 1744–1757.

62. LeCun, Y., Y. Bengio, and G. Hinton, Deep learning. *Nature*, 2015. **521**(7553): pp. 436–444.

63. Bengio, Y., A. Courville, and P. Vincent, Representation learning: a review and new perspectives. *IEEE Transactions on Pattern Analysis and Machine Intelligence*, 2013. **35**(8): pp. 1798–1828.

64. Litjens, G., et al., A survey on deep learning in medical image analysis. *Medical Image Analysis*, 2017. **42**: pp. 60–88.

65. Gong, K., et al., The evolution of image reconstruction in PET: from filtered back-projection to artificial intelligence. *PET Clinics*, 2021. **16**(4): pp. 533–542.

66. Han, X., MR-based synthetic CT generation using a deep convolutional neural network method. *Medical Physics*, 2017. **44**(4): pp. 1408–1419.

67. Liu, F., et al., Deep learning MR imaging-based attenuation correction for PET/MR imaging. *Radiology*, 2017: p. 170700.

68. Gong, K., et al., Attenuation correction for brain PET imaging using deep neural network based on Dixon and ZTE MR images. *Physics in Medicine & Biology*, 2018. **63**(12): p. 125011.

69. Ladefoged, C.N., et al., Deep learning based attenuation correction of PET/MRI in pediatric brain tumor patients: evaluation in a clinical setting. *Frontiers in Neuroscience*, 2018. **12**: p. 1005.

70. Spuhler, K.D., et al., Synthesis of patient-specific transmission data for PET attenuation correction for PET/MRI neuroimaging using a convolutional neural network. *Journal of Nuclear Medicine*, 2019. **60**(4): pp. 555–560.

71. Hwang, D., et al., Improving the accuracy of simultaneously reconstructed activity and attenuation maps using deep learning. *Journal of Nuclear Medicine*, 2018. **59**(10): pp. 1624–1629.

72. Hashimoto, F., et al., Deep learning-based attenuation correction for brain PET with various radiotracers. *Annals of Nuclear Medicine*, 2021. **35**(6): pp. 691–701.

73. Torrado-Carvajal, A., et al., Dixon-VIBE deep learning (DIVIDE) pseudo-CT synthesis for pelvis PET/MR attenuation correction. *Journal of Nuclear Medicine*, 2019. **60**(3): pp. 429–435.

74. Ladefoged, C.N., et al., AI-driven attenuation correction for brain PET/MRI: clinical evaluation of a dementia cohort and importance of the training group size. *Neuroimage*, 2020. **222**: p. 117221.

75. Puig, O., et al., Deep-learning-based attenuation correction in dynamic [15O] H_2O studies using PET/MRI in healthy volunteers. *Journal of Cerebral Blood Flow & Metabolism*, 2021. **41**(12): pp. 3314–3323.

76. Hwang, D., et al., Generation of PET attenuation map for whole-body time-of-flight 18F-FDG PET/MRI using a deep neural network trained with simultaneously reconstructed activity and attenuation maps. *Journal of Nuclear Medicine*, 2019. **60**: pp. 1183–1189.

77. Hwang, D., et al., Data-driven respiratory phase-matched PET attenuation correction without CT. *Physics in Medicine & Biology*, 2021. **66**(11): p. 115009.

78. Hwang, D., et al., Comparison of deep learning-based emission-only attenuation correction methods for positron emission tomography. *European Journal of Nuclear Medicine and Molecular Imaging*, 2022. **49**: pp. 1833–1842.

79. Yang, J., et al., Joint correction of attenuation and scatter in image space using deep convolutional neural networks for dedicated brain 18 F-FDG PET. *Physics in Medicine & Biology*, 2019. **64**: p. 075019.

80. Shiri, I., et al., Direct attenuation correction of brain PET images using only emission data via a deep convolutional encoder-decoder (Deep-DAC). *European Radiology*, 2019. **29**: pp. 6867–6879.

81. Van Hemmen, H., et al., A deep learning-based approach for direct whole-body PET attenuation correction. *Journal of Nuclear Medicine*, 2019. **60**(supplement 1): p. 569.

82. Liu, F., et al., A deep learning approach for 18 F-FDG PET attenuation correction. *EJNMMI Physics*, 2018. **5**(1): pp. 1–15.

83. Goodfellow, I., et al., Generative adversarial nets. *Advances in Neural Information Processing Systems*, 2014. **27**: pp. 1–9.

84. Yi, X., E. Walia, and P. Babyn, Generative adversarial network in medical imaging: a review. *Medical Image Analysis*, 2019. **58**: p. 101552.

85. Pozaruk, A., et al., Augmented deep learning model for improved quantitative accuracy of MR-based PET attenuation correction in PSMA PET-MRI prostate imaging. *European Journal of Nuclear Medicine and Molecular Imaging*, 2021. **48**(1): pp. 9–20.

86. Armanious, K., et al., Independent attenuation correction of whole body [18 F] FDG-PET using a deep learning approach with Generative Adversarial Networks. *EJNMMI Research*, 2020. **10**(1): pp. 1–9.

87. Gong, K., et al., MR-based attenuation correction for brain PET using 3D cycle-consistent adversarial network. *IEEE Transactions on Radiation and Plasma Medical Sciences*, 2020. **5**: pp. 185–192.

88. Dong, X., et al., Synthetic CT generation from non-attenuation corrected PET images for whole-body PET imaging. *Physics in Medicine & Biology*, 2019. **64**(21): p. 215016.

89. Zhu, J.-Y., et al. Unpaired image-to-image translation using cycle-consistent adversarial networks. In *Proceedings of the IEEE International Conference on Computer Vision*. 2017.

9

High-Resolution Image Estimation using Deep Learning

Xianjin Dai

Stanford University, Stanford, CA, USA

Xiaofeng Yang

Emory University, Atlanta, GA, USA

CONTENTS

9.1 Introduction

Medical imaging plays a critical role in daily clinical practice including diagnosis, intervention, and treatment. X-ray computed tomography (CT), magnetic resonance imaging (MRI), and ultrasound imaging (US) are the most commonly used imaging modalities for qualitative and quantitative evaluation of diseases in the human body. High-resolution images are always highly desired for better assessment of diseases, image-guided intervention, treatment planning, and so forth. For example, in radiation therapy, CT is a primary imaging modality which is critical in treatment planning, dose calculation, and target and organs-at-risk (OARs) delineation. Higher spatial resolution CT images offer more accurate anatomical information, which typically leads to more accurate target delineation and dose calculation. Especially for those small organs such as lenses, optic nerves, and chiasm, higher spatial resolution is necessary to delineate the organs accurately (Luo et al. 2018; Budida and Sankar 2019). Ultrasound imaging has been commonly used for real-time guidance of applicator insertion (Watkins et al. 2011; White et al. 2013) or radioactive seed

DOI: 10.1201/9781003243458-12

implantation (Wang et al. 2009; Alnaghy et al. 2017) in brachytherapy. However, it typically requires a relatively long data acquisition time for obtaining high-resolution images, which may be difficult under some circumstances due to the limited hardware capability, patient tolerance, and so forth. For instance, in ultrasound imaging, three-dimensional (3D) volumetric images are typically obtained by reconstructing from two-dimensional (2D) images that are acquired by mechanical scanning (Guo et al. 2018). However, the speed of the mechanical scan is always slow even though ultrasound imaging systems typically can reach a very high frame rate of acquiring 2D images, which may not be able to satisfy the real-time image guidance for intervention. Meanwhile, variations among different operators are always unavoidable for a mechanical scan with handheld probes.

Efforts have been continuously made to solve the challenges of simultaneously achieving both high spatial and temporal resolution in medical imaging. For example, compressed sensing (CS)-based methods have been introduced for the reconstruction of 3D images from sub-sampled data in either ultra-fast US systems (Liebgott, Prost, and Friboulet 2013; Schretter et al. 2017; Mitrovic et al. 2020) or MRI systems (Lustig et al. 2008; Jaspan, Fleysher, and Lipton 2015). As advances in modern machine learning techniques, especially deep learning algorithms (Goodfellow et al. 2016; Akhtar and Mian 2018; Kendall and Gal 2017; Voulodimos et al. 2018), deep-learning-based methods have been extensively investigated for medical image reconstruction (Kim et al. 2018; Shen et al. 2019; Wang et al. 2018; Wang, Ye, and De Man 2020a), analysis (Dai, Lei, Zhang, et al. 2020d; Dai, Lei, Liu, et al. 2020b; Dai, Lei, Wang, et al. 2020c; Fu et al. 2020; Ibragimov et al. 2017; Lei et al. 2020), synthesis (Wang, Lei, Fu et al. 2020b; Dai, Lei, Fu, et al. 2020a), and so forth. Deep-learning-based image reconstruction algorithms have shown very promising results over traditional CS-based methods (Jin et al. 2017; Han and Ye 2018; Hammernik et al. 2018; Zhu et al. 2018). While encouraging, current deep-learning-based algorithms are commonly trained using a large amount of paired (raw data plus ground truth) datasets. However, in clinical practice, these paired raw data with ground truth are difficult to obtain. Sometimes these images may not even be obtainable because of the restricted clinical situations.

In this chapter, we present a self-supervised learning framework using cycle consistent generative adversarial network (cycleGAN) for estimating high-resolution 3D images from sparsely acquired 2D images. We validated our proposed method for US, CT, and MRI high-resolution image estimation. Without any extra paired training dataset, our model can estimate high-resolution images by using only the input image itself.

9.2 Methods

9.2.1 Self-supervised Learning Framework

The self-supervised learning framework is shown in Figure 9.1(Dai et al. 2021). Our proposed model consists of two independent cycleGAN modules, which are respectively trained with paired 2D high- and low-resolution images. Instead of collecting paired 2D high- and low-resolution images, the following process is applied to obtain these pairs from the input images themselves. Consider a 3D image $I(x, y, z)$, in medical imaging, the spatial resolution along two directions (assuming along x and y axes) are generally much higher than that along the third direction (assuming along z axis). To distinguish the different spatial resolution along different directions. We can denote the original input image $I(x, y, z)$ as $I(x_H, y_H, z_L)$. Here, H represents high resolution while L indicates low resolution.

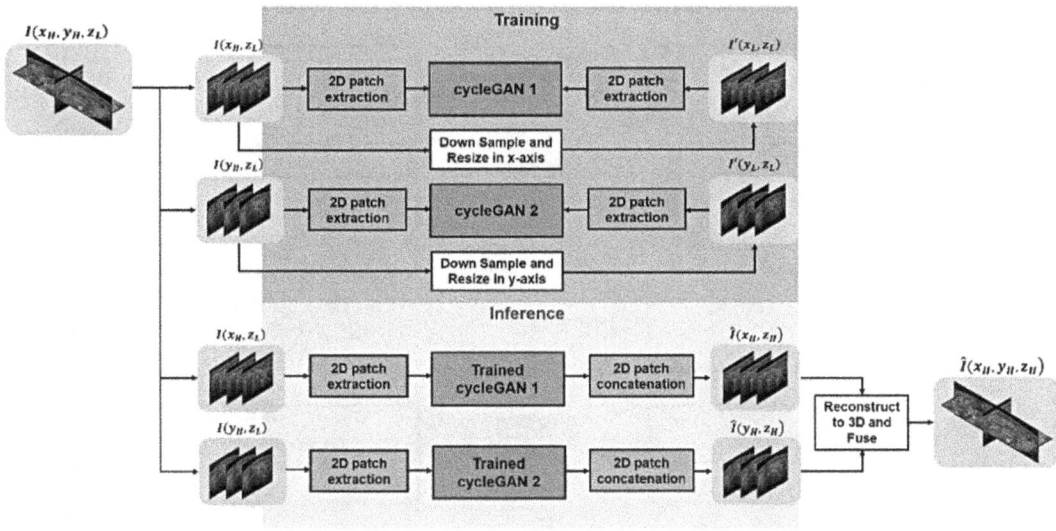

FIGURE 9.1
The schematic diagram of the proposed self-supervised learning framework. cycleGAN: cycle consistent generative adversarial network. (Figures adapted with permission from Dai et al. 2021.)

First, two sets of 2D images are collected from the original 3D image, namely, $I(x_H, z_L)$ and $I(y_H, z_L)$. Then, downsampling with a factor of n is applied along x and y axes respectively to obtain $I'(x_L, z_L)$ and $I'(y_L, z_L)$. Finally, using the paired $I'(x_L, z_L)$ and $I(x_H, z_L)$, and $I'(y_L, z_L)$ and $I(y_H, z_L)$ to respectively train the two cycleGANs. Once the two cycleGANs were trained, then, in the inference, high-resolution 2D images $\hat{I}(x_H, z_H)$ and $\hat{I}(y_H, z_H)$ can be respectively estimated by the cycleGANs taking $I(x_H, z_L)$ and $I(y_H, z_L)$ as inputs.

9.2.2 The cycleGAN

Generative adversarial net (GAN) is an advanced machine learning framework, which was originally introduced in 2014 (Goodfellow et al. 2014). Since then, extensive investigations have been carried out including building new models, exploiting their applications in computer vision, medical image analysis, and so forth. GAN and its variations primarily rely on the competition between a generator and a discriminator by using characteristic adversarial loss. The generator produces fake currency, while the discriminator detects the counterfeit currency. The cycleGAN, one of the GAN variations, was introduced in 2017 (Zhu et al. 2017) to tackle the challenges in image-to-image translation while the paired source and target images are not available. In our proposed model, cycleGAN was adopted and modified to generate high-resolution images from low-resolution images. As described in the previous section, here, cycleGAN actually is trained using paired high- and low-resolution images, which is different from the original cycleGAN (Zhu et al. 2017). The diagram of the cycleGAN architecture used in our model is shown in Figure 9.2 (Dai et al. 2021). In the training phase, the cycleGAN takes the paired high- and low-resolution images as input. A synthetic high-resolution image (sHR) is produced from the input low-resolution image (LR). The sHR is used for prompting the discriminator to identify real high-resolution image (HR) from sHR. Meanwhile, a synthetic low-resolution image (sLR) is generated by the second generator taking HR as input. The sLR is then utilized for prompting the second discriminator to recognize real LR from sLR. The generators and discriminators are pitted

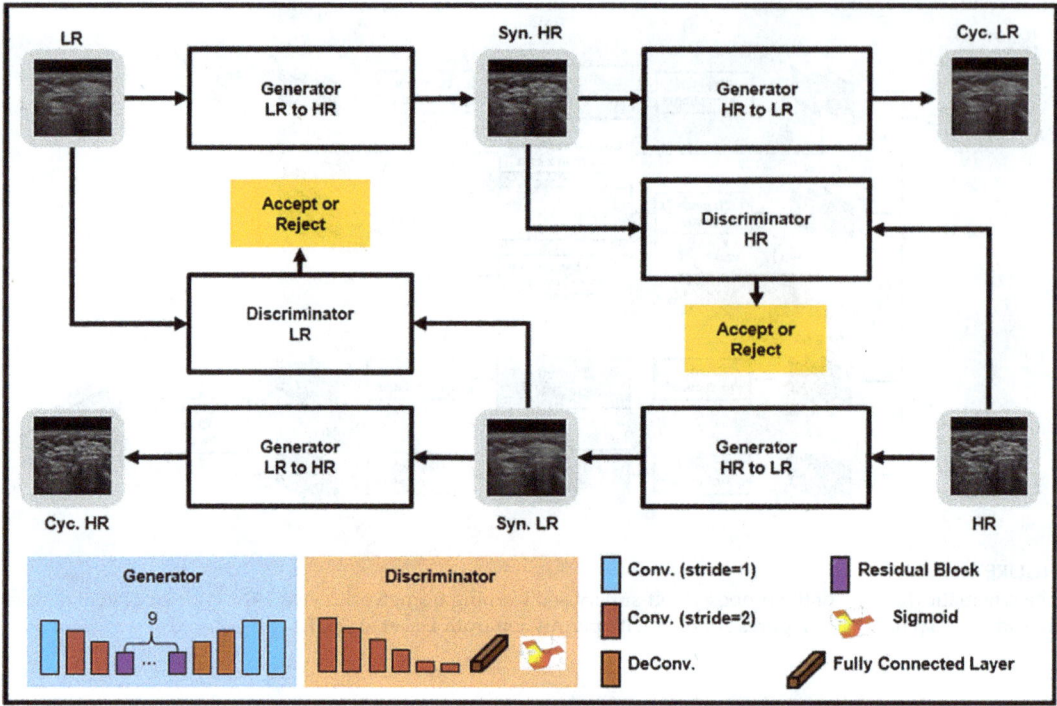

FIGURE 9.2
The network architecture of cycleGAN. LR, low-resolution image; HR, high-resolution image; Syn., synthetic; Cyc., cycle; Conv., convolution; DeConv., deconvolution. (Figures adapted with permission from Dai et al. 2021.)

against each other to obtain accurate sHR generation. Moreover, cycle consistency supervision is adopted to force the mapping from LR to HR to be close to a one-to-one mapping due to the nature of ill-posed problem mapping from LR to HR, which is implemented through adding two additional reversed order cycles to loop the feed forward paths. In our implementation, as shown in Figure 9.2, each generator consists of one convolutional layer with a stride size of 1×1, two convolutional layers with a stride size of 2×2, nine residual blocks (ResNet), two deconvolution layers and two convolutional layers with a stride size of 1×1, while each discriminator is constructed by six convolutional layers with a stride size of 2×2, one fully connected layer and a sigmoid operation layer. Meanwhile, residual learning is incorporated into the cycleGAN network because the differences between these LR and HR images are minimized in our study. Residual learning will improve the convergent efficiency.

9.3 Results

To validate the proposed method, we collected datasets from three different modalities, ultrasound imaging, CT, and MRI under the approval by the local institutional review board (IRB). The experimental results are shown in the following sections.

9.3.1 Ultrasound High-resolution Image Estimation

Automatic breast ultrasound (ABUS) images from 70 breast cancer patients and transrectal ultrasound (TRUS) images from 45 prostate cancer patients were collected. The ABUS images are 8-bit images (intensity levels 0–255) with a voxel size of $0.27 \times 0.27 \times 0.5$ mm^3, while TRUS images are 8-bit images with a voxel size of $0.12 \times 0.12 \times 2.0$ mm^3.

9.3.1.1 Breast US

ABUS images have a relatively high resolution in each dimension (0.27 mm $\times$ 0.27 mm $\times$ 0.5 mm), so, we use these original images as the ground truth to validate the proposed method. To obtain low-resolution images, we downsampled the original images by a factor of n along y axis, which yields images with a voxel size of 0.27 mm $\times$ (0.27 $\times$ n) mm $\times$ 0.5 mm. Figure 9.3 shows an illustrative example with the factor of $n = 5$. Axial-, sagittal-,

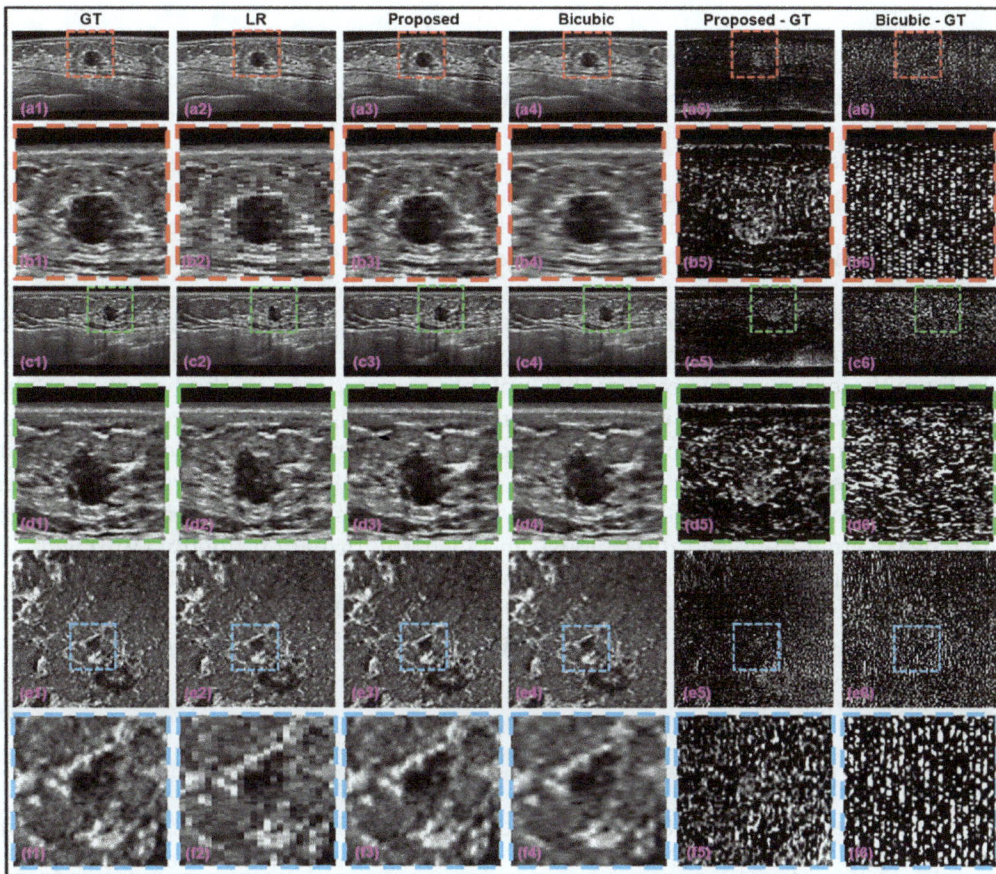

FIGURE 9.3

Synthesis of high-resolution breast ultrasound images. (a1–a6) are the axial-view images with the red-box-area zoom-in images (b1–b6), (c1–c6) show the sagittal-view images and their green-box-area zoom-in images (d1–d6), and (e1–e6) are the coronal views with the blue-box-are zoom-in images (f1–f6). GT, ground truth; LR, downsampled low-resolution images; Proposed, images predicted by the proposed method; Bicubic, images obtained through bicubic interpolation; Proposed – GT, the differences between the proposed method predictions and the ground truth; Bicubic – GT, the differences between the images by bicubic interpolation and the ground truth. (Figures adapted with permission from Dai et al. 2021.)

TABLE 9.1

Overall quantitative results achieved by the proposed method compared to bicubic interpolation and deepGAN for breast cases with a factor of 3 and 5.

Method	Factor of $n = 3$			Factor of $n = 5$		
	MAE	PSNR (dB)	VIF	MAE	PSNR (dB)	VIF
Proposed	0.90 ± 0.15	37.88 ± 0.88	0.69 ± 0.01	1.39 ± 0.23	34.88 ± 0.86	0.65 ± 0.01
deepGAN	1.25 ± 0.13	36.76 ± 0.63	0.68 ± 0.01	1.61 ± 0.31	33.32 ± 1.17	0.64 ± 0.01
Bicubic	4.65 ± 0.48	31.83 ± 0.63	0.60 ± 0.01	5.92 ± 1.18	31.68 ± 1.23	0.46 ± 0.01
p-value (Proposed vs. deepGAN)	<0.01	<0.01	<0.01	<0.01	<0.01	<0.01
p-value (Proposed vs. Bicubic)	<0.01	<0.01	<0.01	<0.01	<0.01	<0.01

Table adapted with permission from Dai et al. 2021.

and coronal-view images are respectively shown in the first, third, and fifth rows, while zoom-in the images within these small red, green, and blue boxes around the breast tumor area are shown second, fourth, and sixth rows. To assess the performance of our proposed method, we also did the same experiment using traditional bicubic interpolation algorithm. In Figure 9.3, ground truth, low-resolution, prediction by our proposed method, and prediction by bicubic interpolation images are respectively shown in the first, second, third and fourth columns. To clarify the improvement by our proposed method, we also calculated the difference between the predictions by our proposed method and bicubic interpolation and the ground truth, which are shown in the fifth and sixth columns respectively. From these images, we can see that, notably our proposed method offers much better image quality than bicubic interpolation method. We also did quantitative assessment of the performance of our proposed method. Same experiments were conducted using our proposed method, bicubic interpolation, and a deep attention generative adversarial network-based method (deepGAN) (He et al. 2020). Metrics including mean absolute error (MAE), peak signal-to-noise ratio (PSNR), and visual information fidelity (VIF) were calculated. Table 9.1 shows the summarized results for image enhancement factors of 3 and 5. Statistical analysis using paired student t-test shows that our proposed method significantly ($p < 0.01$) outperforms the bicubic interpolation and deepGAN methods approach in all the metrics.

9.3.1.2 Prostate US

For TRUS images from prostate cancer patients, the voxel size is $0.12 \times 0.12 \times 2.0$ mm^3, which is already very low resolution in one direction (2.0 mm). Thus, we did a prospective study. Figure 9.4 shows one example of resolution improvement by a factor of $n = 5$. For comparison, bicubic interpolation was also used to estimate the high-resolution images. In Figure 9.4, the first column is the original low-resolution images, the second column is the estimation by bicubic interpolation, and the last column is the prediction by our proposed method. From either axial-view (first row), sagittal-view (second row), or coronal-view (last row) images, we can see the improvements and advantages of our proposed method over the bicubic interpolation method clearly.

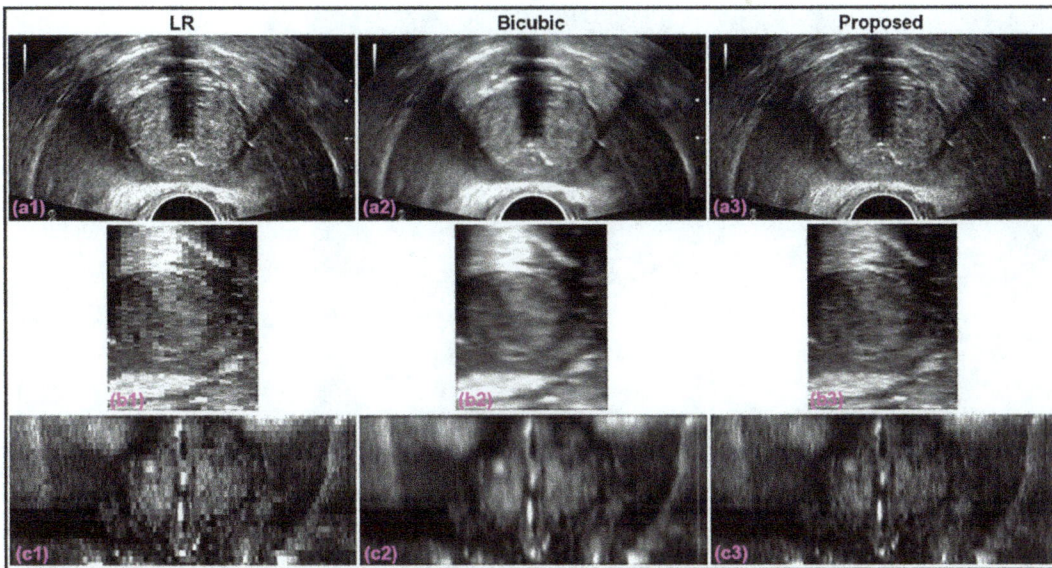

FIGURE 9.4
One example of generating high-resolution prostate ultrasound images with a factor of 5. (a1–a3) are the axial-view images, (b1–b3) show the sagittal-view images, and (c1–c3) are the coronal views. LR, low-resolution input images; Bicubic, images obtained through bicubic interpolation; Proposed, images predicted by the proposed method. (Figures adapted with permission from Dai et al. 2021.)

9.3.2 CT High-resolution Image Estimation

CT images of 75 head and neck (HN) cancer patients undergoing radiotherapy were collected to assess the performance of our proposed framework (Xie et al. 2021) from a Siemens SOMATOM Definition Edge (Erlangen, Germany) scanner with 120 kVp and exposure from 273 to 478 mAs. These images have a voxel size of $1 \times 1 \times 1$ mm³. Like the case of breast US in the previous section, these CT images have relatively high resolution. To evaluate our proposed method, for validation purposes, we averaged the CT images by every three consecutive slices to obtain low-resolution images with a voxel size of $1 \times 1 \times 3$ mm³. Figure 9.5 (Xie et al. 2021) shows one example of generating high-resolution CT images with a factor of three. For comparison, we also did the same experiments using a supervised deep convolutional neural network method (Park's method) (Park et al. 2018). The first column in Figure 9.5 shows the original high-resolution images ($1 \times 1 \times 1$ mm³), the second column shows the low-resolution images ($1 \times 1 \times 3$ mm³), the third column is the prediction by Park's method ($1 \times 1 \times 1$ mm³), while the fourth column is the prediction by our proposed method ($1 \times 1 \times 1$ mm³). Compared to the ground-truth images, the low-resolution images have stair-like and blurry appearances for the bony and soft tissue structures, especially from these sagittal and coronal views. Compared to the low-resolution images, both Park's method prediction and our proposed prediction can restore the high-resolution images. Closely investigate on these estimated images by Park's and our proposed methods, we can see that these regions indicated by the yellow and red arrows in the estimated images by Park's method show more blurry and bulky bone structures while being compared to those in the ground truth and estimated images by our proposed method. However, regions indicated by the green arrows in the images estimated by Park's method show better results than those in the estimated images by our proposed method.

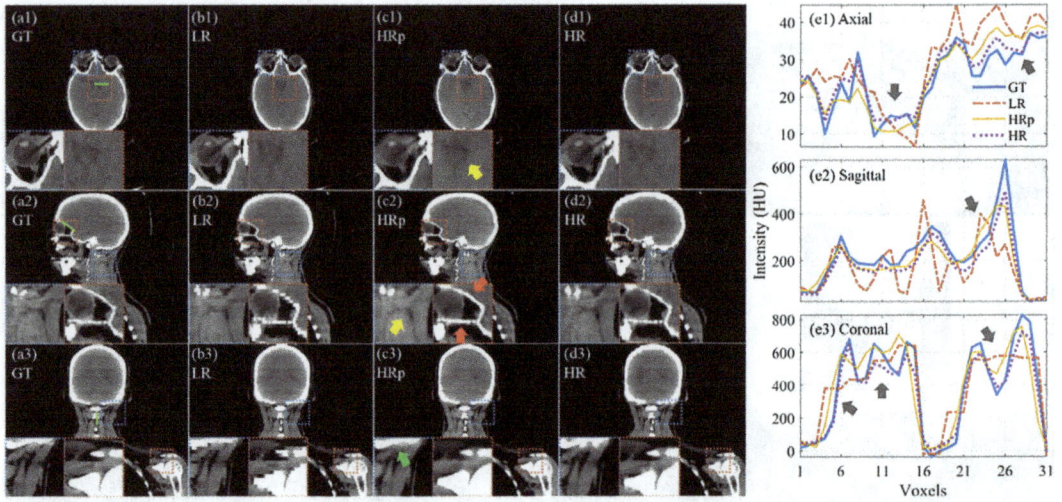

FIGURE 9.5

Comparison on the images and profiles of high-resolution CT image generation. The axial (a1, b1, c1 and d1), sagittal (a2, b2, c2, and d2) and coronal (a3, b3, c3, and d3) images of a head and neck patient are shown in the first to third rows, respectively. The images of ground truth (a1, a2, and a3), LR with 3 mm slice thickness (b1, b2, and b3), synthetic HRp3 with the compared Park's method (c1, c2, and c3) and synthetic HR with the proposed self-supervised CycleGAN method (d1, d2, and d3) are respectively shown in the first to fourth columns. Two regions of interest (ROIs), which mainly contain low- and high-contrast tissues, are selected and enlarged for each slice direction. The CT image intensity profiles along the solid green lines depicted in the ground truth images are also accordingly shown in the last column (e1, e2, and e3). (Figures adapted with permission from Xie et al. 2021.)

To further evaluate our proposed method, the Hounsfield unit (HU) intensity profiles of these images along the solid lines in the ground-truth images are shown in the last column of the Figure 9.5. The profiles in the axial plane primarily show the soft tissue comparisons while those in the sagittal and coronal planes show the bony structures. Through carefully comparing between those estimations (both Park's and our proposed methods) and the ground truth, we can see that the differences between the estimations by our proposed method and the ground truth are smaller than that by Park's method.

To quantitatively assess the performance of our proposed method, metrics including the mean absolute errors (MAEs), the edge keeping index (EKI) (Bhadauria and Dewal 2013), the structural similarity index measurement (SSIM) (Wang et al. 2004; Thung and Raveendran 2009; Xie et al. 2018), the information fidelity criterion (IFC) (Sheikh, Bovik, and de Veciana 2005) and the visual information fidelity in pixel domain (VIFP) (Gu et al. 2012) were calculated. Figure 9.6 shows the boxplot of all the 75 HN patients. The median value of each metric for each patient has also been computed. From the boxplot and t-test, we can see that our proposed method significantly improved all the investigated metrics. The summarized results are detailed in Table 9.2. By our proposed method, the improvements can be seen from Table 9.2 range from 7% for SSIM to ~130% for EKI. However, with the Park's method, the improvements range from 1% for SSIM to 48% for EKI.

9.3.3 MRI High-resolution Image Estimation

To further exploit the application of our proposed for MRI high-resolution image estimation, public dataset from the multimodal brain tumor segmentation challenge 2020 (BraTS2020) (Menze et al. 2015; Bakas et al. 2017; Bakas et al. 2018) were adopted (Xie et al. 2022). There

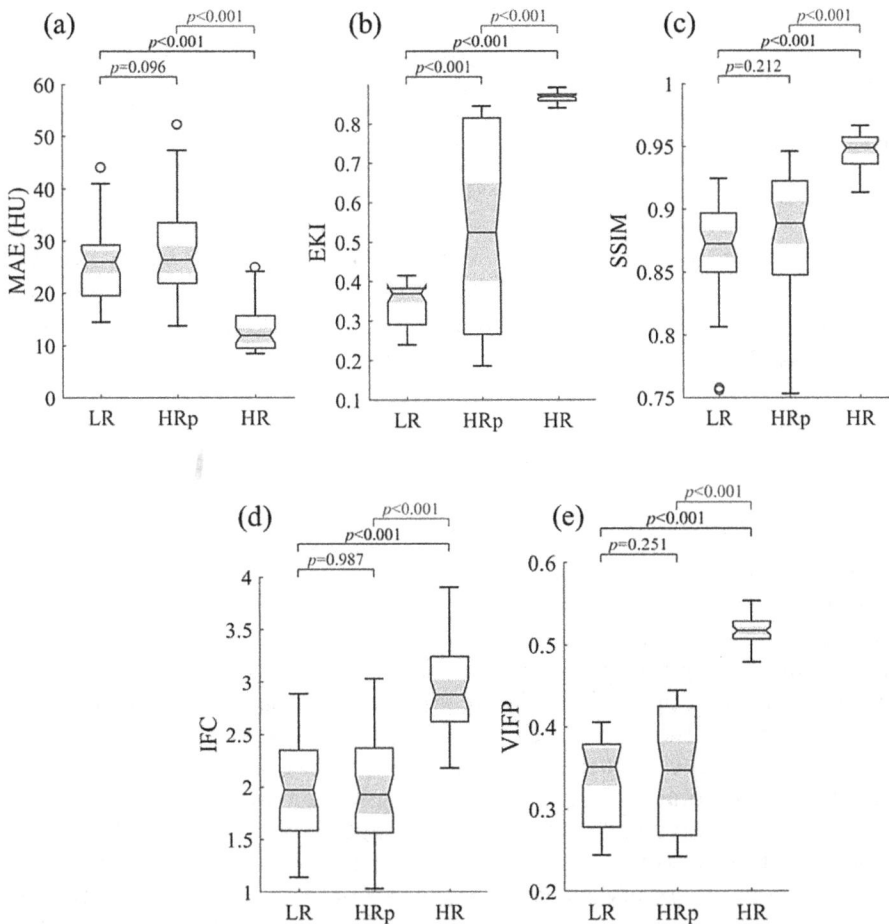

FIGURE 9.6

Boxplots and two-sample t-test results of the median values of the investigated image quality metrics for all 75 head and neck patients. The median value of each metric for each patient is calculated from the metric measurements of the sagittal and coronal slice images. All HRp and HR images are generated from LR images with 3mm slice thickness. (Figures adapted with permission from Xie et al. 2021.)

are multiple sequence MR images exist in the BraTS2020 dataset including T1-weighted, contrast enhance T1-weighted, T2-weighted, and T2 Fluid Attenuated Inversion Recovery (FLAIR). These MR images were acquired from multiple (n=19) institutions using different clinical protocols and various scanners. For validation purposes, we manually excluded 94 patients out of the total 369 patients in the BraTS2020 dataset because these MR images are either with low resolution or strong artifacts. The MR images of selected 275 patients all have the same resolution with a voxel size of $1 \times 1 \times 1$ mm^3. To exploit the feasibility of our proposed framework for MR high-resolution image estimate, these selected MR images in BraTS2020 dataset were downsampled to low-resolution images (LR^{CC}) along cranial-caudal (CC) direction. The downsampling method is simple, essentially extracting one slice every three slices. After downsampling, the low-resolution images have a voxel size of $1 \times 1 \times 3$ mm^3. Figure 9.7 shows one example of generating high-resolution

TABLE 9.2

Statistics and percentage improvements, compared to LR images, of the mean values of the investigated image quality metrics for all 75 head and neck patients (the 3 mm slice thickness case). The mean and standard deviation values of each metric for each patient is calculated from the metric measurements of the sagittal and coronal slice images.

	LR	HRp		HR	
	Mean ± Std	Mean ± Std	% Improvements Compared to LR	Mean ± Std	% Improvements Compared to LR
MAE (HU)	23.26 ± 5.75	27.71 ± 7.84	−19.11%	14.85 ± 3.28	36.15%
EKI	0.37 ± 0.06	0.55 ± 0.26	48.44%	0.85 ± 0.03	129.13%
SSIM	0.88 ± 0.04	0.88 ± 0.04	0.23%	0.94 ± 0.01	6.65%
IFC	2.11 ± 0.51	1.86 ± 0.42	−11.83%	2.77 ± 0.38	31.40%
VIFP	0.36 ± 0.06	0.35 ± 0.08	−1.89%	0.50 ± 0.03	39.84%

Table adapted with permission from Xie et al. 2021.

images from low-resolution images. For the purpose of comparison, we conducted the same experiments using bicubic interpolation, Zhao's self-supervised method with deep convolutional neural networks (DCNN) (Zhao et al. 2018; Zhao et al. 2021; Lim et al. 2017), supervised learning using the same networks as the proposed method, and our proposed self-supervised learning approach. In Figure 9.7, the axial-, sagittal-, and coronal-view are shown in the first to the third rows, respectively. Closely investigate on these images, we can see that there are stripe artifacts inside and sawtooth artifacts at the edges of brain area in these sagittal and coronal LR^{CC} images. By using bicubic interpolation ($Cubic^{CC}$), these artifacts can be reduced to some extent. But the images were blurred and lack of detailed information. On the contrary, these deep-learning-based methods including Zhao's self-supervised method, supervised learning with our network, and our proposed self-supervised approach can generate high-resolution images close to the ground truth. Our proposed self-supervised method outperforms Zhao's self-supervised method but inferior to the supervised learning using our network. Table 9.3 shows the quantitative results, from which we can see that our proposed method outperforms both bicubic interpolation and Zhao's self-supervised methods.

9.4 Discussion

The experiments of estimating high-resolution images for ultrasound imaging, CT, and MRI have shown that our proposed method outperforms traditional bicubic interpolation methods as well as some conventional deep-learning-based algorithms. Meanwhile, it shows the generic of our proposed method, which is not restricted to one specific medical imaging modality, instead, it can handle all the commonly used medical imaging modalities. One of the most favorable advantages by using our proposed method is that no extra training dataset is required. Deep-learning-based algorithms are typically considered as data-driven methods which require a relatively large amount of 'labeled' data for training usable models. In clinical practice, these data might not be available or rather difficult to obtain due to the limited capability of facilities, patient specifications, and so forth. Under

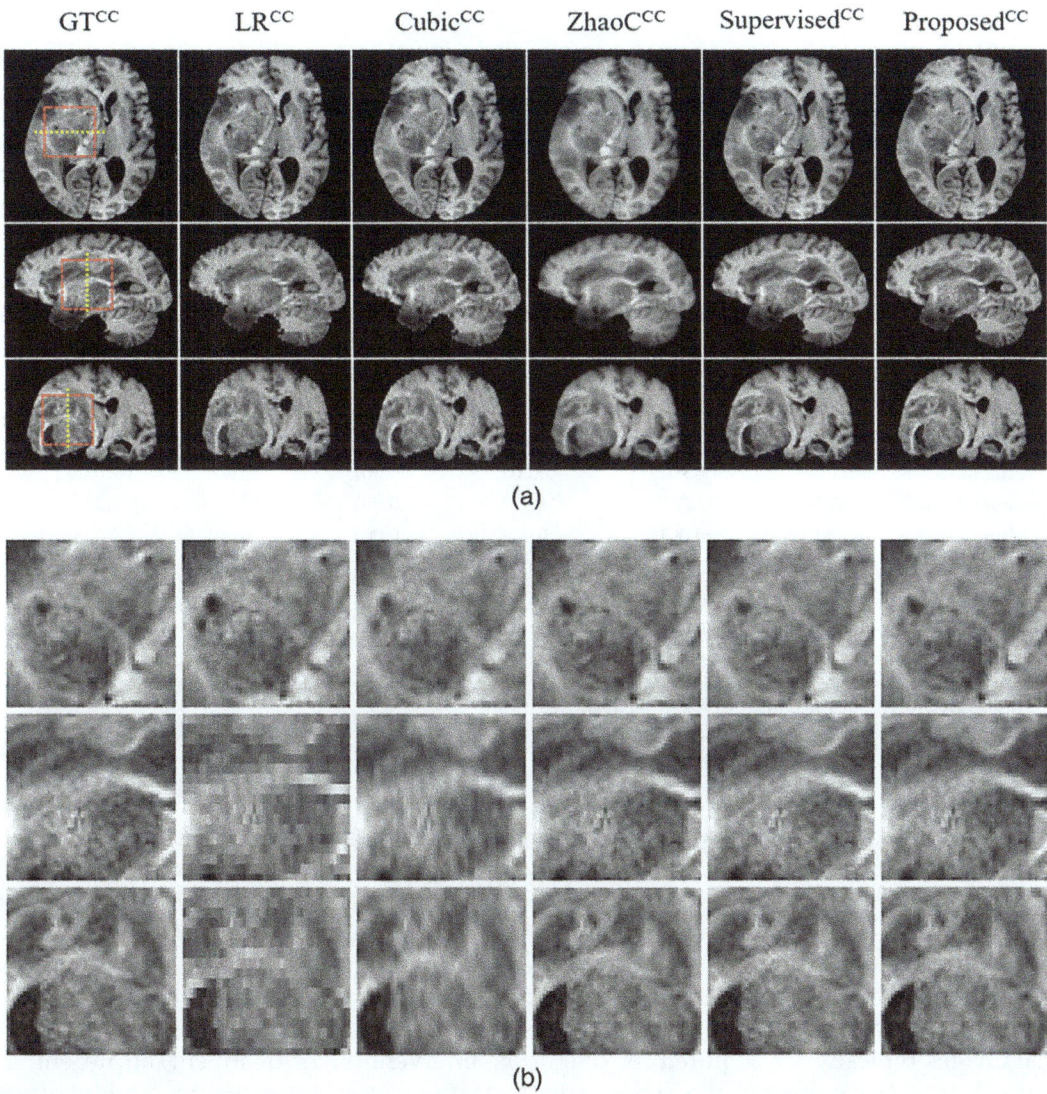

(a)

(b)

FIGURE 9.7

T1-weighted MR high-resolution image estimate. The T1 MR images (a) and zoom-in views of the red dot-dash box regions with the same subfigure layouts (b) of a brain cancer patient in the BraTS2020 dataset. In each subfigure, shown in columns one through six are the images of ground truth (GT^{CC}, 1 mm × 1 mm × 1 mm), low-resolution in the cranial-caudal direction (LR^{CC}, 1 mm × 1 mm × 3 mm) with 3 mm slice step, bicubic interpolation (Cubic^{CC}, 1 mm × 1 mm × 1 mm) and high-resolution predicted by the Zhao's method (ZhaoC^{CC}, 1 mm × 1 mm × 1 mm), the proposed method with supervised training (Supervised^{CC}, 1 mm × 1 mm × 1 mm) and the proposed method (Proposed^{CC}, 1 mm × 1 mm × 1 mm), respectively; the images of the axial, sagittal and coronal planes are shown in the first to the third rows, respectively. (Figures adapted with permission from Xie et al. 2022.)

the manner of self-supervised learning, our proposed method take the input images for both training and inference to estimate high-resolution images.

Certainly, some limitations still exist in our current model, which have to be resolved in our future investigations. Our proposed method by learning the mapping from low resolution to high resolution in two dimensions to predict the high resolution in the third

TABLE 9.3

Quantitative results of the T1-weighted images of all the investigated patients in the BraTS2020 dataset. Labeling of the compared methods follows the term definitions of Figure 9.7. The metrics of normalized mean absolute error (NMAE), peak signal-to-noise ratios (PSNR), structural similarity index measure (SSIM), edge keeping index (EKI), information fidelity criterion (IFC) and visual information fidelity in pixel domain (VIFP) were calculated while taking the GTCC images as references. Metric numbers with the first, second and third best values are shown in bold, underscore and italic fonts, respectively.

	NMAE	PSNR	SSIM	EKI	IFC	VIFP
LR^{CC} (*)	0.155 ± 0.041	24.575 ± 1.775	0.841 ± 0.055	0.336 ± 0.088	1.579 ± 0.295	0.322 ± 0.071
Cubic^{CC} (‡)	*0.098 ± 0.034*	*28.953 ± 2.313*	0.927 ± 0.038	0.546 ± 0.093	2.830 ± 0.864	*0.522 ± 0.123*
ZhaoC^{CC} (⚶)	0.116 ± 0.032	28.488 ± 1.909	*0.932 ± 0.023*	*0.924 ± 0.062*	*2.938 ± 0.595*	0.496 ± 0.056
Supervised^{CC}(∴)	**0.045 ± 0.015**	**37.578 ± 1.933**	**0.992 ± 0.004**	**0.990 ± 0.010**	**6.456 ± 1.085**	**0.810 ± 0.036**
Proposed^{CC} (∷)	<u>0.055</u> ± 0.015	<u>34.244</u> ± 1.889	<u>0.983</u> ± 0.006	<u>0.929</u> ± 0.035	<u>4.133</u> ± 0.698	<u>0.694</u> ± 0.041
p-value (‡ vs. *)	<0.001	<0.001	<0.001	<0.001	<0.001	<0.001
p-value (⚶ vs. ‡)	<0.001	0.070	0.062	<0.001	0.088	0.002
p-value (∴ vs. ‡)	<0.001	<0.001	<0.001	<0.001	<0.001	<0.001
p-value (∷ vs. ‡)	<0.001	<0.001	<0.001	<0.001	<0.001	<0.001
p-value (⚶ vs. ‡)	<0.001	<0.001	<0.001	<0.001	<0.001	<0.001
p-value (⚶ vs. ∴)	<0.001	<0.001	<0.001	0.292	<0.001	<0.001
p-value (∴ vs. ∷)	<0.001	<0.001	<0.001	<0.001	<0.001	<0.001

Table adapted with permission from Xie et al. 2022.

dimension from low-resolution images assumes the isotropy of tissue feature. However, due to the heterogeneity of human body, this assumption might be debatable. Therefore, further careful investigations have to be conducted to understand the isotropic nature of image features in different medical imaging modalities with respective to different body sites or different tissue types. Another notable limitation is that the proposed method relies on the high resolution in two dimensions. In other words, if the input images are low resolution in all three dimensions, then the proposed method is not applicable. Moreover, medical images typically are acquired for diagnosis, intervention, or treatment, at present, in our validation experiments, the disease sites or features were ignored, which might affect the performance of our proposed method.

9.5 Conclusion

In summary, in this chapter, we introduce a self-supervised learning-based framework that can be used to estimate high-resolution images, which was validated for three different medical imaging modalities including ultrasound imaging, CT, and MRI. In particular, without using any extra paired dataset, the proposed method has the capability of estimating high-resolution images from sparsely acquired 2D images, which is typically the case in medical imaging for diagnosis, intervention, and therapeutics.

References

Akhtar, N., and A. Mian. 2018. Threat of adversarial attacks on deep learning in computer vision: A survey. *IEEE Access* 6:14410–14430.

Alnaghy, S., D. L. Cutajar, J. A. Bucci, et al. 2017. BrachyView: Combining LDR seed positions with transrectal ultrasound imaging in a prostate gel phantom. *Physica Medica* 34:55–64.

Bakas, S., H. Akbari, A. Sotiras, et al. 2017. Advancing the cancer genome atlas glioma MRI collections with expert segmentation labels and radiomic features. *Scientific Data* 4 (1):170117.

Bakas, S., M. Reyes, A. Jakab, et al. 2018. Identifying the Best Machine Learning Algorithms for Brain Tumor Segmentation, Progression Assessment, and Overall Survival Prediction in the BRATS Challenge. https://ui.adsabs.harvard.edu/abs/2018arXiv181102629B

Bhadauria, H. S., and M. L. Dewal. 2013. Medical image denoising using adaptive fusion of curvelet transform and total variation. *Computers and Electrical Engineering* 39 (5):1451–1460.

Budida, A., and A. B. Sankar. 2019. Enhanced protocols for CT slice thickness on clinical target volume for 3d radiation therapy. *International Journal of Innovative Technology and Exploring Engineering* 8 (12):1365–1370.

Dai, X., Y. Lei, Y. Fu, et al. 2020a. Multimodal MRI synthesis using unified generative adversarial networks. *Medical Physics* 47 (12):6343–6354.

Dai, X., Y. Lei, Y. Liu, et al. 2020b. Intensity non-uniformity correction in MR imaging using residual cycle generative adversarial network. *Physics in Medicine and Biology* 65 (21):215025.

Dai, X., Y. Lei, T. Wang, et al. 2021. Self-supervised learning for accelerated 3D high-resolution ultrasound imaging. *Medical Physics* 48 (7):3916–3926.

Dai, X., Y. Lei, T. Wang, et al. 2020c. Synthetic MRI-aided Head-and-Neck Organs-at-Risk Auto-Delineation for CBCT-guided Adaptive Radiotherapy. *arXiv preprint*, arXiv:04275.

Dai, X., Y. Lei, Y. Zhang, et al. 2020d. Automatic multi-catheter detection using deeply supervised convolutional neural network in MRI-guided HDR prostate brachytherapy. *Medical Physics* 47 (9):4115–4124.

Fu, Y., Y. Lei, T. Wang, W. J. Curran, T. Liu, and X. Yang. 2020. Deep learning in medical image registration: a review. *Physics in Medicine and Biology* 65 (20):20TR01.

Goodfellow, I., Y. Bengio, A. Courville, and Y. Bengio. 2016. *Deep learning*. Vol. 1: MIT Press, Cambridge.

Goodfellow, I., J. Pouget-Abadie, M. Mirza, et al. 2014. Generative adversarial nets. *Paper Read at Advances in Neural Information Processing Systems*.

Gu, K., G. Zhai, X. Yang, and W. Zhang. 2012. An improved full-reference image quality metric based on structure compensation. *Paper read at Proceedings of The 2012 Asia Pacific Signal and Information Processing Association Annual Summit and Conference*, 3–6 Dec. 2012.

Guo, R., G. Lu, B. Qin, and B. Fei. 2018. Ultrasound imaging technologies for breast cancer detection and management: A review. *Ultrasound in Medicine & Biology* 44 (1):37–70.

Hammernik, K., T. Klatzer, E. Kobler, et al. 2018. Learning a variational network for reconstruction of accelerated MRI data. *Magnetic Resonance in Medicine* 79 (6):3055–3071.

Han, Y., and J. C. Ye. 2018. Framing U-net via deep convolutional framelets: Application to sparse-view CT. *IEEE Transactions on Medical Imaging* 37 (6):1418–1429.

He, X., Yang L., Y. Liu, et al. 2020. Deep attentional GAN-based high-resolution ultrasound imaging. *Paper Read at Medical Imaging 2020: Ultrasonic Imaging and Tomography*.

Ibragimov, Bulat, and L. Xing. 2017. Segmentation of organs-at-risks in head and neck CT images using convolutional neural networks. *Medical physics* 44 (2):547–557.

Jaspan, O. N., R. Fleysher, and M. L. Lipton. 2015. Compressed sensing MRI: a review of the clinical literature. *The British Journal of Radiology* 88 (1056):20150487.

Jin, K. H., M. T. McCann, E. Froustey, and M. Unser. 2017. Deep convolutional neural network for inverse problems in imaging. *IEEE Transactions on Image Processing* 26 (9):4509–4522.

Kendall, A., and Y. Gal. 2017. What uncertainties do we need in bayesian deep learning for computer vision? *Paper Read at Advances in Neural Information Processing Systems*.

Kim, K., D. Wu, K. Gong, et al. 2018. Penalized PET reconstruction using deep learning prior and local linear fitting. *IEEE Transactions on Medical Imaging* 37 (6):1478–1487.

Lei, Y., Y. Fu, T. Wang, et al. 2020. Deep Learning in Multi-organ Segmentation. *arXiv preprint*, arXiv:10619.

Liebgott, H., R. Prost, and D. Friboulet. 2013. Pre-beamformed RF signal reconstruction in medical ultrasound using compressive sensing. *Ultrasonics* 53 (2):525–533.

Lim, B., S. Son, H. Kim, S. Nah, and K. M. Lee. 2017. Enhanced Deep Residual Networks for Single Image Super-Resolution. https://ui.adsabs.harvard.edu/abs/2017arXiv170702921L

Luo, H., Y. He, F. Jin, et al. 2018. Impact of CT slice thickness on volume and dose evaluation during thoracic cancer radiotherapy. *Cancer Management and Research* 10:3679–3686.

Lustig, M., D. L. Donoho, J. M. Santos, and J. M. Pauly. 2008. Compressed sensing MRI. *IEEE Signal Processing Magazine* 25 (2):72–82.

Menze, B. H., A. Jakab, S. Bauer, et al. 2015. The multimodal brain tumor image segmentation benchmark (BRATS). *IEEE Transactions on Medical Imaging* 34 (10):1993–2024.

Mitrovic, J., Z. Ignjatovic, L. Pietra, W. J. Sehnert, and V. Dogra. 2020. Compressed sensing for reduced hardware footprint in medical ultrasound. *Ultrasonics* 108:106214.

Park, J., D. Hwang, K. Y. Kim, S. K. Kang, Y. K. Kim, and J. S. Lee. 2018. Computed tomography super-resolution using deep convolutional neural network. *Physics in Medicine and Biology* 63 (14):145011.

Schretter, Colas, Shaun Bundervoet, David Blinder, Ann Dooms, Jan D'hooge, and Peter Schelkens. 2017. Ultrasound imaging from sparse RF samples using system point spread functions. *IEEE Transactions on Ultrasonics, Ferroelectrics, and Frequency Control* 65 (3):316–326.

Sheikh, H. R., A. C. Bovik, and G. de Veciana. 2005. An information fidelity criterion for image quality assessment using natural scene statistics. *IEEE Transactions on Image Processing* 14 (12):2117–2128.

Shen, L., W. Zhao, and L. Xing. 2019. Patient-specific reconstruction of volumetric computed tomography images from a single projection view via deep learning. *Nat Biomed Eng* 3:880–888. https://doi.org/10.1038/s41551-019-0466-4.

Thung, K., and P. Raveendran. 2009. A survey of image quality measures. *Paper Read at 2009 International Conference for Technical Postgraduates (TECHPOS)*, 14–15 Dec. 2009.

Voulodimos, A., N. Doulamis, A. Doulamis, and E. Protopapadakis. 2018. Deep learning for computer vision: A brief review. *Computational Intelligence and Neuroscience* 2018:7068349.

Wang, Ge, Jong Chu Ye, Klaus Mueller, and Jeffrey A Fessler. 2018. Image reconstruction is a new frontier of machine learning. *IEEE Transactions on Medical Imaging* 37 (6):1289–1296.

Wang, Ge, Jong Chul Ye, and Bruno De Man. 2020a. Deep learning for tomographic image reconstruction. *Nature Machine Intelligence* 2 (12):737–748.

Wang, J., Y. Jiang, J. Li, S. Tian, W. Ran, and D. Xiu. 2009. Intraoperative ultrasound-guided iodine-125 seed implantation for unresectable pancreatic carcinoma. *Journal of Experimental & Clinical Cancer Research* 28 (1):88.

Wang, Tonghe, Yang Lei, Yabo Fu, et al. 2020b. A review on medical imaging synthesis using deep learning and its clinical applications. *Journal of Applied Clinical Medical Physics* 22:11–36.

Wang, Z., A. C. Bovik, H. R. Sheikh, and E. P. Simoncelli. 2004. Image quality assessment: from error visibility to structural similarity. *IEEE Transactions on Image Processing* 13 (4):600–612.

Watkins, J. M., P. L. Kearney, K. J. Opfermann, S. J. Ackerman, J. M. Jenrette, and M. F. Kohler. 2011. Ultrasound-guided tandem placement for low-dose-rate brachytherapy in advanced cervical cancer minimizes risk of intraoperative uterine perforation. *Ultrasound in Obstetrics & Gynecology* 37 (2):241–244.

White, E. C., M. R. Kamrava, J. Demarco, et al. 2013. High-dose-rate prostate brachytherapy consistently results in high quality dosimetry. *International Journal of Radiation Oncology, Biology, Physics* 85 (2):543–548.

Xie, H., T. Niu, S. Tang, X. Yang, N. Kadom, and X. Tang. 2018. Content-oriented sparse representation (COSR) for CT denoising with preservation of texture and edge. *Medical Physics* 45 (11):4942–4954.

Xie, Huiqiao, Yang Lei, Tonghe Wang, et al. 2022. Synthesizing high-resolution magnetic resonance imaging using parallel cycle-consistent generative adversarial networks for fast magnetic resonance imaging. *Medical Physics* 49 (1):357–369.

Xie, Huiqiao, Yang Lei, Tonghe Wang, et al. 2021. High through-plane resolution CT imaging with self-supervised deep learning. *Physics in Medicine & Biology* 66 (14):145013.

Yang, Guang, Simiao Yu, Hao Dong, et al. 2017. DAGAN: Deep de-aliasing generative adversarial networks for fast compressed sensing MRI reconstruction. *IEEE Transactions on Medical Imaging* 37 (6):1310–1321.

Zhao, C., A. Carass, B. E. Dewey, J. L. Prince, and Ieee. 2018. Self super-resolution for magnetic resonance images using deep networks. *2018 IEEE 15th International Symposium on Biomedical Imaging*. New York: IEEE.

Zhao, C., B. E. Dewey, D. L. Pham, P. A. Calabresi, D. S. Reich, and J. L. Prince. 2021. SMORE: A self-supervised anti-aliasing and super-resolution algorithm for MRI using deep learning. *IEEE Transactions on Medical Imaging* 40 (3):805–817.

Zhu, B., J. Z. Liu, S. F. Cauley, B. R. Rosen, and M. S. Rosen. 2018. Image reconstruction by domain-transform manifold learning. *Nature* 555 (7697):487–492.

Zhu, Jun-Yan, Taesung Park, Phillip Isola, and Alexei A. Efros. 2017. Unpaired image-to-image translation using cycle-consistent adversarial networks. *Paper Read at Proceedings of the IEEE International Conference on Computer Vision*.

10

2D–3D Transformation for 3D Volumetric Imaging

Zhen Tian

The University of Chicago, Chicago, IL, USA

Xiaofeng Yang

Emory University, Atlanta, GA, USA

CONTENTS

10.1 Introduction

Respiratory motion causes the movement of tumors and organs in the thoracic and abdominal sites, such as lung, liver, pancreas[1–3]. Moreover, the magnitude, baseline, period, and regularity of respiratory motion can vary from time to time[4]. For those motion-related treatment sites, the intra-fractional and inter-fractional variations of respiratory motion can lead to significant dosimetric uncertainties in radiotherapy[2,5,6], particularly for stereotactic body radiation therapy (SBRT) that delivers very high doses in a few treatment fractions[7,8]. 2D kV imaging (*e.g.*, kV fluoroscopy) can provide continuous 2D images during treatment to monitor the tumor motion in real time. However, as the patient's 3D anatomy is projected onto the 2D imager plane, 2D imaging usually has poor tumor visibility[9]. Hence, it is highly desired to provide real-time volumetric imaging during the treatment delivery of radiotherapy for more accurate in-treatment motion monitoring and more active motion management.

Traditional computed tomography approaches reconstruct a 3D image from a set of 2D projections acquired at many different angles, via mathematical methods in the form of filtered back-projection. These traditional reconstruction methods require that the number of the 2D angular projections be sufficient to satisfy the classic Shannon–Nyquist sampling theorem, which poses a practically achievable limit on imaging time and hence makes it impossible to realize the real-time volumetric imaging via the traditional methods[10]. Although compressed sensing techniques have been successfully employed in iterative image reconstruction to reduce the required projection number[11–16], the resultant sparsity

DOI: 10.1201/9781003243458-13

is still far away from meeting the extremely high sparsity requirement of real-time volumetric imaging[10]. Efforts have been devoted to pushing the sparse sampling to the limit of a single projection by leveraging machine learning and patient-specific prior knowledge[17-20]. The main idea of these machine learning methods is to build a patient-specific lung motion model from the patient's own four dimensional computed tomography (4D CT) images as prior knowledge using deformable registration and principle component analysis (PCA), and then estimate the PCA coefficients of the lung motion model from a single 2D projection using either optimization, linear regression or sparse learning to derive the 3D image of the patient's new anatomy. However, the robustness of these machine learning methods to handle the variations in patient's positioning, respiratory motion, and anatomy change is still a concern for routine clinical use[9].

Deep learning techniques have attracted much attention for their ability to learn complex relationships and have found widespread applications across different disciplines. Recently, deep learning techniques have been successfully employed to learn the 2D–3D transformation in the feature domain to generate a synthetic 3D CT image from a single 2D kV projection by Shen et al.[10] and Lei et al.[21], and have shown to be robust to handle the variations in patient positioning, respiratory motion, and anatomical change and hence promising for clinical implementation to achieve real-time volumetric imaging. This chapter will present these two deep-learning-based 2D–3D transformation methods in detail.

10.2 Methods

10.2.1 Overview of the Deep-Learning-Based 2D–3D Transformation Methods

The framework of the deep-learning-based 2D–3D transformation methods consists of two major stages: a training stage and an inference stage. At the training stage, a deep-learning network is built using deep neural networks, and trained under one or multiple supervision mechanisms that perform learnable parameter optimization to learn the 2D-to-3D mapping. The network takes a single 2D kV projection image as input, and outputs a generated synthetic 3D CT image. The training data is a set of paired 2D projection and 3D CT image, with the 3D CT image used as ground truth to supervise the network training. The two networks developed by Shen et al. and Lei et al. have a similar network structure[10,21]. Both networks consist of three modules, that is, an encoding module, a transformation module, and a decoding module. The encoding module is to learn the feature representation by extracting semantic features from the input 2D projection image. By reshaping the 2D feature map extracted by the encoding module to a 3D feature map, the transformation module bridges the 2D and 3D feature spaces. The decoding module is to learn the transformation from the 3D feature domain to the 3D image domain. Figure 10.1 shows the detailed architecture of the deep-learning network named TransNet built by Lei *et al*. The well-trained network will be used in the inference stage following the same feed-forward path, that is, extracting 2D features from the input single 2D projection and feeding it into the network to generate the synthetic 3D CT image.

10.2.2 Network Training

Unlike deep learning applications in computer vision and other fields, data availability is usually a limiting factor in a medical setting, making it a fundamental challenge to achieve

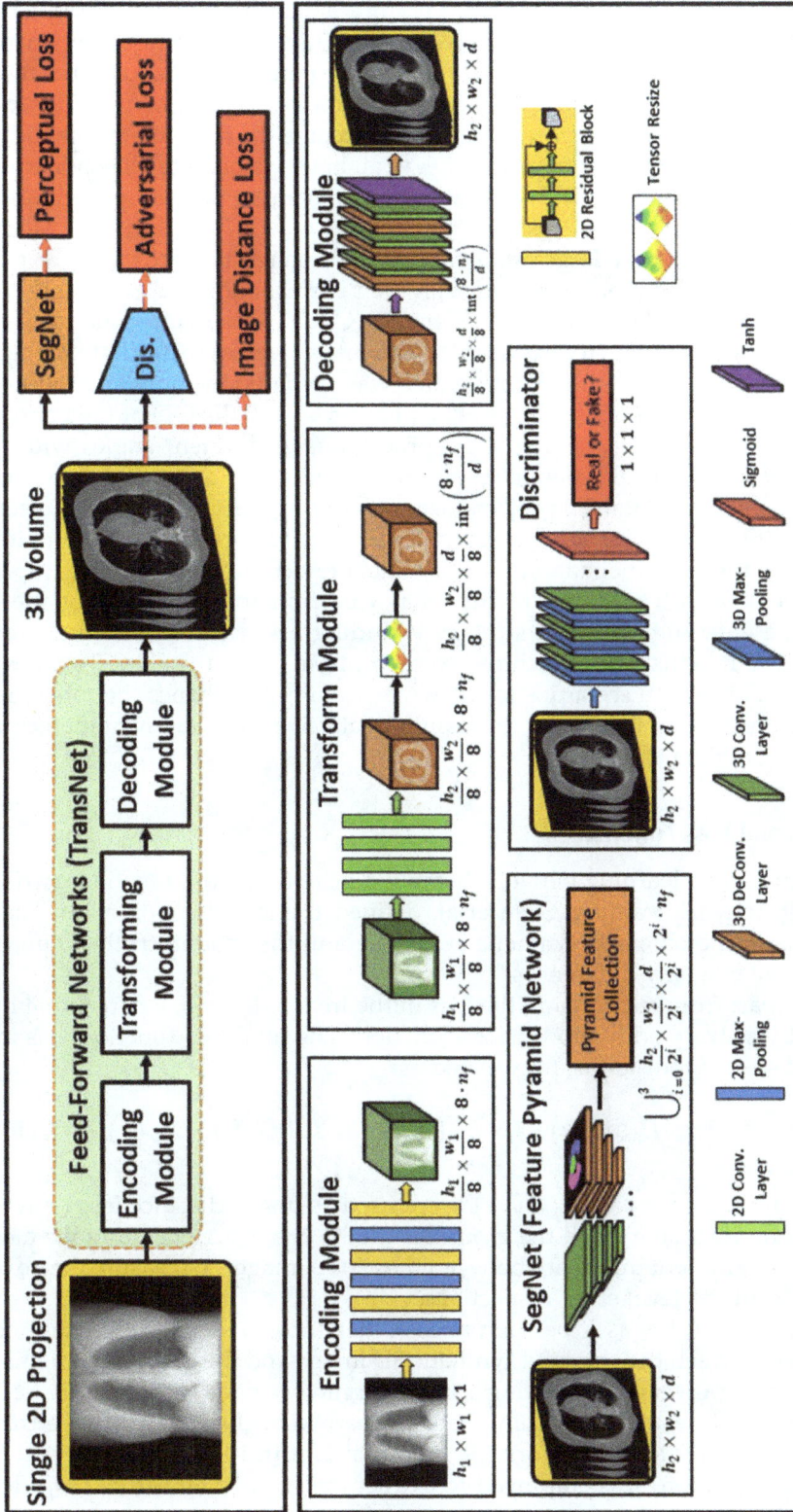

FIGURE 10.1

Schematic flow chart of a deep-learning-based 2D–3D transformation network named TransNet, built by *Lei et al.* The top box shows the framework of the network. The bottom one shows the detailed architecture of each network module. The SegNet and the discriminator show here are used for perceptual supervision and adversarial supervision, which will be present in Section 10.2.3. (Reprinted from "Deep learning-based real-time volumetric imaging for lung stereotactic body radiation therapy: a proof of concept study," by Yang Lei, Zhen Tian, Tonghe Wang, Kristen Higgins, Jeffery D Bradley, Walter J Curran, Tian Liu and Xiaofeng Yang, *Physics in Medicine & Biology*, 65 (2020) 235003, published on 20 November 2020. Reprinted with permission.)

the generalizability of deep learning models for a large patient population of big variations in tumor shape, size, location, and motion. On the other hand, despite the patient's anatomy, motion, and treatment setup changes, the same patient still holds a relatively high anatomical similarity between CT simulation and radiation treatment, which provides a patient-specific prior and can be leveraged to overcome the fundamental challenge of generalizability. Hence, both Shen et al. and Lei et al. have trained their deep-learning networks to be patient-specific to enable scenarios most relevant to the very patient under treatment.

The network training requires a large number of 3D CT images, that are free of respiratory motion or with minimum motion, paired with their corresponding 2D kV X-ray projections acquired by the clinical onboard cone-beam CT (CBCT) system integrated into the Linacs. In order to reduce unnecessary exposure to patients as well as to facilitate the clinical workflow of using the networks, the training data was prepared by digitally producing 2D projection images from each phase CT (that is, phase-resolved 3D CT) of the patient's own 4D CT, instead of acquiring the actual 2D projections at different angles with the patient holding his or her breath at different respiratory amplitude. Ray-tracing algorithms were used to produce these digital 2D projections by following the CBCT imaging geometry. However, the 4D CT images only represent one treatment scenario, as the patient anatomy, treatment setup, and the magnitude, baseline, period, and regularity of respiratory motion at the time of each treatment fraction may vary from those captured on the 4D CT images. Hence, a data augmentation strategy including a series of translations, rotations, and organ deformations was introduced to each phase CT of the 4D CT images to mimic various potential clinical scenarios for network training to enhance the robustness of the network. Meanwhile, by increasing the training data size, data augmentation could also avoid overfitting of the network.

10.2.3 Supervision and Loss Function

The performance of a deep learning network heavily relies on the loss function used for supervision during network training. Shen et al. defined the loss function as the mean squared error between the obtained synthetic 3D image and the ground truth 3D image, focusing on the image intensity differences[10].

In Lei's work[21], apart from supervising the loss in the image domain, a perceptual loss and an adversarial loss were added to the loss function. The total loss function was in a form of a weighted sum of four terms, formulated as:

$$L_{\text{total}}\left(\hat{y}, y\right) = \lambda_{\text{mae}} \cdot L_{\text{mae}}\left(\hat{y}, y\right) + \lambda_{gd} \cdot L_{gd}\left(\hat{y}, y\right) + \lambda_p \cdot L_p\left(f_{\hat{y}}, f_y\right) + L_{\text{adv}}\left(\hat{y}\right) \qquad (10.1)$$

The first two terms, $L_{\text{mae}}\left(\hat{y}, y\right)$ and $L_{gd}\left(\hat{y}, y\right)$, were used as image distance loss between the obtained synthetic 3D image $\hat{y}$ and the ground truth 3D image y, calculating the mean absolute error and the gradient difference between these two images to measure the voxel-level error and the gradient-level error, respectively.

The third term, $L_p\left(f_{\hat{y}}, f_y\right)$, was a perceptual loss that measured the perceptual and semantic differences between the obtained synthetic 3D image and the ground truth image in order to enforce more power on generating accurate organ boundary in the synthetic 3D images. To calculate this loss, a deep learning network named SegNet, which was a traditional end-to-end feature pyramid network (as shown in Figure 10.1) and pretrained for image semantic segmentation, was employed to extract the high-level feature map from the generated synthetic 3D image (*i.e.*, $f_{\hat{y}} = \cup_{i=1}^{N} F_s^i(\hat{y})$) and the one from the ground truth image (*i.e.*, $f_y = \cup_{i=1}^{N} F_s^i(y)$). Here, N denotes the number of pyramid levels of the feature map. The perceptual loss was defined as the Euclidean distance between the two feature maps, calculated as

$$L_p\left(f_{\hat{y}}, f_y\right) = \sum_{i=1}^{N} \frac{\omega_i}{C_i \cdot H_i \cdot W_i \cdot D_i} \left\| F_s^i(\hat{y}) - F_s^i(y) \right\|_2^2 \tag{10.2}$$

where C_i denotes the number of the feature map channels at ith pyramid level, and H_i, W_i and D_i denotes the height, width, and depth of that feature map. As the feature map at a higher pyramid level is much coarser, the weight for that level's perceptual loss ω_i should be enlarged to compensate for that. Lei et al. used $\omega_i = 1.2^{i-1}$ in their study.

The fourth term in the total loss function, $L_{\text{adv}}(\hat{y})$, was an adversarial loss that tried to improve the realism of the generated synthetic 3D image to be comparable to the ground truth 3D image. The main idea of this adversarial loss was very similar to the previous conditional generative adversarial network studies[22,23], in which a discriminator was employed to compete against the generator during network training in order to improve the realism of the generator's output. In this 2D–3D transformation application, the 2D–3D transformation network served as the generator. As shown in Figure 10.1, a traditional fully convolution network was used as a discriminator to discriminate the generated synthetic 3D image from the ground truth image. It had a binary output, denoted as $F_{\text{dis}}(\cdot)$, with 0 denoting that the discriminator thinks the input 3D image is synthetic image and 1 denoting that the discriminator thinks the input 3D image is real. The discriminator's loss was to measure its recognition error of differentiating synthetic images from the real images, calculated as

$$L_{\text{dis}}\left(\hat{y}, y\right) = SCE\left(F_{\text{dis}}(\hat{y}), 0\right) + SCE\left(F_{\text{dis}}(y), 1\right) \tag{10.3}$$

Here, the function $SCE(\cdot, \cdot)$ denotes sigmoid cross entropy between two distributions. A smaller value of the sigmoid cross entropy means more similarity between the two distributions and hence a smaller recognition error. The adversarial loss $L_{\text{adv}}(\hat{y})$ was to measure the recognition accuracy of the discriminator for the generated synthetic 3D images, that is,

$$L_{\text{adv}}\left(\hat{y}\right) = SCE\left(F_{\text{dis}}(\hat{y}), 1\right) \tag{10.4}$$

A smaller value of $L_{\text{adv}}(\hat{y})$ means a higher recognition error by thinking the generated synthetic 3D images as the real image. It can be seen that the discriminator loss and the adversarial loss compete against each other. During network training, by minimizing the discriminator's loss to improve the discriminator's accuracy of differentiating synthetic 3D images from the real images while minimizing the adversarial loss of the generator to maximize the recognition error for the high-quality synthetic 3D images that can fool the discriminator, the generator could yield a synthetic 3D image of a realistic level comparable to the ground truth 3D image.

A rule of thumb to determine the values of the weighting factors in a total loss function of multiple terms is that the initial values of different loss items should be in the same order of magnitude numerically. Then, the values of these weighting factors need to be adjusted via experiments in order to reflect the desired optimization priority and loss importance. The final network performance should be generally not sensitive to small per-turbation of the weighting factors, given that the network converged after enough number of iterations.

10.3 Evaluation Results

As the real-time volumetric imaging does not exist in current clinics, there is no ground truth of the patient's instantaneous 3D image during treatment delivery. Hence, both Shen's and Lei's studies used the paired digitally reconstructed projection and the phase CT image of the 4D CT image that were not seen during network training to evaluate the trained network as a proof-of-concept study on real-time volumetric imaging[10,21]. Multiple metrics, such as mean absolute error (MAE), peak signal-to-noise-ratio (PSNR), and structural similarity index metric (SSIM), were used to quantitatively evaluate the image accuracy and quality of the generated synthetic 3D images. Lei et al. observed that adding perceptual supervision to the network can significantly enhance its sensitivity for accurate organ boundary regression, and adversarial supervision significantly improves the realism of the generated synthetic 3D images in relative to the ground truth images[21]. Lei et al. reported that thanks to these two extra supervisions, their final network outperforms Shen's method, yielding less blurry images, smaller MAE, and larger PSNR and SSIM (as shown in Figure 10.2 and Table 10.1)[21].

Lei et al. has also performed an experiment to test the robustness of their final network[21]. In the experiment, the original 4D CT images of a patient case were deformed to mimic a challenging testing scenario that had doubled motion amplitude and 50% tumor shrinkage in relative to those in the original 4D CT. 2D projections were then generated from the deformed images at different projection angles and input into the final network to generate the synthetic 3D images. As shown in Figure 10.3 and Table 10.2, 3D images with acceptable accuracy and image quality are obtained at this challenging scenario. The tumors in the generated synthetic image and the corresponding ground truth image were manually delineated by physicians, and the center of mass distance (COMD) between these two images was less than 2.3 mm at this challenging scenario.

10.4 Discussion

Shen et al. proposed to employ a deep neural network to learn the mapping relationship between a single 2D projection image to the corresponding 3D volume, for the first time[10]. Inspired by this work, Lei et al. employed a more advanced GAN network for this ill-posted 2D–3D transformation problem, and integrated perceptual supervision and adversarial supervision into the network training[21].

Shen et al. compared their deep learning method with the previous PCA lung motion model-based 2D–3D registration method developed by Li et al.[17,18]. It was found that although both methods produced similar results in an ideal scenario, the deep learning method outperformed the PCA-based method in more realistic scenarios when the patient positioning slightly deviated from that of the 4D CT scan[10]. The robustness of the deep learning methods

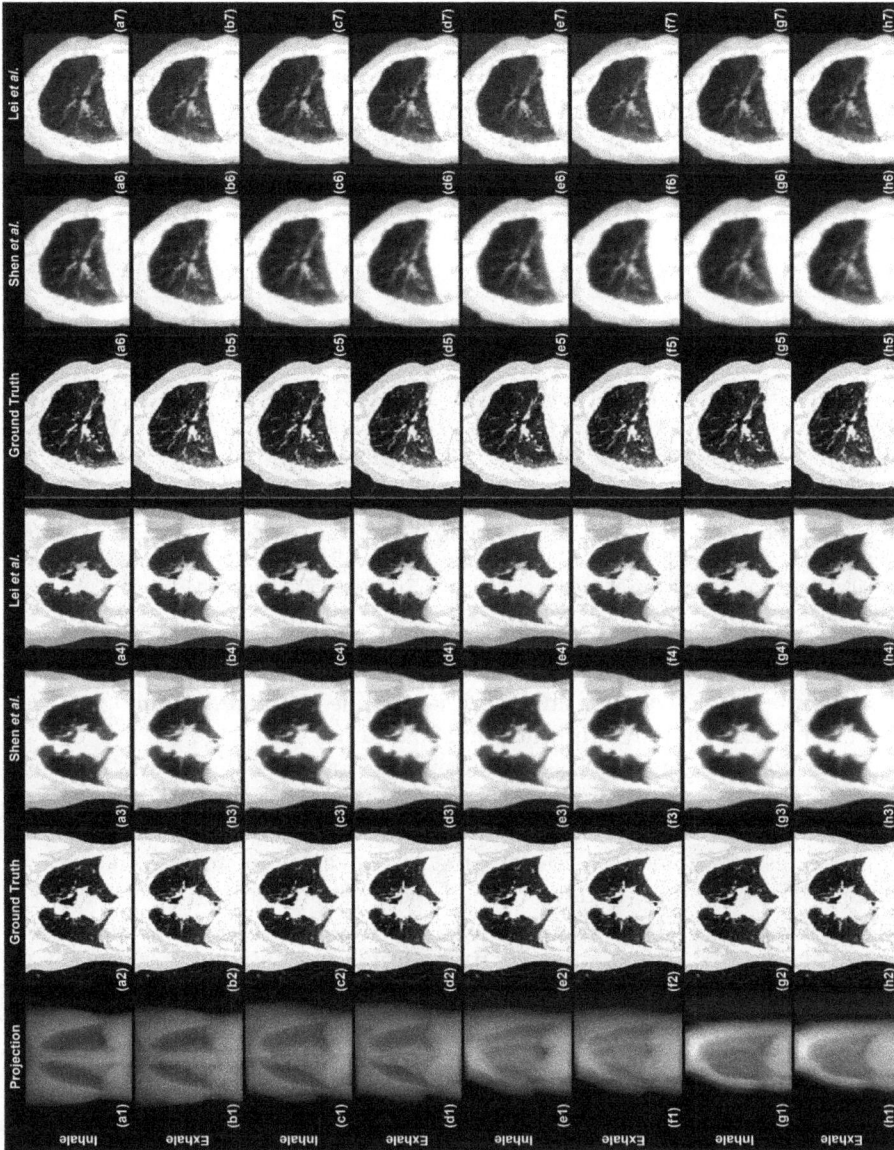

FIGURE 10.2

The 3D CT images generated for a patient case at inhale (phase 0) and exhale (phase 5) phases by Shen's method[10] and Lei's method[21], displayed in coronal and sagittal views. (a), (b) correspond to the projection angle 0°; (c) (d) (e), (f) and (g), (h) correspond to projection angle 30°, 60°, 90°, respectively. The window of [−1000, 0] HU was used to display all the CT images. (Reprinted from "Deep learning-based real-time volumetric imaging for lung stereotactic body radiation therapy: a proof of concept study," by Yang Lei, Zhen Tian, Tonghe Wang, Kristen Higgins, Jeffery D Bradley, Walter J Curran, Tian Liu and Xiaofeng Yang, Physics in Medicine & Biology, 65 (2020) 235003, published on 20 November 2020. Reprinted with permission.)

TABLE 10.1

Numerical comparison result of Shen's method and Lei's method for the inhale and exhale phases at four projection angles on a same patient dataset of 20 patients. For each patient, 40 experiments were performed using different combinations of the testing phases and projection angles. Specifically, at each experiment, the 2D projection of one breathing phase at one projection angle was selected for testing, and the pairs of the phase CT at the other breathing phases and their 2D projection at the same projection angle were used as the raining data. The metric values are averaged over all the patient cases. MAE is calculated within whole volume. PSNR and SSIM are calculated within tumor ROIs

Inhale	MAE (HU)		PSNR (dB)		SSIM	
(phase 0)	Shen et al.	Lei *et al.*	Shen et al.	Lei *et al.*	Shen et al.	Lei *et al.*
0°	115.7±23.2	90.3±15.8	14.5±2.2	15.6±2.2	0.804±0.100	0.849±0.081
30°	127.7±16.9	105±14.2	14.1±2.3	15.1±2.3	0.779±0.115	0.829±0.088
60°	121.2±14.3	97.7±9.7	14.5±2.6	15.5±2.6	0.812±0.081	0.854±0.068
90°	123.5±15.3	103.9±12.1	14.2±2.8	15.2±2.8	0.780±0.136	0.832±0.099
exhale	MAE (HU)		PSNR (dB)		SSIM	
(phase 5)	Shen et al.	Lei et al.	Shen et al.	Lei et al.	Shen et al.	Lei et al.
0°	116.1±22.3	91.3±15.2	14.6±2.4	15.8±2.4	0.801±0.107	0.847±0.085
30°	129.8±16.9	107.3±14.3	14.1±2.3	15.1±2.3	0.737±0.235	0.798±0.17
60°	121.9±13.6	99.1±9.3	14.7±2.5	15.8±2.5	0.815±0.075	0.856±0.062
90°	124.5±14.5	105.4±10.7	14.4±2.7	15.4±2.8	0.758±0.231	0.816±0.16

Reprinted from "Deep learning-based real-time volumetric imaging for lung stereotactic body radiation therapy: a proof of concept study," by Yang Lei, Zhen Tian, Tonghe Wang, Kristen Higgins, Jeffery D Bradley, Walter J Curran, Tian Liu and Xiaofeng Yang, Physics in Medicine & Biology, 65 (2020) 235003, published on 20 November 2020. Reprinted with permission.

has also been demonstrated by Lei *et al.* in an experiment of the simulated challenging scenario with doubled respiratory motion amplitude and 50% tumor shrinkage[21].

Common deep-learning methods train networks on one group of patients and apply to other patients, requiring the training group to be representative enough. This 2D–3D mapping problem is very ill-posed and has a high requirement on the similarity between the training data and the testing data. Considering the large variation among individual patients, both Shen et al. and Lei et al. have trained their deep learning networks to be patient-specific using the patient's own 4D CT images and digitally reconstructed 2D projections. Data augmentation (e.g., translation, rotation, flipping, scaling, etc.) is employed to crease the data size of the patients-specific training data to mimic various scenarios, including daily treatment setup variation and the change of respiratory motion and patient anatomy. Hence, the representativeness of the training dataset should usually be fulfilled for each specific patient, as long as the patient's anatomy does not change dramatically between 4D CT scan and treatment (such as the patient loses too much weight, or one side of the lung got collapsed during the treatment course).

The proof-of-concept studies performed by Shen et al. and Lei et al. have well demonstrated the feasibility of deriving a 3D volumetric image from a single 2D projection, paving the way for in-treatment real-time volumetric imaging. While for the clinical implementation, the 2D projections that are digitally reconstructed from 4D CT images for network training need to be improved to have similar image quality compared to the actual in-treatment 2D kV projections that will be used to drive the instantaneous 3D

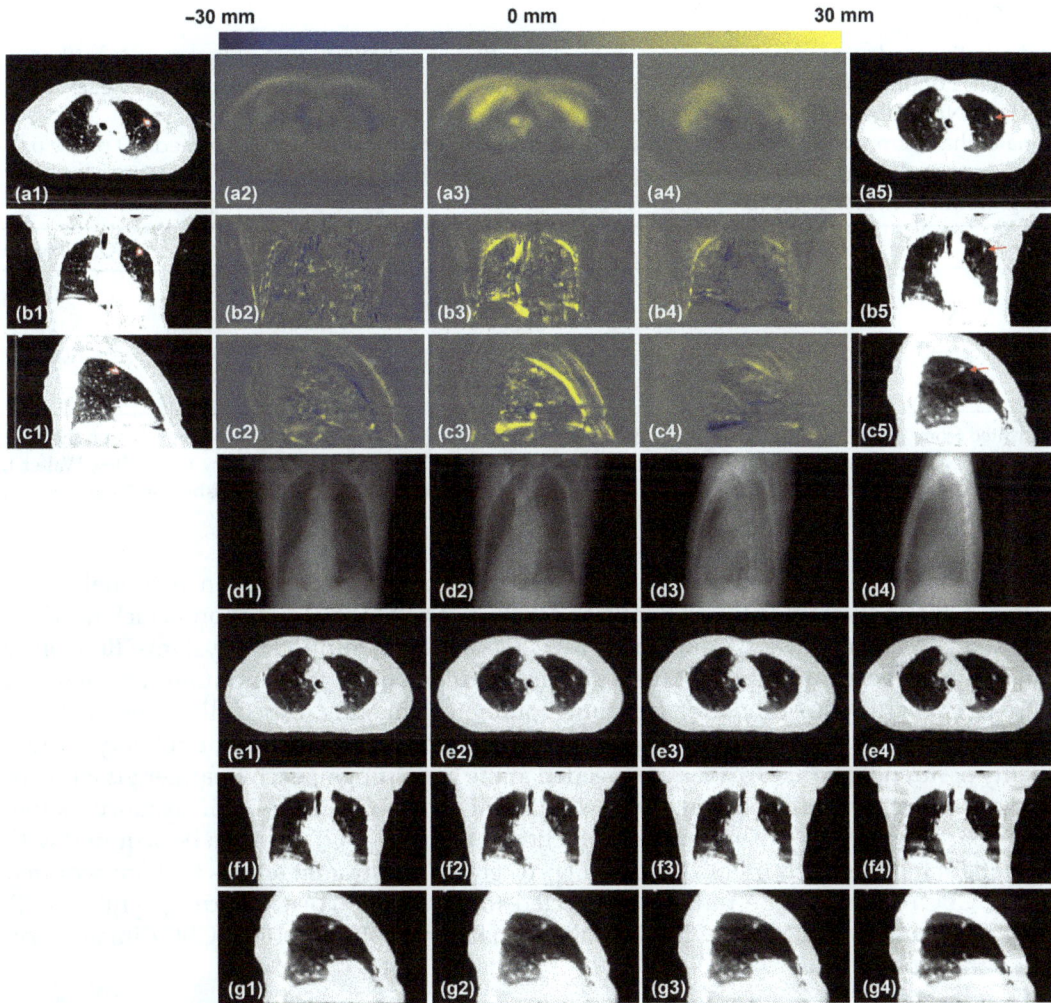

FIGURE 10.3
Visual results of a simulation of a challenging testing scenario with doubled motion amplitude and ~50% tumor shrinkage in relative to the 4D CT images, performed by *Lei* et al. (a1)–(c1) shows the original CT at phase 0 (inhale phase) in axial, sagittal and coronal views, where tumor contour (red) is overlaid; (a2) –(a4), (b2) –(b4) and (c2) –(c4) show simulated deformation vector field (DVF) in x, y, and z axis to simulate the doubled motion amplitude and ~50% tumor shrinkage; (a5–c5) shows the deformed CT transformed CTs; (d1–4) shows the projection data of the deformed CT image at 0°, 30°, 60°, 90°, respectively; (e1–4), (f1–4), and (g1–4) show the corresponding synthetic 3D images generated by Lei's method from the projection data in the three views. The window level for all CT images is set to [−1000, 0] HU. The window level for DVF is set to [−30, 30] mm (since the maximum motion is set to 30 mm). (Reprinted from "Deep learning-based real-time volumetric imaging for lung stereotactic body radiation therapy: a proof of concept study," by Yang Lei, Zhen Tian, Tonghe Wang, Kristen Higgins, Jeffery D. Bradley, Walter J. Curran, Tian Liu, and Xiaofeng Yang, *Physics in Medicine & Biology*, 65 (2020) 235003, published on November 20, 2020. Reprinted with permission.)

image. Compared to the ideal projections generated by the conventional ray-tracing methods, the actual kV projections acquired by the flat panel detector are usually contaminated with scatters and noises. This disparity can be addressed by Monte Carlo simulations to simulate the physics events of every imaging photon to generate more realistic projections

TABLE 10.2

Quantitative evaluation results obtained for a simulation of a challenging testing scenario with doubled motion amplitude and ~50% tumor shrinkage for a patient case, performed by Lei *et al.* MAE, PSNR, SSIM, and COMD were calculated for the 3D images generated at phase 0 (inhale phase) from a 2D projection at different projection angles, using the deformed 3D phase 0 CT images as ground truth. MAE is calculated within the whole volume. PSNR and SSIM are calculated within tumor ROIs.

Projection	MAE (HU)	PSNR (dB)	SSIM	COMD (mm)
0°	79.0	15.9	0.825	0.80
30°	94.7	14.3	0.816	0.61
60°	90.7	13.9	0.768	1.08
90°	101.8	12.8	0.751	2.29

Reprinted from "Deep learning-based real-time volumetric imaging for lung stereotactic body radiation therapy: a proof of concept study," by Yang Lei, Zhen Tian, Tonghe Wang, Kristen Higgins, Jeffery D. Bradley, Walter J. Curran, Tian Liu, and Xiaofeng Yang, *Physics in Medicine & Biology*, 65 (2020) 235003, published on November 20, 2020. Reprinted with permission.

for network training. Moreover, real-time volumetric imaging has an extremely high requirement on the computational efficiency of the 2D–3D transformation, which needs to be further accelerated using more powerful GPU cards and multi-GPU systems. Reliability and safety are very important for future clinical implementation and routine clinical use of these deep learning methods for real-time volumetric imaging, especially if the synthetic 3D images will be used for active motion management during treatment delivery. Proper quality assurance procedures may be needed at the beginning of each treatment fraction or every few fractions or when an obvious anatomical change occurs. For instance, before treatment delivery of a given treatment fraction, a regular CBCT scan can be acquired with the patient holding their breath. A few 2D projections of different angles that are acquired by the CBCT scan, will be randomly selected to derive the corresponding synthetic 3D images for quality assurance, with the patient anatomy captured on the CBCT image serving as the ground truth.

References

1. Shirato H, Seppenwoolde Y, Kitamura K, Onimura R, Shimizu S. Intrafractional tumor motion: lung and liver. *Paper presented at*: Seminars in Radiation Oncology 2004.
2. Keall PJ, Mageras GS, Balter JM, et al. The management of respiratory motion in radiation oncology report of AAPM Task Group 76 a. *Medical Physics* 2006;33(10):3874–3900.
3. Mori S, Hara R, Yanagi T, et al. Four-dimensional measurement of intrafractional respiratory motion of pancreatic tumors using a 256 multi-slice CT scanner. *Radiotherapy and Oncology* 2009;92(2):231–237.
4. Zhao B, Yang Y, Li T, et al. Statistical analysis of target motion in gated lung stereotactic body radiation therapy. *Physics in Medicine & Biology* 2011;56(5):1385.
5. Schmidt ML, Hoffmann L, Kandi M, Møller DS, Poulsen PRJAO. Dosimetric impact of respiratory motion, interfraction baseline shifts, and anatomical changes in radiotherapy of non-small cell lung cancer. *Acta Oncologica* 2013;52(7):1490–1496.
6. Mutaf Y, Scicutella C, Michalski D, et al. A simulation study of irregular respiratory motion and its dosimetric impact on lung tumors. *Physics in Medicine & Biology* 2011;56(3):845.

7. Van den Begin R, Engels B, Gevaert T, et al. Impact of inadequate respiratory motion management in SBRT for oligometastatic colorectal cancer. *Radiotherapy and Oncology* 2014;113(2):235–239.

8. Liu G, Hu F, Ding X, et al. Simulation of dosimetry impact of 4DCT uncertainty in 4D dose calculation for lung SBRT. *Radiaton Oncology* 2019;14(1):1–12.

9. Yan H, Tian Z, Shao Y, Jiang SB, Jia X. A new scheme for real-time high-contrast imaging in lung cancer radiotherapy: a proof-of-concept study. *Physics in Medicine & Biology* 2016;61(6):2372.

10. Shen L, Zhao W, Xing L. Patient-specific reconstruction of volumetric computed tomography images from a single projection view via deep learning. *Nature Biomedical Engineering* 2019;3(11):880–888.

11. Tang J, Nett BE, Chen G-H. Performance comparison between total variation (TV)-based compressed sensing and statistical iterative reconstruction algorithms. *Physics in Medicine & Biology* 2009;54(19):5781.

12. Jorgensen JS, Sidky EY, Pan XJI. Quantifying admissible undersampling for sparsity-exploiting iterative image reconstruction in X-ray CT. *IEEE Transactions on Medical Imaging* 2012;32(2):460–473.

13. Choi K, Wang J, Zhu L, Suh TS, Boyd S, Xing LJM. Compressed sensing based cone-beam computed tomography reconstruction with a first-order method a. *Medical Physics* 2010;37(9):5113–5125.

14. Lee H, Xing L, Davidi R, et al. Improved compressed sensing-based cone-beam CT reconstruction using adaptive prior image constraints. *Physics in Medicine & Biology* 2012;57(8):2287.

15. Xu Q, Yu H, Mou X, Zhang L, Hsieh J, Wang GJI. Low-dose X-ray CT reconstruction via dictionary learning. *IEEE Transactions on Medical Imaging* 2012;31(9):1682–1697.

16. Tian Z, Jia X, Yuan K, Pan T, Jiang SBJ. Low-dose CT reconstruction via edge-preserving total variation regularization. *Physics in Medicine & Biology* 2011;56(18):5949.

17. Li R, Jia X, Lewis JH, et al. Real-time volumetric image reconstruction and 3D tumor localization based on a single x-ray projection image for lung cancer radiotherapy. *Medical Physics* 2010;37(6Part1):2822–2826.

18. Li R, Lewis JH, Jia X, et al. On a PCA-based lung motion model. *Physics in Medicine & Biology* 2011;56(18):6009.

19. Zhao Q, Chou C-R, Mageras G, Pizer SJI. Local metric learning in 2D/3D deformable registration with application in the abdomen. *IEEE Transactions on Medical Imaging* 2014;33(8):1592–1600.

20. Xu Y, Yan H, Ouyang L, et al. A method for volumetric imaging in radiotherapy using single x-ray projection. *Medical Physics* 2015;42(5):2498–2509.

21. Lei Y, Tian Z, Wang T, et al. Deep learning-based real-time volumetric imaging for lung stereotactic body radiation therapy: a proof of concept study. *Physics in Medicine & Biology* 2020;65(23):235003.

22. Nie D, Trullo R, Lian J, et al. Medical image synthesis with deep convolutional adversarial networks. *IEEE Transactions on Biomedical Engineering* 2018;65(12):2720–2730.

23. Harms J, Lei Y, Wang T, et al. Paired cycle-GAN-based image correction for quantitative cone-beam computed tomography. *Medical Physics* 2019;46(9):3998–4009.

11

Multimodality MRI Synthesis

Liangqiong Qu

The University of Hong Kong, Pok Fu Lam, Hong Kong SAR, China

Yongqin Zhang

Northwest University, Xi'an, China

Zhiming Cheng

Hangzhou Dianzi University, Hangzhou, China

Shuang Zeng

The University of Hong Kong, Pok Fu Lam, Hong Kong SAR, China

Xiaodan Zhang

Beijing University of Technology, Beijing, China

Yuyin Zhou

University of California, Santa Cruz, CA, USA

CONTENTS

DOI: 10.1201/9781003243458-14

11.1 Introduction

Magnetic resonance imaging (MRI) is a noninvasive and versatile medical imaging technique that generates images of the body with strong magnetic fields, magnetic field gradients, and radio waves (McRobbie et al., 2017). MRI scanner consists of four components, a major magnet for polarizing the sample, a shim coil for inhomogeneity shift correction, a gradient system to localize the regions to be scanned, and a radio frequency system to excite samples and receive the signal. MRI differs from CT and PET scans that do not use X-rays or ionizing radiation, and usually provides better contrast than regular CT, PET and X-rays, thus has been extensively used for disease diagnosis, staging, and follow-ups (Gosain et al., 2005).

The commercial magnetic field for MRI ranges from 0.2 to 7.0T, with the majority of the MRI system operating at 1.5 T. High (or ultra-high) field MRI scanners produce images with exceptional anatomical details, potentially improving the accuracy of treatment and prognosis of diseases (Liu et al., 2021), see Figure 11.1 for an example of MR images captured with magnetic field 3T and 7T respectively (Qu et al., 2020). However, high-field MRI scanners are often cost-prohibitive and not always accessible in the clinic. According to the use of different radio frequency pulses and gradients, MRI can also be divided into different MRI modalities (also known as different MRI sequences) (Preston, 2006), see Figure 11.2 for an example of several common MRI sequences. Since each MRI modality captures

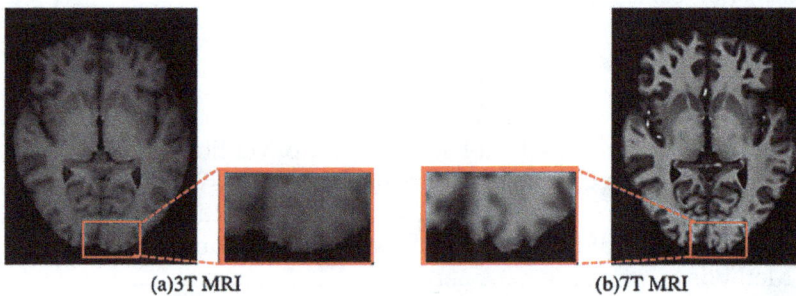

(a)3T MRI   (b)7T MRI

FIGURE 11.1
An example of MR images with different magnetic fields.

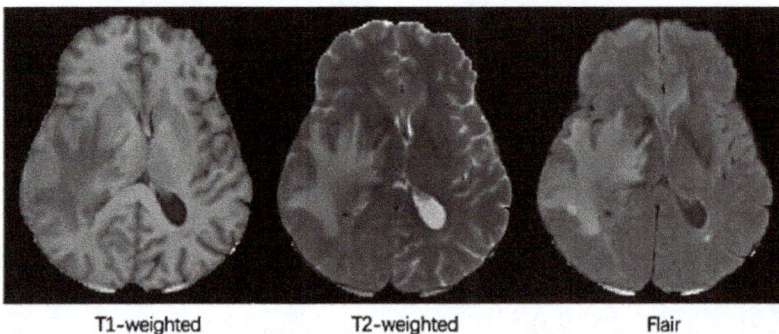

T1-weighted   T2-weighted   Flair

FIGURE 11.2
An example image of the most common MRI sequences. From left to right: T1-weighted, T2-weighted, and FLAIR.

specific characters of the underlying anatomical information, the complementary multiple modalities are often utilized to aid clinical analysis (Zhou et al., 2020), such as brain lesion segmentation (Kamnitsas et al., 2016), therapy planning (Kaus et al., 2007), etc. However, acquiring sufficient multimodality MR images in clinical applications is often hindered by missing imaging protocols or unacceptable long scanning time. In addition, the imperfections of the data acquisition process often cause MR images with intensity inhomogeneity, which produces images with non-anatomic intensity variations and thus adversely affects the clinical diagnosis (Venkatesh et al., 2020).

To overcome the above-mentioned challenges, MRI synthesis, which is defined as a mapping between target images and the given source images in MRI, has gained wide popularity in recent years, such as MRI synthesis across different modalities (e.g., T1-weighted to T2-weighted, T1 to FLAIR (Olut et al., 2018, Yu et al., 2019), MRI synthesis between different quality (e.g., 3T MRI to 7T MRI) (Chaudhari et al., 2018, Qu et al., 2020), and MRI inhomogeneity correction (Vovk et al., 2007, Xiang et al., 2020, Venkatesh et al., 2020). Despite the different application scopes, they share a similar synthesis principle, which aims at synthesizing the missing target-modality image from any given source-modality image. A timeline of the development on MRI synthesis methods is shown in Figure 11.3. Earlier work for MRI synthesis was initially an atlas-based approach, which uses the co-registered atlases from source- and target-modalities to build the atlas-to-modality deformation field in source modality, and then applies the deformation field to the target-modality atlases to synthesize the target-modality like image (Catana et al., 2010, Berker et al., 2012, Burgos et al., 2014). However, atlas-based approaches are sensitive to registration accuracy and are usually limited to healthy subjects since they rely on geometric transformations, and the majority of the atlas is usually built upon healthy subjects (Nie et al., 2018).

Different from atlas-based approaches that rely on strict geometric transformations, another popular type of MRI synthesis method is the learning-based approach, which

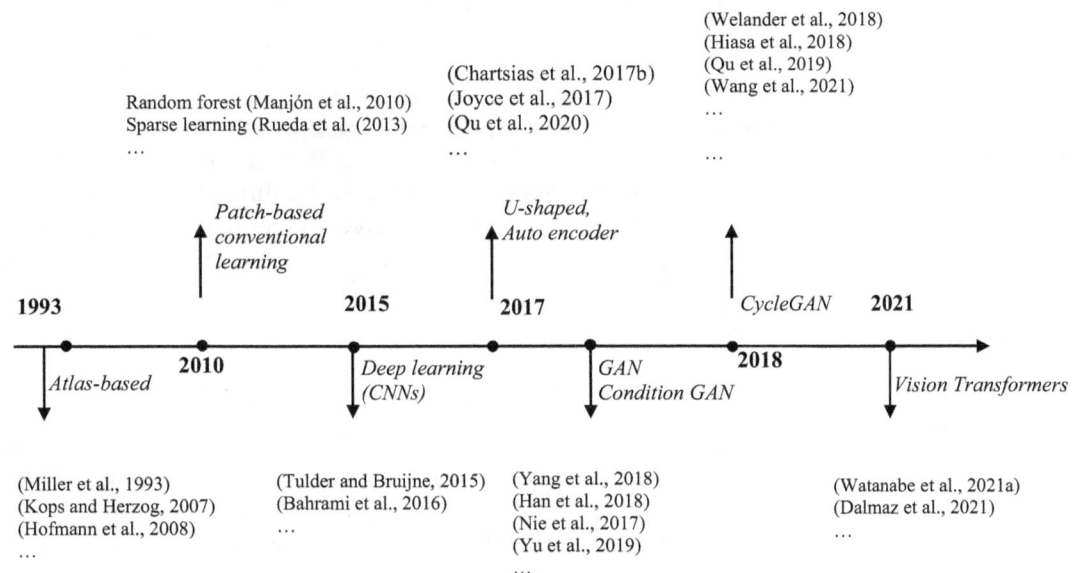

FIGURE 11.3
The development of models for multimodality MRI synthesis.

directly learns the mapping between source- and target-modality images with data-driven approaches such as principal component analysis, random forests, sparse representation, etc. (Manjón et al., 2010, Jog et al., 2014, Huynh et al., 2015). For example, Rueda et al. (2013) generated the high-resolution (HR) MR image from its corresponding low-resolution (LR) MR image by coupling the low- and high-frequency information with a sparsity-based learning method. Ye et al. (2013) proposed a patch-based searching method for multimodality MR image synthesis.

While effective, these patch-based conventional learning methods may not be optimal due to the limited capability of handcrafted feature representation (Chartsias et al., 2017b). More recently, deep learning-based methods with rich feature description capability have demonstrated state-of-the-art performance in multimodality MRI synthesis. Various deep learning methods have been proposed to tackle the multimodality MRI synthesis problems, ranging from the earlier deep convolutional neural networks (CNNs), and generative adversarial networks (GANs) (Chartsias et al., 2017a, Welander et al., 2018, Zhang et al., 2018b, Wang et al., 2021, Sun et al., 2022) to the more recent vision transformers (Watanabe et al., 2021, Dalmaz et al., 2022).

For example, Van Nguyen et al. (2015) proposed a CNN-based framework to integrate the intensity features of image voxels and spatial information for synthesizing MRI T2 from MRI T1. To explore multimodality inputs, a more complex U-shaped CNN architecture with an encoder- and decoder-based architecture was proposed for synthesizing FLAIR from T1, T2, and Diffusion-weighted imaging (DWI) (Chartsias et al., 2017b). In order to generate more realistic target images, a fully connected convolutional neural (FCN) based generative adversarial network was proposed for generating 7T MR from 3T MR images (Nie et al., 2018). Hiasa et al. (2018) applied the cycle generative adversarial network (CycleGAN) (Zhu et al., 2017) approach for cross-modality image synthesis and introduced a gradient consistency loss to encourage edge alignment between different modalities. Dalmaz et al. (2021) introduced a transformer-based generator with several aggregated residual transformer blocks for multimodality MRI synthesis.

In the remainder of the chapter, we mainly focus on the evolution of patch-based conventional learning methods and deep learning methods for MRI synthesis. Specifically, we will introduce our work for 7T MRI prediction from 3T MRI using one of the conventional learning-based methods in Section 2, one fully supervised deep learning method with CNNs in Section 3.1, and one semi-supervised deep learning method with GANs in Section 3.2. Finally, a brief conclusion, including a discussion of possible future research, is given in Section 4. We hope this chapter could promote a better understanding of the principle of multimodality MRI synthesis and the recent progress of technical methods in this field.

11.2 Multimodality MRI Synthesis via Patch-Based Conventional Learning

Patch-based conventional learning methods, also called example-based methods, aim at training a model that maps the patches extracted from the source modality to those from the target-modality. Figure 11.4 shows a general flowchart for multimodality MRI synthesis with a patch-based conventional learning method. It usually consists of three stages: 1) Source- and target- dictionaries building based on the extracted patches from the original source- and target-modality images with methods such as dictionary learning, local

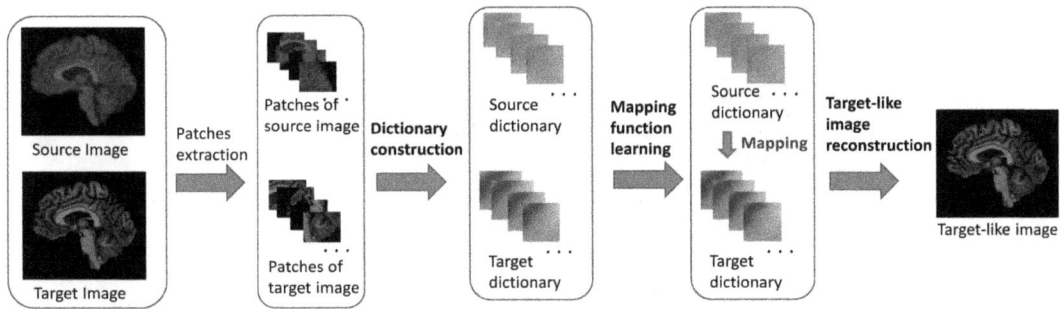

FIGURE 11.4
A general flowchart for patch-based conventional learning methods for multimodality MRI synthesis.

dictionary, global dictionary, etc.; 2) Learning of the mapping function between source- and target- dictionaries representation with machine learning methods (such as random forest, dictionary learning, sparse learning, etc.); and 3) Target-modality image reconstruction from the learned mapping function and the dictionary representation of the target-modality. A number of conventional learning methods have been proposed in recent years. For example, Roy et al. (2013) presented an example-based super-resolution framework to synthesize high-resolution MR images from multi-contrast atlases. Bahrami et al. (2015) proposed a hierarchical sparse representation method with multi-level canonical correlation analysis (MCCA) for the reconstruction of 7T-like MR images from 3T MR images.

In this section, we present our recent work on 7T MRI prediction from 3T MRI to illustrate the general procedure of multimodality MRI synthesis with patch-based conventional learning methods (Zhang et al., 2018a). Specifically, we introduce a dual-domain cascaded regression (DDCR) framework with two parallel and interactive multi-stage regression streams based on spatial and frequency domains to synthesize 7T-like MR images from the routine 3T MR images.

11.2.1 A Dual-Domain Cascaded Regression for 7T MRI Prediction from 3T MRI

The first 7T MRI scanner was approved for clinical use by the United States Food and Drug Administration (FDA) in 2017. Compared with 3T MRI, 7T MRI typically affords greater anatomical details and faster image reconstruction, which may benefit the diagnosis of diseases (van der Zwaag et al., 2016). However, 7T MRI scanners are significantly more expensive and hence less common at hospitals and clinical institutions. Accordingly, this motivates the research on 7T-like MR image synthesis using low-field images (e.g., 3T MR images).

The basic goal of 7T MR image synthesis is to map low-resolution (LR) 3T MR images to high-resolution (HR) 7T MR images. Following the patch-based machine learning methods, we formulate the mapping between the 7T MR and 3T MR images as a regression problem. Specifically, we regress the local patches of 7T MR images from local patches of 3T MR images and use intensity and spectral transformations to improve the quality of 7T-like MR image synthesis, see Figure 11.5.

Dictionary Building. Given an input (source) 3T MR image X, Q pairs of well-aligned 3T MR and 7T MR exemplar images $\{\mathbf{Z}_{3T}, \mathbf{Z}_{7T}\}$ for training, we divide the input image into patches $\mathbf{x}$ of size $p \times p \times p$ for patch regression. For each input patch, we collect L_1 most similar patches $\{\mathbf{z}_{3T, l} | l = 1, \cdots, L_1\}$ from the 3T exemplar images with the block-matching

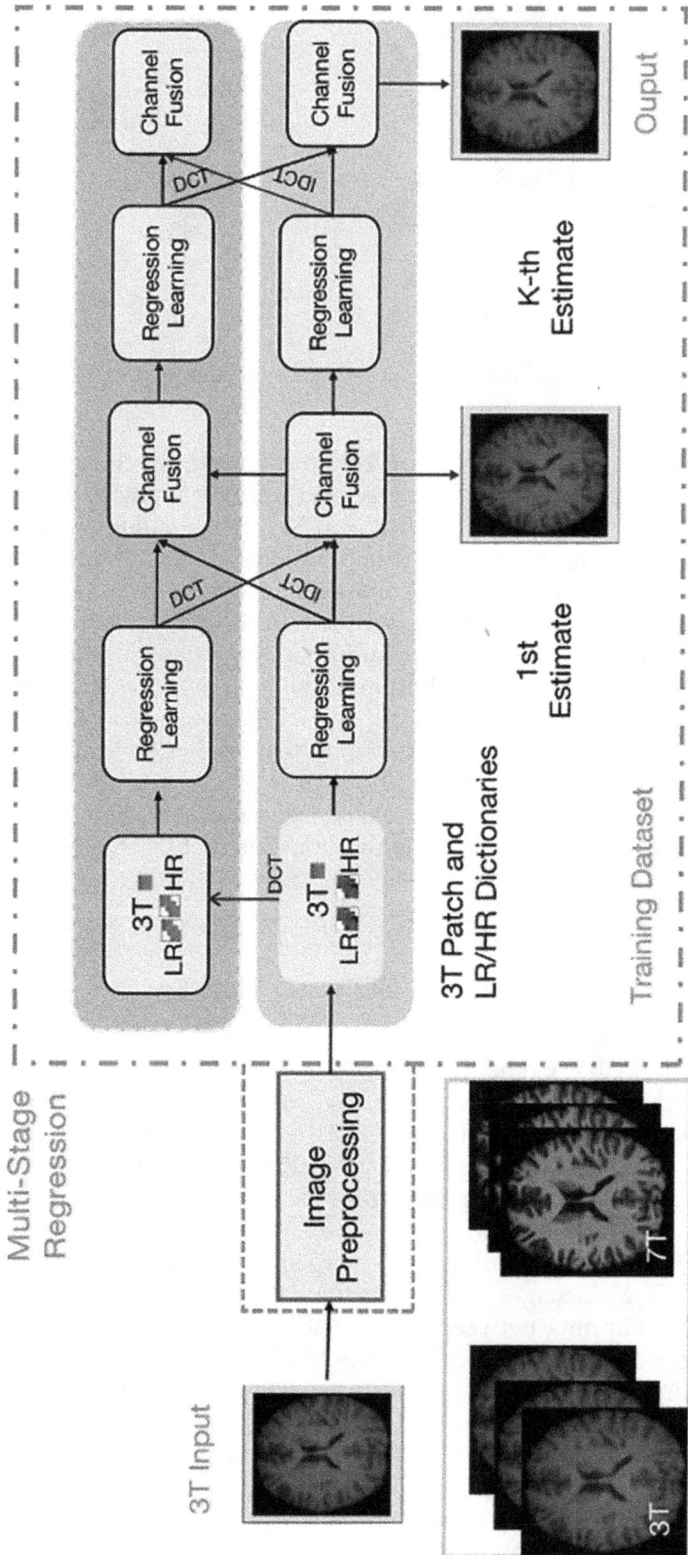

FIGURE 11.5

A framework of the dual-domain cascaded regression (DDCR) on spatial and frequency domains with two parallel and interactive multi-stage regression streams. DCT and IDCT are the forward discrete cosine transforms (DCT) and inverse discrete cosine transforms (IDCT), respectively.

method (Brunig and Niehsen, 2001). The 7T patches $\{z_{7T, l} | l = 1, \cdots, L_1\}$ with the same locations of 3T exemplar patches are also collected from the corresponding 7T exemplar images. These 3T and 7T patch pairs are employed for the construction of the LR and HR dictionaries: $D_{LR} = [z_{3T, 1}, \cdots, z_{3T, L1}]$ and $D_{HR} = [z_{7T, 1}, \cdots, z_{7T, L1}]$ in the spatial domain, respectively.

Dual-domain Cascaded Regression. To represent the mapping from the LR dictionary to the HR dictionary, a linear regression model is defined as

$$D_{HR} = B_s D_{LR} + \varepsilon, \tag{11.1}$$

where B_s is the projection matrix, and ε is the error. The ridge regression (Zhang et al., 2015) is used to solve the inverse problem in an optimization form:

$$\hat{B}_s = \min_{B_s} \| D_{HR} - B_s D_{LR} \|_2^2 + \lambda \| B_s \|_2^2, \tag{11.2}$$

where λ is the regularization parameter. By taking the first derivative of (2) with respect to the variable B_s and setting it to zero, the projection matrix is expressed in a closed-form solution:

$$\hat{B}_s = D_{HR} D'_{LR} \left(D_{LR} D'_{LR} + \lambda E \right)^{-1}, \tag{11.3}$$

where D'_{LR} denotes the transpose of the matrix D_{LR}, and E is an identity matrix. According to the inverse matrix identity (Petersen and Pedersen, 2008), the estimated projection matrix $\hat{B}_s$ in the spatial domain can be rewritten in another form with low computational complexity:

$$\hat{B}_s = D_{HR} \left(D'_{LR} D_{LR} + \lambda E \right)^{-1} D'_{LR}. \tag{11.4}$$

The corresponding synthesized HR dictionary $\hat{D}_{se}$ from the LR dictionary D_{LR} in the following form:

$$\hat{D}_{se} = \hat{B}_s D_{LR}, \tag{11.6}$$

where $\hat{D}_{se}$ is referenced for the construction of the LR dictionary for the next regression stage.

For the first regression stage in the frequency domain, let $\hat{D}_{se}$, U_{LR} and U_{HR} stand for the respective DCT coefficients of x, D_{LR} and D_{HR}. Similar to the regression in the spatial domain, the synthesized 7T component $\hat{\beta}_{te}$ and HR dictionary $\hat{U}_{te}$ in the frequency domain can be separately computed as

$$\hat{\beta}_{te} = U_{HR} \left(U'_{LR} U_{LR} + \lambda E \right)^{-1} U'_{LR} \alpha, \tag{11.7}$$

and

$$\hat{U}_{te} = U_{HR} \left(U'_{LR} U_{LR} + \lambda E \right)^{-1} U'_{LR} U_{LR}. \tag{11.8}$$

These temporary results are referenced as the input and the constructed LR dictionary for the next stage.

7T-Like MR Image Reconstruction. After obtaining $\hat{\mathbf{B}}_s$, the preliminary synthesized 7T MR patch $\hat{\mathbf{y}}_{se}$ from the input 3T patch $\mathbf{x}$ is a simple matrix projection:

$$\hat{\mathbf{y}}_{se} = \hat{\mathbf{B}}_s\mathbf{x}. \tag{11.9}$$

After regression on both spatial and frequency domains, we further fuse the regression results using the square root of mean square (RMS). Further, the cascaded stages of regression are carried out in the streams of respective domains. Specifically, for the cascaded regression in the spatial domain, the synthesized HR patch and dictionary at the current stage are taken as the input and the LR dictionary of the next stage, respectively. Similarly, in the frequency domain, the synthesized HR component and dictionary at the current stage are treated as the input LR component and dictionary of the next stage. On the other hand, the HR dictionaries from the current stage in the spatial and frequency domains are treated as the HR dictionaries of the next stage, respectively. With the setup of input, LR and HR dictionaries on both domains, the regression results at the current stage are computed and then fused as the input for the next stage. With K stages of regression, all synthesized 7T patches are collected to construct the final result.

11.2.2 An Evaluation of Different Patch-Based Conventional Learning Methods

In this subsection, we present an evaluation of DDCR with different patch-based conventional learning methods, including histogram matching (HMAT), and MCCA (Bahrami et al., 2015). Meanwhile, to further illustrate the benefit of dual-domain strategy for the cascaded regression, DDCR was also compared with single spatial-domain cascaded regression, denoted as SDCR. All these methods were evaluated on a 7T MRI prediction from a 3T MRI task with 15 recruited 3T and 7T brain MR images (Zhang et al., 2018a). All synthesized images were compared with the real 7T MR images for the computation of peak signal-to-noise ratio (PSNR) and structural similarity (SSIM).

Dataset and Pre-Processing. With the local institutional review board, 15 adults were recruited for MR data acquisition in this study. The 3T and 7T brain MR images of all subjects were acquired with Siemens Magnetom Trio 3T and 7T MRI scanners, respectively. Specifically, for 3T MR images, T1 images of 224 coronal slices were obtained with the 3D magnetization-prepared rapid gradient-echo (MP-RAGE) sequence. The imaging parameters of 3D MP-RAGE sequence were as follows: repetition time (TR) = 1900 ms, echo time (TE) = 2.16 ms, inversion time (TI) = 900 ms, flip angle (FA) = 9°, and voxel size = 1 × 1 × 1 mm³. For 7T MR images, T1 images of 191 sagittal slices were also obtained with the 3D MP2-RAGE sequence. The imaging parameters of 3D MP2-RAGE sequence were as follows: TR = 6000 ms, TE = 2.95 ms, TI = 800/2700 ms, FA = 4°/4°, and voxel size = 0.65 × 0.65 × 0.65 mm³. As the gradient-echo pulse sequences were used for image acquisition, there was only little distortion between the obtained 3T and 7T MR images, which ensures imaging consistency across magnetic fields.

The source 3T MR image and all the pairs of 3T and 7T exemplar MR images {J$_{3T}$, J$_{7T}$} were linearly registered to MNI standard space (Holmes et al., 1998) using FLIRT (Jenkinson et al., 2002) to remove pose differences. Specifically, all 7T exemplar MR images were linearly registered to the MNI standard space with an individual template (Holmes et al., 1998). The 3T exemplar MR image was then rigidly aligned to its corresponding 7T MR image. After registration, bias correction (Sled et al., 1998) and skull stripping (Shi et al., 2010) were performed. The image intensity values were normalized using histogram

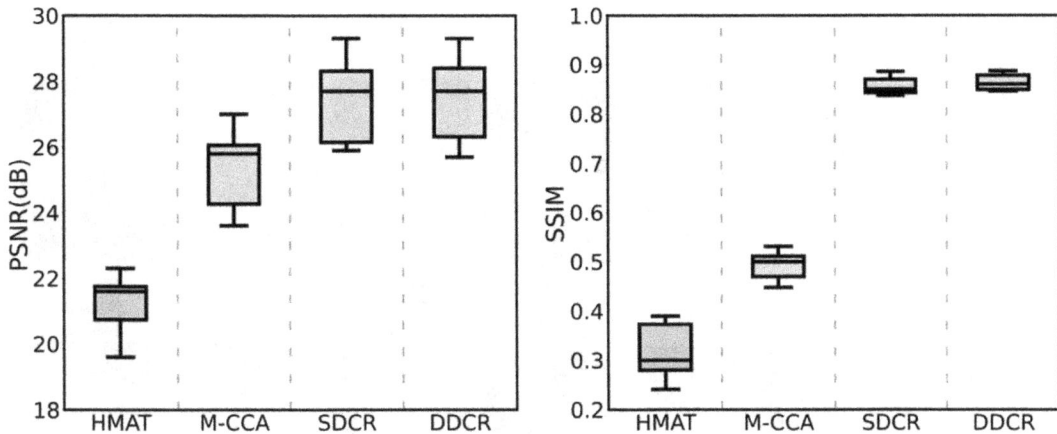

FIGURE 11.6
Boxplots for PSNR and SSIM values of different patch-based conventional learning methods. The middle line of each box is the median, the edges mark the 25th and 75th percentiles, and the whiskers extend to the minimum and the maximum. For all methods, the respective medians of PSNR and SSIM values are as follows: (a) HMAT (PSNR = 21.6 dB, SSIM = 0.30), (b) MCCA (PSNR = 25.8 dB, SSIM = 0.50), (c) SDCR (PSNR =27.7 dB, SSIM = 0.85), and (e) DDCR (PSNR = 27.7 dB, SSIM = 0.86).

matching and scaled to the range of [0, 1]. Histogram matching was performed separately for 3T and 7T MR images. For 3T MR images, the histograms of all normalized 3T exemplar MR images were matched to the histogram of the normalized input 3T MR image. Following that, the normalized 7T exemplar MR image, whose corresponding 3T exemplar MR image was nearest to the input 3T MR image in Euclidean distance, was chosen as referenced 7T MR image for the histogram matching of all remaining 7T MR images.

Performance Evaluation. Figure 11.6 shows the boxplots of PSNR and SSIM values of 15 synthesized 7T MR images from different patch-based conventional learning methods. As can be observed, DDCR generally achieves higher PSNR and SSIM than the other baseline methods. Even though SDCR almost achieves the same PSNR values as DDCR, SDCR has distinctly lower SSIM values than DDCR, thus validating the effectiveness of the dual-domain strategy. Figure 11.7 shows the axial, coronal, and sagittal views of synthesized 7T MR images for one randomly selected subject. It can also be found from Figure 11.7 that the synthesized 7T image by DDCR has better image quality and less distortion.

11.3 Multimodality MRI Synthesis via Deep Learning

Patch-based conventional learning methods for multimodality MRI synthesis rely on the quality of handcrafted medical image features, which usually have limited representation power and may deteriorate the final performance. Deep learning techniques alleviate the need for handcrafted features and have demonstrated state-of-the-art performance on various multimodality MRI synthesis tasks. Dong et al. (2014), in their seminal work, show how a three-layer full CNN can be used to learn LR-to-HR mappings in an end-to-end manner. Various neural network architectures have since then been proposed to improve the multimodality MRI synthesis performance, e.g., residual learning (He et al.,

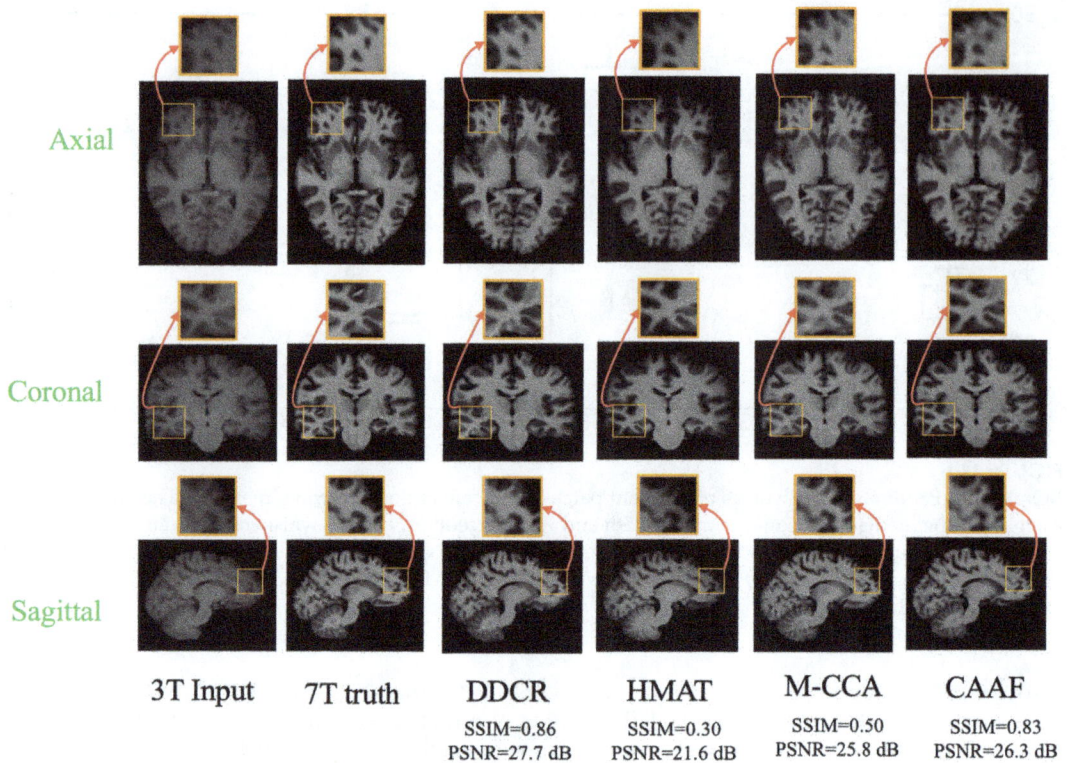

FIGURE 11.7
Visual comparison of different patch-based conventional learning methods. The axial, coronal, and sagittal views of synthesized 7T MR images with close-up views of specific regions for one subject are shown.

2016, Oktay et al., 2016), dense connections (Huang et al. 2017a, Chen et al. 2018), U-shaped architecture Unet (Ronneberger et al., 2015, Qu et al., 2020), etc.

In addition to the above CNN-based methods that apply mean squared error (MSE) loss or mean absolute error (MAE) for optimization in pixel space, several efforts have been devoted to improving perceptual quality with perceptual loss via GANs (Nie et al., 2017, Nie et al. 2018, Yu et al. 2019). For example, a generative adversarial network was proposed for generating a more realistic 7T MRI from 3T MRI images (Nie et al., 2018). Dar et al. (2019) employed a conditional GAN-based approach for MRI synthesis to alleviate model collapse and exploding gradients issues in GANs training. While effective, these fully supervised deep learning methods usually require a large amount of paired data of the same subject for training. However, acquiring paired data can be challenging, and they are hence not always available in large quantities. In light of this challenge, numerous efforts have been devoted to leveraging the unpaired source- and target-modality images with the advanced neural network architectures or unsupervised/semi-supervised/ weakly-supervised strategies, to learn the desired mapping when the paired data are scarce (Wang et al., 2021, Zhang et al., 2018b, Sun et al., 2022, Chartsias et al., 2017a, Welander et al., 2018). For example, Hiasa et al. (2018) applied an unsupervised GAN model, CycleGAN (Zhu et al., 2017) approach for cross-modality image synthesis and introduced with an extra gradient consistency loss to encourage edge alignment between different modalities.

In the following subsection, we will first introduce our recent work (Qu et al., 2020) for 7T MRI prediction from 3T MRI using one of the fully supervised CNN models in Section 3.1. In this work, we introduce a U-shaped like CNN model that leverages the wavelet domain as a prior to facilitate effective reconstruction of multi-frequency image details. We then introduce a wavelet-based semi-supervised adversarial learning framework to synthesize 7T MR images from their 3T counterparts by applying a cycle-consistent loss to leverage unpaired 3T and 7T MR images to learn the 3T-to-7T mapping when 3T-7T paired data are scarce (Qu et al., 2019).

11.3.1 A Fully Supervised 7T MRI Prediction from 3T MRI with CNNs

In this subsection, we present our work of 7T MRI prediction from 3T MRI via a fully supervised CNN model (Qu et al., 2020). Compared to routine 3T MRI, 7T MRI provides images with higher resolution and high signal-to-noise ratio, potentially improving diagnostic and prognostic value (van der Zwaag et al., 2016). Good image synthesis should preserve both global, low-frequency contrast and local, high-frequency details. We introduce a U-shaped deep learning network that fuses complementary information from spatial and wavelet domains to synthesize 7T MR images from their 3T counterparts. We leverage wavelet transformation to facilitate effective multi-scale reconstruction, taking into account both low-frequency tissue contrast and high-frequency anatomical details. We design a wavelet-based affine transformation (WAT) layer to effectively encode the wavelet domain prior into our network. Our network (see Figure 11.8), called WATNet, consists of a feature extraction branch and an image reconstruction branch. The feature extraction branch learns the complex 3T-to-7T mappings for different frequency components with several flexible WAT layers. The image reconstruction branch synthesizes 7T MR images from the wavelet modulated spatial information.

11.3.1.1 Method

Given a 3T MR image I, our goal is to synthesize a 7T-like MR image $\hat{O}$ that is as similar as possible to the ground-truth 7T MR image O. Synthesis of a high-quality 7T MR image from its 3T counterpart is challenging since 3T and 7T MR images not only differ in high-frequency details but also low-frequency contrast. We argue that the wavelet transformation, which can depict multi-level frequency components of an image, is beneficial for the synthesis of high-quality 7T MR images. To this end, instead of directly learning the mapping function, we incorporate wavelet domain as a prior to guide the prediction of 7T images with multi-level frequency components:

$$\hat{O} = R_\theta \left(I \,|\, \{W_l\} \right), \tag{11.10}$$

where W_l denotes the wavelet coefficients of I at level l. In our implementation, we set $l = 1, 2, 3$.

Wavelet-based affine transformation (WAT) layer. We further introduce a parameter-efficient and flexible WAT layer for endowing the network with information from the wavelet domain. The WAT layer is inspired by conditional normalization (CN) (Huang and Belongie, 2017) and performs spatial element-wise affine transform, allowing greater adaptivity to local spatial details. The WAT layer, shown in Figure 11.8, learns a set of affine

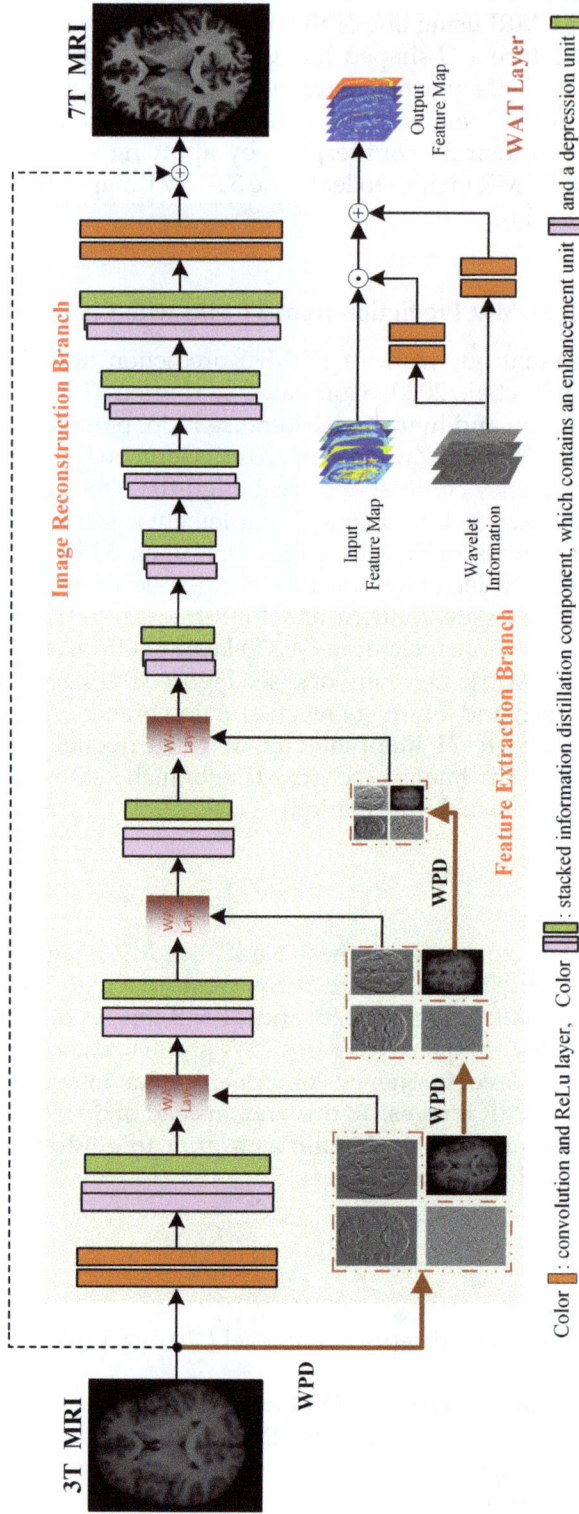

FIGURE 11.8
A fully supervised deep learning network predicts the 7T MR image using a 3T MR image and its wavelet coefficients.

parameters $\{\gamma_l, \beta_l\}$ based on the wavelet coefficients W_l, and performs element-wise affine transformation spatially on intermediate feature maps F as follows:

$$\text{WAT}(F \mid \gamma_l, \beta_l) = \gamma_l \odot F + \beta_l, \tag{11.11}$$

where the affine parameters γ_l and β_l are functions of wavelet coefficients W_l with $\gamma_l = U(W_l)$ and $\beta_l = U(W_l)$, and they have the same dimensions as F. The mapping functions U and V can be arbitrary. In our implementation, they are implemented as two convolution layers that can be optimized as part of the whole network. The WAT layer modulates the feature maps F by scaling them linearly with γ_l and then shifting with β_l, conditioned on wavelet coefficients.

Network architecture. Our network, calls WATNet (see Figure 11.8), takes a 3T MR image and its wavelet coefficients and predicts the 7T MR image. It consists of two branches: a feature extraction branch and an image reconstruction branch.

Feature extraction branch. The feature extraction branch consists of (1) two convolution layers for generating feature maps from the 3T MR image, (2) four stacked information distillation blocks for extraction of hierarchical features, and (3) three WAT layers for modulating the intermediate feature maps based on the wavelet coefficients.

For conciseness, we denote the convolution layer as conv(N, K, S, P) with N outputs, kernel size K, stride S, and pad size P. ReLu denotes the nonlinear rectified linear unit function layer. The feature extraction block is constructed with two conv(64, 3, 1, 1) layers. Each of the convolution layers is followed by a ReLu layer.

The stacked information distillation block (DBlock), originally proposed by Hui et al. (2018), is a variation of densely connected layers (Huang et al., 2017a). As suggested in Hui et al. (2018), employing feature recalibration in a network instead of using all feature maps without distinction can help suppress useless information. DBlock consists of an enhancement unit and a compression unit, which help gain more effective information and improve the representation power of the network. As illustrated in Figure 11.9, *the* enhancement unit consists of six convolution layers, each followed by a ReLu layer, which is omitted here for simplicity. The enhancement unit consists of slice and concatenation operations. The compression unit consists of a 1×1 convolution layer and a ReLu layer, and distills relevant and useful information for the subsequent block.

We denote the three WAT layers in the feature extraction branch as WAT_l, where $l \in \{1, 2, 3\}$. We do not add the WAT layer in the fourth DBlock since the wavelet coefficients are small. For each WAT layer, we perform three levels of wavelet transformation on the 3T image, obtaining the $\{W_l\}$, $l = 1, 2, 3$. WAT_l takes W_l as input and generates a set of affine parameters $\{\gamma_l, \beta_l\}$. The affine transformation is then applied spatially on the intermediate feature maps F_l generated by DBlock_l.

Image reconstruction branch. This branch takes the wavelet-modulated feature maps and generates the residual image that is added to the 3T MR image for generating the 7T MR image. As illustrated in Figure 11.8, the architecture of this branch is almost a mirror of the feature extraction branch, without the WAT layers and with conv (64, 1, 2, 0) − ReLu in the compression unit replaced by conv (64, 1, 1, 0)−ReLu−deconv, where deconv denotes deconvolution with upsampling.

Loss function. We use the MAE as the loss function since it provides better convergence than the widely used MSE loss. Given a batch of N predicted and ground-truth images $\{\hat{O}^i, O^i\}_{i=1}^N$, the MAE loss function is defined as,

$$L = \sum_{i}^{N} \left\| \hat{O}^i - O^i \right\|_1. \tag{11.12}$$

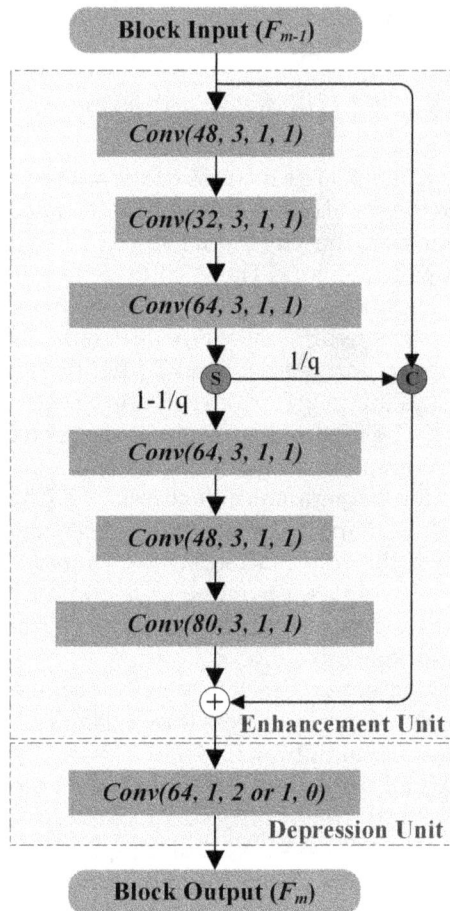

FIGURE 11.9
Stacked information distillation block (DBlock). For conciseness, we omit the ReLu layer that follow each convolution layer. denotes channel concatenation and denotes slice operation, slicing the feature maps into two segments, $1/q$ part of the feature maps via the short path and the remaining $1 - 1/q$ via the long path. q is set to 4.

11.3.1.2 Performance Evaluation

In this subsection, we compared WATNet with several state-of-the-art 7T MR image synthesis methods, including conventional learning methods with multi-level canonical correlation analysis (MCCA (Bahrami et al., 2015)), random forest (RF (Bahrami et al., 2017)), the dual-domain cascaded regression (DDCR presented in Section 2), and one deep learning method with three CNNs (CAAF (Bahrami et al., 2016)). Following Section 2.2, we used the same 15 pairs of 3T and 7T brain MR images in our experiments, applied the same pre-processing steps, and adopted leave-one-out cross-validation for performance evaluation. One pair of 3T and 7T images were used for testing, four pairs for validation, and the remaining ten pairs for training.

WATNet was implemented with Caffe (Jia et al., 2014) and optimized using Adam (Kingma and Ba, 2014). The filter weights were initialized according to (Hui et al., 2018). We set the batch size to 32, the weight decay to 0.0001, and the learning rate to 0.002 (halved every 10 epochs). WATNet was trained with the image patches extracted from 3T and 7T

MR images. The patch size was 64 × 64 × 3, covering three consecutive axial slices to promote inter-slice continuity. In the testing phase, the input 3T image was processed slice by slice to synthesize the final 7T-like MR image. Values resulting from patch overlap are averaged. To prevent overfitting, the training data were augmented in three ways: (1) Random rotation with 90°, 180°, and 270°; (2) Random scaling with a factor in the range [0.55, 0.9]; (3) Left- right Flipping. Training and testing were performed using a single GeForce GTX 1080 Ti. Training took around 18 hours.

Following Section 2, we evaluate the performance of different 7T MR image synthesis methods with two commonly used image quality metrics PSNR and SSIM. For fair comparisons, we use either the qualitative/quantitative results reported in the papers or the source code provided by the authors.

Figure 11.10 compares the results of different methods and shows the residuals of the predicted images with respect to the ground-truth 7T MR image. The respective PSNR and SSIM are also provided. It can be observed that WATNet yields results that are significantly closer to the ground truth. This can be attributed to the multi-frequency learning capability of WATNet. In contrast, MCCA (Bahrami et al., 2015), RF (Bahrami et al., 2017), and CAAF (Bahrami et al., 2016) do not consider the multi-scale nature of the problem, leading to unsatisfactory results. For example, MCCA synthesizes a 7T MR image with unsatisfactory tissue contrast, as shown in Figure 11.10 (c). Similar to our WATNet, DDCR (Zhang et al., 2018a) aims to synthesize the 7T MR image using information from both spatial and frequency domains, resulting in performance superior to methods based only on the spatial domain. However, limited by the linear mapping function, the results of DDCR (Zhang et al., 2018a) are still unsatisfactory. For example, this is shown by the loss of gray matter in the second row of Figure 11.10 (f) and also the blurry tissue boundaries in the fifth row of Figure 11.10(f).

Figure 11.11 shows the boxplots of PSNR and SSIM values computed based on the predicted images with respect to the ground truth. WATNet again demonstrates the best performance among all the compared methods. Specifically, WATNet improves the state-of-the-art PSNR/SSIM performance from 27.51/0.8580 given by DDCR (Zhang et al., 2018a) to 28.27/0.8782.

11.3.2 A Semi-Supervised Adversarial Learning for 7T MRI Prediction from 3T MRI

This subsection describes a wavelet-based semi-supervised adversarial learning framework to synthesize 7T-like MR images from their 3T counterparts. Unlike the deep learning method in Section 3.1 which relies on supervision requiring a significant amount of 3T-to-7T paired MR data, this method applies a semi-supervised learning mechanism to leverage unpaired 3T and 7T MR images to learn the 3T-to-7T mapping when 3T-to-7T paired data are scarce. This is achieved via a cycle generative adversarial network that operates in the joint spatial-wavelet domain for the synthesis of multi-frequency details.

11.3.2.1 Method

The goal is to learn the mapping from 3T MR images in domain X to 7T MR images in domain Y, given a set of training samples $\{x_i\}_{i=1}^{N} \subset X$ and $\{y_i\}_{i=1}^{M} \subset Y$, where $X = X' \cup X''$ and $Y = Y' \cup Y''$. The training samples in X' and Y' are paired 3T-7T MR images, whereas the samples in X'' and Y'' are unpaired. To learn from unpaired data, we couple the primal 3T-to-7T mapping task with a dual 7T-to-3T mapping task. As shown in Figure 11.12, the wavelet-based semi-supervised adversarial network (denoted as SemiWave) consists

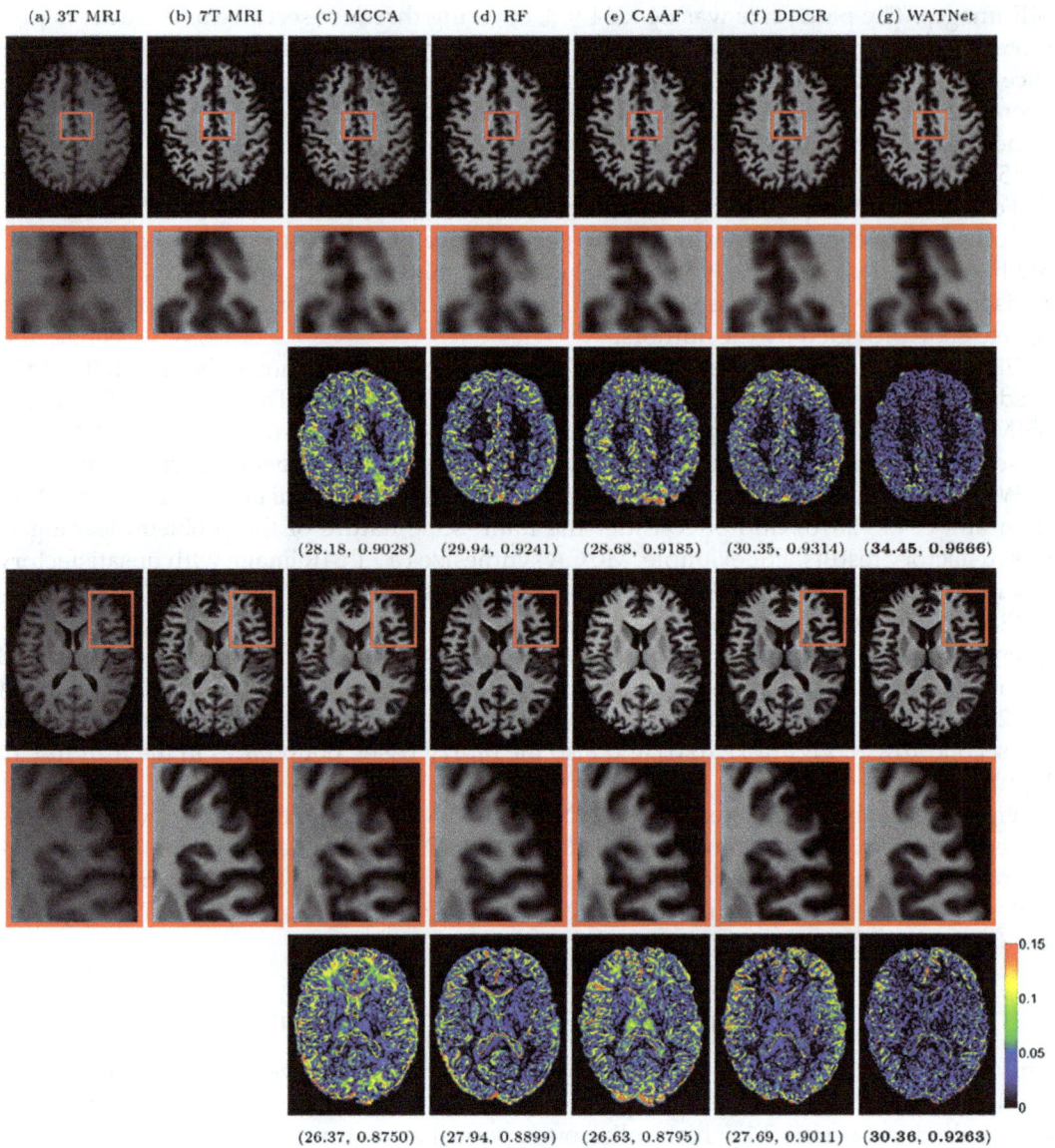

FIGURE 11.10

7T-like MR images synthesized via WATNet and other methods are shown in axial view along with the prediction error maps. (PSNR, SSIM) values are shown at the bottom.

of two mapping functions: a primal 3T-to-7T mapping ($G : x \rightarrow y$) and a dual 7T-to-3T mapping ($F: y \rightarrow x$), two associated adversarial discriminators (D_{3T} and D_{7T}), and one wavelet coefficient extractor (E_w). Discriminator D_{7T} tries to distinguish between the real 7T MR image y and the synthesized 7T MR image $G(x)$. Similarly, D_{3T} tries to distinguish between x and $F(y)$. The wavelet coefficient extractor E_w computes the wavelet coefficients of four frequency bands. Four losses are involved in regularizing these two mappings: adversarial loss, pair-wise reconstruction loss, cycle consistency loss, and wavelet loss.

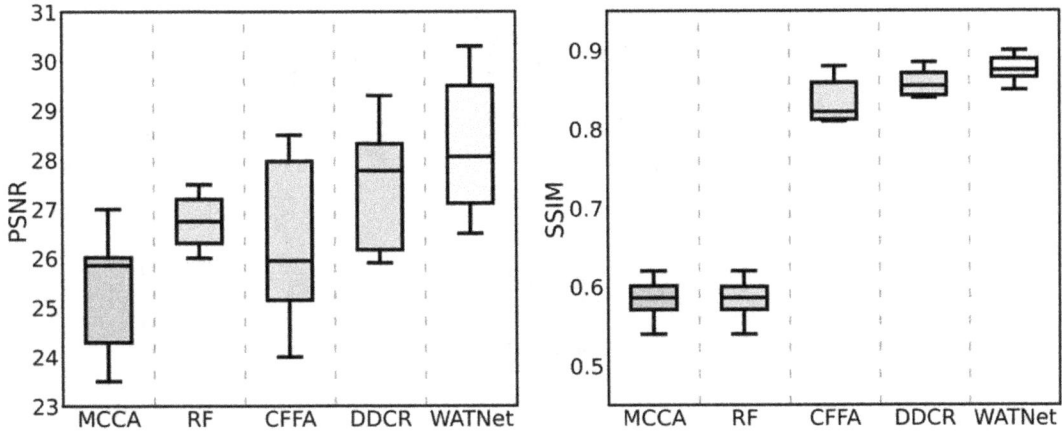

FIGURE 11.11
Boxplots of PSNR and SSIM values for five different 7T MR image synthesis methods. The average PSNR and SSIM values for all the methods are: (25.52, 0.4840) for MCCA (26.55, 0.5728) for RF, (27.05, 0.8406) for CFFA, (27.51, 0.8580) for DDCR, and (28.27, 0.8782) for WATNet.

Adversarial loss. Adversarial loss is applied to both mapping functions, aiming to match the distribution of the synthetic images to the distribution of the target images. For generator G and its corresponding discriminator D_{7T}, the adversarial loss is defined as

$$\mathcal{L}_{\text{gan}}\left(G, D_{7T}\right) = E_{y \in Y}\left[\log D_{7T}\left(y\right)\right] + E_{x \in X}\left[1 - \log D_{7T}\left(G(x)\right)\right], \tag{11.13}$$

where G tries to generate a synthetic 7T MR image $G(x)$ that resembles a real 7T MR image y in Y, whereas D_{7T} aims to distinguish between a synthetic 7T MR image $G(x)$ and a real image y. G tries to minimize this objective function, whereas D_{7T} aims to maximize it, i.e., arg $\min_G \max_{D7T} \mathcal{L}_{\text{gan}}(G, D_{7T})$. Similarly, an adversarial loss is also applied to generator F and its discriminator D_{3T} as

$$\mathcal{L}_{\text{gan}}\left(F, D_{3T}\right) = E_{x \in X}\left[\log D_{3T}\left(x\right)\right] + E_{y \in Y}\left[1 - \log D_{3T}\left(F(y)\right)\right]. \tag{11.14}$$

Pair-wise reconstruction loss. Networks with adversarial loss are often affected by model collapse and training instability. We incorporate a pair-wise reconstruction loss to impose additional constraints on the generators with a small number of paired images. Specifically, the generator is tasked not only to fool the discriminator but also to minimize the pixel-wise intensity difference between synthetic and real images. Given a set of paired 3T and 7T MR images$\{X', Y'\}$ the pair-wise reconstruction loss is defined as

$$\mathcal{L}_{p\_s}\left(G, F\right) = E_{x \in X', y \in Y'}\left[\left\|y - G(x)\right\|_1 + \left\|x - F(y)\right\|_1\right]. \tag{11.15}$$

We choose L1 distance rather than L2 to encourage image sharpness.

Cycle consistency loss. Training an effective 7T MR image synthesis model with adversarial loss and pair-wise reconstruction loss often requires a significant amount of 3T-to-7T paired data. However, acquiring paired data is often challenging due to factors such as prolonged scanning time, patient discomfort, and costs. Acquisition of unpaired data is relatively straightforward. In light of this, we design a semi-supervised learning mechanism to

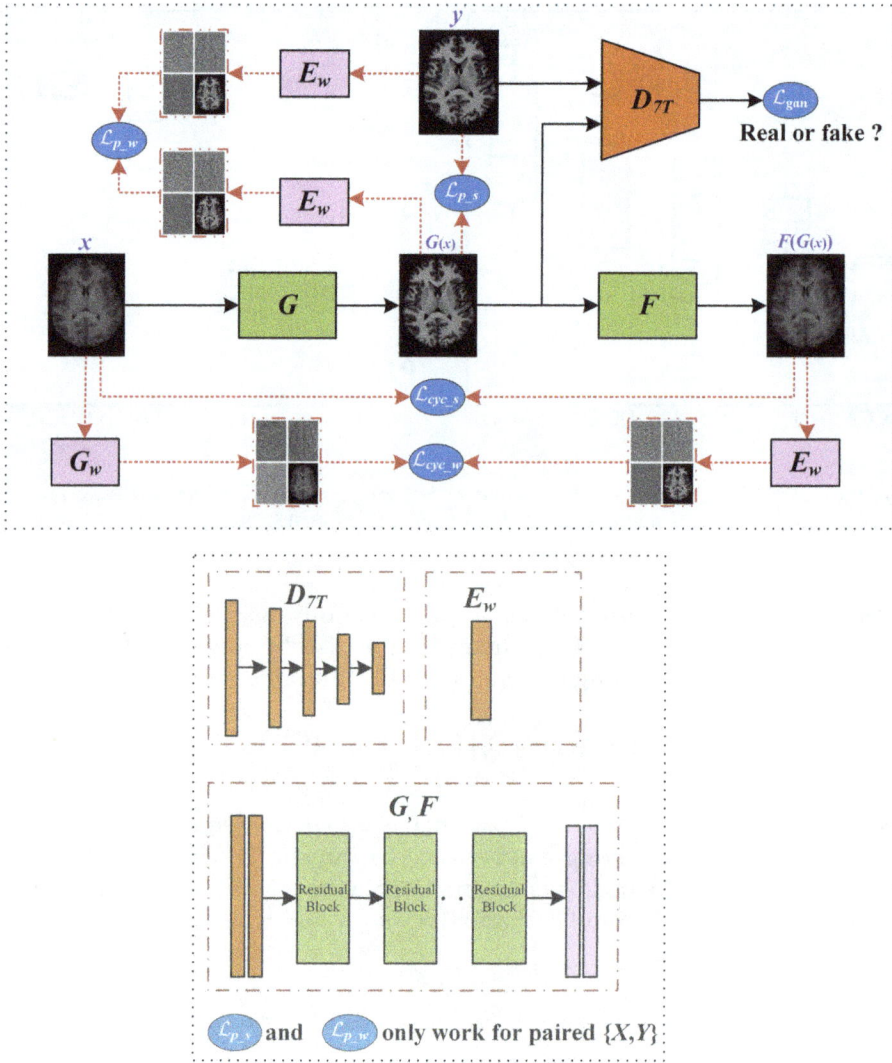

FIGURE 11.12
Framework of SemiWave. Only the flow of x → G(x) →F(G(x)) is shown for simplicity. In this flow, SemiWave consists of two mapping functions: G : x → y and F : y → x, one associated adversarial discriminator D_{7T} and one wavelet coefficient extractor Ew. Four types of loss modules are used to regularize these two mappings: adversarial loss module $\mathcal{L}_{gan}$, pair-wise reconstruction loss module $\mathcal{L}_{p\_s}$, cycle consistency loss module $\mathcal{L}_{cyc\_s}$, and wavelet loss module $\mathcal{L}_{p\_w}$ and $\mathcal{L}_{cyc\_w}$. $\mathcal{L}_{p\_s}$ and $\mathcal{L}_{p\_w}$ work for paired data, whereas $\mathcal{L}_{gan}$, $\mathcal{L}_{cyc\_s}$ and $\mathcal{L}_{cyc\_w}$ work for both paired and unpaired data.

leverage both the paired and unpaired data to learn the 3T-to-7T mapping. Specifically, to learn from extra unpaired data, a cycle consistency constraint is enforced on G and F. Following (Zhu et al., 2017), the cycle consistency constraint encourages similarity between each synthetic 7T MR image G(x) generated by the 3T-to-7T mapping G with the original 3T image x when mapped back via the 7T-to-3T mapping F, i.e., F(G(x)) ≈ x. Similarly, G(F(y)) ≈ y. The cycle consistency loss is defined as

$$\mathcal{L}_{cyc\_s}(G,F) = E_{x \in X}\left[\left\|x - F\big(G(x)\big)\right\|_1\right] + E_{y \in Y}\left[\left\|y - G\big(F(y)\big)\right\|_1\right]. \qquad (11.16)$$

Wavelet loss. The pair-wise reconstruction loss and cycle consistency loss aim at keeping the pixel/voxel-wise intensity consistency between synthetic and real images. These L1/L2 driven pixel-wise losses tend to produce over-smoothed outputs and fail to capture anatomical details (Huang et al., 2017b). We thus introduce a wavelet loss to encourage the network to generate images with realistic details. Specifically, a wavelet coefficient extractor E_w is first utilized to decompose the image into its wavelet coefficients in four bands. Given a 7T MR image y, we denote its wavelet coefficients in four bands as $\{y_w^A, y_w^H, y_w^V, y_w^D\}$. The approximation coefficients y_w^A capture the global topology information, and the detail coefficients $\{y_w^H, y_w^V, y_w^D\}$ store the high-frequency details. The wavelet loss $\mathcal{L}_{p\_w}(G)$ between the generated image $G(x)$ and real image y is defined as

$$\mathcal{L}_{p\_w}(G) = E_{x \in X', y \in Y'} \left[\lambda_1 \left\| y_w^A - \left(G(x)\right)_w^A \right\|_1 + \lambda_2 \left(\left\| y_w^H - \left(G(x)\right)_w^H \right\|_1 \right. \right.$$
$$\left. \left. + \left\| y_w^V - \left(G(x)\right)_w^V \right\|_1 \right) + \lambda_3 \left\| y_w^D - \left(G(x)\right)_w^D \right\|_1 \right], \tag{11.17}$$

where $\lambda_1, \lambda_2, \lambda_3$ are the weights to balance the importance of wavelet coefficients of different bands. Emphasis on the reconstruction of high-frequency wavelet coefficients helps recover better anatomical details, whereas emphasis on the prediction of low-frequency wavelet coefficients enforces consistency in global information. During training, we gradually increased the weights λ_2 and λ_3 in order to generate more realistic 7T MR images with greater anatomical details.

Similarly, we also apply wavelet loss $\mathcal{L}_{p\_w}(F)$ between the synthetic image $F(y)$ and real image x, and the wavelet loss $\mathcal{L}_{cyc\_w}(G, F)$ between real images $\{x, y\}$ and reconstructed images $\{F(G(x)), G(F(y))\}$.

Overall Objective Function. The overall objective function is defined as

$$\mathcal{L}(G, F, D_{3T}, D_{7T}) = \mathcal{L}_{gan} + \alpha \mathcal{L}_{p\_s} + \beta \mathcal{L}_{cyc\_s} + \gamma \mathcal{L}_{p\_w} + \delta \mathcal{L}_{cyc\_w}, \tag{11.18}$$

where $\mathcal{L}_{p\_s}$ and $\mathcal{L}_{p\_w}$ are only valid for paired data $\{X', Y'\}$, and the others are valid for all the training data $\{X, Y\}$. Parameters $\alpha, \beta, \gamma, \delta$ balance the contributions of different loss modules.

Network architecture. The detailed network architecture of SemiWave is shown in Figure 11.12. Similar to Zhu et al. (2017), the generators G and F contain two stride-2 convolutions, six residual blocks, and two fractionally-strided convolutions with a stride of 1/2. Instance normalization is applied after each convolutional layers. For discriminators D_{7T} and D_{3T}, a 70 × 70 PatchGans scheme (Isola et al., 2017, Zhu et al., 2017) is applied to classify whether 70 × 70 overlapping patches are real or fake. The wavelet coefficient extractor E_w is based on one-level Haar wavelet packet decomposition (Akansu et al., 2001). It is composed of a stride-2 convolution with its filter weights determined by Haar-based filters.

11.3.2.2 Performance Evaluation

In this subsection, we present an evaluation of SemiWave with several state-of-the-art fully supervised 7T MR image synthesis methods, including four fully supervised patch-based conventional learning methods (MCCA (Bahrami et al., 2015), RF (Bahrami et al., 2017), DDCR introduced in Section 2), and two fully supervised deep learning methods (CAAF (Bahrami et al., 2016) and WATNet in Section 3.1). Following Section 2.2, we evaluated on

TABLE 11.1

Mean PSNR and SSIM for different 7T MR image synthesis methods. The competing methods CAAF, DDCR, and our previous WATNet were trained with the fully paired data. SemiWave was trained with around 28.5% paired data and 71.5% unpaired data

	MCCA	RF	CFFA	DDCR	WATNet	SemiWave
PSNR	25.52	26.55	27.05	27.51	**28.27**	27.85
SSIM	0.4840	0.5728	0.8406	0.8580	0.8782	**0.8740**

the same 15 pre-processed pairs of 3T and 7T T1 weighted MR brain images with the same commonly accepted metrics PSNR, and SSIM. Leave-one-out cross-validation was used for performance evaluation. One pair of 3T and 7T MR images were used for testing, and the remaining 14 pairs were used for training. Specifically, the training set was divided into two subsets for SemiWave: four pairs for paired training and the remaining ten pairs were randomly shuffled for unpaired training.

The network was implemented with Pytorch and optimized with Adam (Kingma and Ba, 2014). The negative log likelihood objectives of adversarial loss $\mathcal{L}_{gan}$ were replaced by a more stable least-square loss (Zhu et al., 2017). In order to stabilize the training, the batch size was set to two with one batch from paired data and another from unpaired data. The learning rate was set to 0.0002 for the first 100 epochs and linearly decayed to zero in the

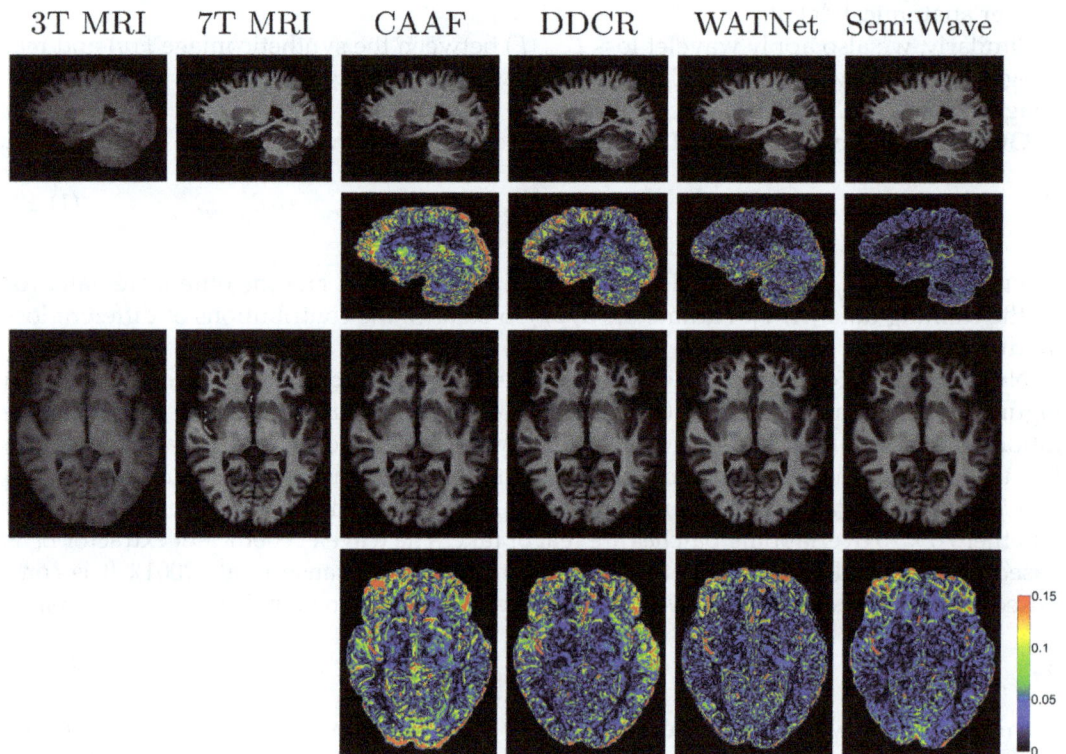

FIGURE 11.13

7T MR images synthesized using SemiWave and three other methods in sagittal and axial views, along with the prediction error maps.

subsequent 100 epochs. Parameters α, γ were set to 30, and parameters β,δ were set to 10. Parameters λ_2 and λ_3 were initially set to be small and were gradually increased for a greater contribution from high-frequency wavelet coefficients. Specifically, parameters λ_1, λ_2, λ_3 were linearly increased from 1,1,1 to 1,3,5 for the first 50 epochs and then fixed for the following epochs. In the training phase, several consecutive axial slices were extracted from the 3D images as the training images. Horizontal flipping, random scaling, and random rotation were applied to augment the training data.

Table 11.1 shows the quantitative comparison results of the different 7T MR image synthesis methods. Even with only 28.5% paired data, SemiWave still achieves superior performance compared with all the conventional learning methods and one deep learning method CFFA, and even comparable results to the deep learning WATNet presented in Section 3.1, where an advanced U-shape architecture with several WAT layers are involved. Noticeably, with the cycle consistency loss and the adversarial loss, SemiWave improves the state-of-the-art SSIM value from 0.8782 given by WATNet to 0.8740. The visual comparison results in Figure 11.13 also demonstrated the capacity of SemiWave in generating realistic 7T MR images with greater anatomical details.

11.4 Conclusion

Multimodality MRI synthesis has demonstrated increasing attention in clinics and has been widely used in various applications, including MRI synthesis across different modalities, MRI synthesis across different qualities, and MRI inhomogeneity correction. In this chapter, we present a thorough review of the multimodality MRI synthesis methods. Specifically, we use the 7T MRI prediction from 3T MRI (MRI synthesis across different quality) as an example task, and introduce the widely used conventional learning methods and modern deep-learning-based methods for multimodality MRI synthesis. These widely used learning methods (both patch-based conventional learning and deep learning methods) share similar principles and aim at learning the mapping function between the source-modality image and the target-modality image, thus can be applied to various types of multimodality MRI synthesis tasks. The patch-based conventional learning methods usually consist of three stages: 1) source- and target-dictionaries construction; 2) learning of the mapping function between source- and target-domain; and 3) target-modality image reconstruction from the learned mapping function and dictionary representation. We then present a dual-domain cascaded regression model with two parallel and interactive multi-stage regression streams for 7T MRI prediction from 3T MRI, to improve the quality of the synthetic 7T MR images. Compared to the patch-based conventional learning methods, the deep-learning-based methods preclude the use of hand-craft features and show significantly better results than conventional learning methods. We further introduce a CNN-based fully supervised method for 7T MRI prediction method by harnessing information from both spatial and wavelet domains. To alleviate the need for large quantities of paired training data for multimodality MRI synthesis, a semi-supervised adversarial learning method, which leverages the unpaired 3T and 7T MR images to learn the 3T-to-7T mapping, is also presented.

Even though deep learning methods have made rapid breakthroughs in multimodality MRI synthesis, there are still some open questions that need to be tackled before it be readily applicable in the real clinic. Creating robust models that generalize across institutions requires a tremendous amount of training cases. Most deep learning models are currently trained using data with small size, which results in the limited representation of populations, algorithm bias, and poor generalizability problems. Besides, most of the existing models are evaluated only on the in-domain distribution dataset without a multi-site assessment. Thus, developing robust deep learning models that work well not only on the in-domain distribution dataset, but also on the out-of-distribution dataset is highly desired and should be potential future research. In addition, once a multimodality MRI synthesis model is developed and is going to be deployed for clinic usage, the evaluation of the model's robustness when faced with the inevitability of natural changes in imaging protocols, patient variance, and physician behavior is the most critical challenge that the researchers should solve. We thus argue that the study of model deployment in clinical settings should be another active future direction in parallel to the current efforts on the model development front.

References

Akansu, A. N., Haddad, R. A., & Haddad, P. A. 2001. *Multiresolution signal decomposition: transforms, subbands, and wavelets*, San Diego: Academic Press.

Bahrami, K., Shi, F., Rekik, I., & Shen, D. 2016. Convolutional neural network for reconstruction of 7T-like images from 3T MRI using appearance and anatomical features. *Deep learning and data labeling for medical applications*, Athens: Springer.

Bahrami, K., Shi, F., Rekik, I., Gao, Y., & Shen, D. 2017. 7T-guided super-resolution of 3T MRI. *Medical Physics*, 44, 1661–1677.

Bahrami, K., Shi, F., Zong, X., Shin, H. W., An, H., & Shen, D. Hierarchical reconstruction of 7T-like images from 3T MRI using multi-level CCA and group sparsity. *International Conference on Medical Image Computing and Computer-Assisted Intervention*, 2015. Springer, 659–666.

Berker, Y., Franke, J., Salomon, A., Palmowski, M., Donker, H. C., Temur, Y., Mottaghy, F. M., Kuhl, C., Izquierdo-Garcia, D., & Fayad, Z. A. 2012. MRI-based attenuation correction for hybrid PET/MRI systems: a 4-class tissue segmentation technique using a combined ultrashort-echo-time/Dixon MRI sequence. *Journal of Nuclear Medicine*, 53, 796–804.

Brunig, M., & Niehsen, W. 2001. Fast full-search block matching. *IEEE Transactions on Circuits and Systems for Video Technology*, 11, 241–247.

Burgos, N., Cardoso, M. J., Thielemans, K., Modat, M., Pedemonte, S., Dickson, J., Barnes, A., Ahmed, R., Mahoney, C. J., Schott, J. M., Duncan, J. S., Atkinson, D., Arridge, S. R., Hutton, B. F., & Ourselin, S. 2014. Attenuation correction synthesis for hybrid PET-MR scanners: application to brain studies. *IEEE Trans Med Imaging*, 33, 2332–2341.

Catana, C., van der Kouwe, A., Benner, T., Michel, C. J., Hamm, M., Fenchel, M., Fischl, B., Rosen, B., Schmand, M., & Sorensen, A. G. 2010. Toward implementing an MRI-based PET attenuation-correction method for neurologic studies on the MR-PET brain prototype. *Journal of Nuclear Medicine*, 51, 1431–1438.

Chartsias, A., Joyce, T., Dharmakumar, R., & Tsaftaris, S. A. Adversarial image synthesis for unpaired multi-modal cardiac data. *International Workshop on Simulation and Synthesis in Medical Imaging*, 2017a. Springer, 3–13.

Chartsias, A., Joyce, T., Giuffrida, M. V., & Tsaftaris, S. 2017b. Multimodal MR synthesis via modality-invariant latent representation. *IEEE Transactions on Medical Imaging*, 37, 803–814.

Chaudhari, A. S., Fang, Z., Kogan, F., et al. 2018. Super-resolution musculoskeletal MRI using deep learning[J]. *Magnetic Resonance in Medicine*, 80(5), 2139–2154.

Chen, Y., Xie, Y., Zhou, Z., Shi, F., Christodoulou, A. G., & Li, D. Brain MRI super resolution using 3D deep densely connected neural networks. *2018 IEEE 15th International Symposium on Biomedical Imaging (ISBI 2018)*, 2018. IEEE, 739–742.

Dalmaz, O., Yurt, M., & Çukur, T. 2021. ResViT: residual vision transformers for multi-modal medical image synthesis. *arXivpreprint*, arXiv:2106.16031.

Dalmaz, O., Yurt, M., & Ukur, T. 2022. ResViT: residual vision transformers for multi-modal medical image synthesis. *IEEE Transactions on Medical Imaging*, 41, 2598–2614.

Dar, S. U., Yurt, M., Karacan, L., Erdem, A., Erdem, E., & Cukur, T. 2019. Image synthesis in multi-contrast MRI with conditional generative adversarial networks. *IEEE Transactions on Medical Imaging*, 38, 2375–2388.

Dong, C., Loy, C. C., He, K., & Tang, X. Learning a deep convolutional network for image super-resolution. *European Conference on Computer Vision*, 2014. Springer, 184–199.

Gosain, A. K., Klein, M. H., Sudhakar, P. V., & Prost, R. W. 2005. A volumetric analysis of soft-tissue changes in the aging midface using high-resolution MRI: implications for facial rejuvenation. *Plastic & Reconstructive Surgery*, 115, 1153–1155.

Han, C., Hayashi, H., Rundo, L., Araki, R., Shimoda, W., Muramatsu, S., Furukawa, Y., Mauri, G., & Nakayama, H. GAN-based synthetic brain MR image generation. *2018 IEEE 15th International Symposium on Biomedical Imaging (ISBI 2018)*, 2018. IEEE, 734–738.

He, K., Zhang, X., Ren, S., & Sun, J. Deep residual learning for image recognition. *Proceedings of the IEEE Conference on Computer Vision and Pattern Recognition*, 2016. 770–778.

Hiasa, Y., Otake, Y., Takao, M., Matsuoka, T., Takashima, K., Carass, A., Prince, J. L., Sugano, N., & Sato, Y. Cross-modality image synthesis from unpaired data using CycleGAN. *International Workshop on Simulation and Synthesis in Medical Imaging*, 2018. Springer, 31–41.

Hofmann, M., Steinke, F., Scheel, V., Charpiat, G., Farquhar, J., Aschoff, P., Brady, M., Schölkopf, B., & Pichler, B. J. 2008. MRI-based attenuation correction for PET/MRI: a novel approach combining pattern recognition and atlas registration. *Journal of Nuclear Medicine*, 49, 1875–1883.

Holmes, C. J., Hoge, R., Collins, L., Woods, R., Toga, A. W., & Evans, A. C. 1998. Enhancement of MR images using registration for signal averaging. *Journal of Computer Assisted Tomography*, 22, 324–333.

Huang, G., Liu, Z., Van Der Maaten, L., & Weinberger, K. Q. Densely connected convolutional networks. *Proceedings of the IEEE Conference on Computer Vision and Pattern Recognition*, 2017a. 4700–4708.

Huang, X., & Belongie, S. Arbitrary style transfer in real-time with adaptive instance normalization. *Proceedings of the IEEE International Conference on Computer Vision*, 2017. 1501–1510.

Huang, H., He, R., Sun, Z., & Tan, T. Wavelet-SRNET: a wavelet-based CNN for multi-scale face super resolution. *Proceedings of the IEEE International Conference on Computer Vision*, 2017b. 1689–1697.

Hui, Z., Wang, X., & Gao, X. Fast and accurate single image super-resolution via information distillation network. *Proceedings of the IEEE Conference on Computer Vision and Pattern Recognition*, 2018. 723–731.

Huynh, T., Gao, Y., Kang, J., Wang, L., Zhang, P., Lian, J., & Shen, D. 2015. Estimating CT image from MRI data using structured random forest and auto-context model. *IEEE Transactions on Medical Imaging*, 35, 174–183.

Isola, P., Zhu, J.-Y., Zhou, T., & Efros, A. A. Image-to-image translation with conditional adversarial networks. *Proceedings of the IEEE Conference on Computer Vision and Pattern Recognition*, 2017. 1125–1134.

Jenkinson, M., Bannister, P., Brady, M., & Smith, S. 2002. Improved optimization for the robust and accurate linear registration and motion correction of brain images. *Neuroimage*, 17, 825–841.

Jia, Y., Shelhamer, E., Donahue, J., Karayev, S., Long, J., Girshick, R., Guadarrama, S., & Darrell, T. Caffe: convolutional architecture for fast feature embedding. *Proceedings of the 22nd ACM international conference on Multimedia*, 2014. 675–678.

Jog, A., Carass, A., & Prince, J. L. Improving magnetic resonance resolution with supervised learning. *2014 IEEE 11th International Symposium on Biomedical Imaging (ISBI)*, 2014. IEEE, 987–990.

Joyce, T., Giuffrida, M. V., & Tsaftaris, S. A. Robust multi-modal MR image synthesis. *International Conference on Medical Image Computing and Computer-Assisted Intervention*, 2017. Springer, 347–355.

Kamnitsas, K., Ledig, C., Newcombe, V. F., Simpson, J. P., Kane, A. D., Menon, D. K., Rueckert, D., & Glocker, B. 2016. Efficient multi-scale 3D CNN with fully connected CRF for accurate brain lesion segmentation. *Medical Image Analysis*, 36, 61.

Kaus, M. R., Brock, K. K., Pekar, V., Dawson, L. A., Nichol, A. M., & Jaffray, D. A. 2007. Assessment of a model-based deformable image registration approach for radiation therapy planning. *International Journal of Radiation Oncology* Biology* Physics*, 68, 572–580.

Kingma, D. P., & Ba, J. 2014. Adam: a method for stochastic optimization. *arXiv preprint*, arXiv:1412.6980.

Kops, E. R., & Herzog, H. Alternative methods for attenuation correction for PET images in MR-PET scanners. *2007 IEEE Nuclear Science Symposium Conference Record*, 2007. IEEE, 4327–4330.

Liu, Y., Leong, A. T., Zhao, Y., Xiao, L., Mak, H. K., Tsang, A. C. O., Lau, G. K., Leung, G. K., & Wu, E. X. 2021. A low-cost and shielding-free ultra-low-field brain MRI scanner. *Nature Communications*, 12, 1–14.

Manjón, J. V., Coupé, P., Concha, L., Buades, A., Collins, D. L., & Robles, M. 2010. MRI superresolution using self-similarity and image priors. *International Journal of Biomedical Imaging*, 2010, 1–11.

McRobbie, D. W., Moore, E. A., Graves, M. J., & Prince, M. R. 2017. *MRI from picture to proton*, Cambridge: Cambridge University Press.

Miller, M. I., Christensen, G. E., Amit, Y., & Grenander, U. 1993. Mathematical textbook of deformable neuroanatomies. *Proceedings of the National Academy of Sciences*, 90, 11944–11948.

Nie, D., Trullo, R., Lian, J., Petitjean, C., Ruan, S., Wang, Q., & Shen, D. Medical image synthesis with context-aware generative adversarial networks. *International Conference on Medical Image Computing and Computer-Assisted Intervention*, 2017. Springer, 417–425.

Nie, D., Trullo, R., Lian, J., Wang, L., Petitjean, C., Ruan, S., Wang, Q., & Shen, D. 2018. Medical image synthesis with deep convolutional adversarial networks. *IEEE Transactions on Biomedical Engineering*, 65, 2720–2730.

Oktay, O., Bai, W., Lee, M., Guerrero, R., Kamnitsas, K., Caballero, J., de Marvao, A., Cook, S., O'Regan, D., & Rueckert, D. Multi-input cardiac image super-resolution using convolutional neural networks. *International Conference on Medical Image Computing and Computer-Assisted Intervention*, 2016. Springer, 246–254.

Olut, S., Sahin, Y. H., Demir, U., et al. Generative adversarial training for MRA image synthesis using multi-contrast MRI[C]//PRedictive Intelligence in MEdicine: First International Workshop, PRIME 2018, Held in Conjunction with MICCAI 2018, Granada, Spain, September 16, 2018, Proceedings 1. Springer International Publishing, 2018. 147–154.

Petersen, K. B., & Pedersen, M. S. 2008. The matrix cookbook. *Technical University of Denmark*, 7, 510.

Preston, D. C. 2006. Magnetic resonance imaging (mri) of the brain and spine: basics. *MRI Basics, Case Med*, 30, 1–6.

Qu, L., Wang, S., Yap, P. T., & Shen, D. Wavelet-based semi-supervised adversarial learning for synthesizing realistic 7T from 3T MRI. *International Conference on Medical Image Computing and Computer-Assisted Intervention*, 2019. Springer, 786–794.

Qu, L., Zhang, Y., Wang, S., Yap, P. T., & Shen, D. 2020. Synthesized 7T MRI from 3T MRI via deep learning in spatial and wavelet domains. *Medical Image Analysis*, 62, 101663.

Ronneberger, O., Fischer, P., & Brox, T. U-net: convolutional networks for biomedical image segmentation. *International Conference on Medical Image Computing and Computer-Assisted Intervention*, 2015. Springer, 234–241.

Roy, S., Carass, A., & Prince, J. L. 2013. Magnetic resonance image example-based contrast synthesis. *IEEE Transactions on Medical Imaging*, 32, 2348–2363.

Rueda, A., Malpica, N., & Romero, E. 2013. Single-image super-resolution of brain MR images using overcomplete dictionaries. *Medical Image Analysis*, 17, 113–132.

Shi, F., Fan, Y., Tang, S., Gilmore, J. H., Lin, W., & Shen, D. 2010. Neonatal brain image segmentation in longitudinal MRI studies. *Neuroimage*, 49, 391–400.

Sled, J. G., Zijdenbos, A. P., & Evans, A. C. 1998. A nonparametric method for automatic correction of intensity nonuniformity in MRI data. *IEEE Transactions on Medical Imaging*, 17, 87–97.

Sun, H., Xi, Q., Fan, R., Sun, J., Xie, K., Ni, X., & Yang, J. 2022. Synthesis of pseudo-CT images from pelvic MRI images based on an MD-CycleGAN model for radiotherapy. *Physics in Medicine & Biology*, 67, 035006.

van der Zwaag, W., Schäfer, A., Marques, J. P., Turner, R., & Trampel, R. 2016. Recent applications of UHF-MRI in the study of human brain function and structure: a review. *NMR in Biomedicine*, 29, 1274–1288.

Van Nguyen, H., Zhou, K., & Vemulapalli, R. Cross-domain synthesis of medical images using efficient location-sensitive deep network. *International Conference on Medical Image Computing and Computer-Assisted Intervention*, 2015. Springer, 677–684.

Venkatesh, V., Sharma, N., & Singh, M. 2020. Intensity inhomogeneity correction of MRI images using InhomoNet. *Computerized Medical Imaging and Graphics*, 84, 101748.

Vovk, U., Pernus, F., & Likar, B. 2007. A review of methods for correction of intensity inhomogeneity in MRI. *IEEE Transactions on Medical Imaging*, 26, 405–421.

Wang, T., Lei, Y., Curran, W. J., Liu, T., & Yang, X. Contrast-enhanced MRI synthesis from non-contrast MRI using attention CycleGAN. *Medical Imaging 2021: Biomedical Applications in Molecular, Structural, and Functional Imaging*, 2021. International Society for Optics and Photonics, 116001L.

Watanabe, S., Ueno, T., Kimura, Y., Mishina, M., & Sugimoto, N. 2021. Generative image transformer (GIT): unsupervised continuous image generative and transformable model for [123I] FP-CIT SPECT images. *Annals of Nuclear Medicine*, 35, 1203–1213.

Welander, P., Karlsson, S., & Eklund, A. 2018. Generative adversarial networks for image-to-image translation on multi-contrast MR images-a comparison of cyclegan and unit. *arXiv preprint*, arXiv:1806.07777.

Xiang, Z. Z., He, T., Zeng, Y. Y., Liu, F., Shao, B. F., Yang, T., Ma, J. C., Wang, X. R., Yu, S. T., & Liu, L. 2020. Intensity non-uniformity correction in MR imaging using residual cycle generative adversarial network. *Physics in Medicine and Biology*, 65, 215025.

Yang, Q., Li, N., Zhao, Z., Fan, X., Chang, E. I., & Xu, Y. 2018. MRI cross-modality neuroimage-to-neuroimage translation. *arXiv preprint*, arXiv:1801.06940.

Ye, D. H., Zikic, D., Glocker, B., Criminisi, A., & Konukoglu, E. Modality propagation: coherent synthesis of subject-specific scans with data-driven regularization. *International Conference on Medical Image Computing and Computer-Assisted Intervention*, 2013. Springer, 606–613.

Yu, B., Zhou, L., Wang, L., Shi, Y., Fripp, J., & Bourgeat, P. 2019. Ea-GANs: edge-aware generative adversarial networks for cross-modality MR image synthesis. *IEEE Transactions on Medical Imaging*, 38, 1750–1762.

Zhang, Y., Cheng, J.-Z., Xiang, L., Yap, P.-T., & Shen, D. Dual-domain cascaded regression for synthesizing 7T from 3T MRI. *International Conference on Medical Image Computing and Computer-Assisted Intervention*, 2018a. Springer, 410–417.

Zhang, Y., Liu, J., Yang, W., & Guo, Z. 2015. Image super-resolution based on structure-modulated sparse representation. *IEEE Transactions on Image Processing*, 24, 2797–2810.

Zhang, Z., Yang, L., & Zheng, Y. Translating and segmenting multimodal medical volumes with cycle-and shape-consistency generative adversarial network. *Proceedings of the IEEE Conference on Computer Vision and Pattern Recognition*, 2018b. 9242–9251.

Zhou, T., Fu, H., Chen, G., Shen, J., & Shao, L. 2020. Hi-net: hybrid-fusion network for multi-modal MR image synthesis. *IEEE Transactions on Medical Imaging*, 39, 2772–2781.

Zhu, J.-Y., Park, T., Isola, P., & Efros, A. A. Unpaired image-to-image translation using cycle-consistent adversarial networks. *Proceedings of the IEEE International Conference on Computer Vision*, 2017. 2223–2232.

12

Multi-Energy CT Transformation and Virtual Monoenergetic Imaging

Wei Zhao

Beihang University, Beijing, China

CONTENTS

12.1 Introduction

While valuable with millions of scans performed around the world each year, classical single-energy CT (SECT) falls short in tissue type differentiation and quantification. The reason for these limitations is that materials with different elemental compositions can be represented by the same CT numbers (i.e., Hounsfield Units (HU)). The CT number of a specific voxel corresponds to the linear attenuation coefficient, and it depends on the effective atomic number (EAN) and the mass density of the voxel, as well as the effective energy of the X-ray photons. Therefore, for a given SECT image, the physical interpretation of the CT number is ambiguous, and the CT number of a voxel is not unique for any given material, and different materials can correspond to the same CT number. Instead of using a single energy spectrum like SECT, multi-energy CT (MECT) performs attenuation measurements using two or more energy spectra.

DOI: 10.1201/9781003243458-15

In the diagnostic energy range, there are mainly two types of interactions of photons with matter, namely the photoelectric effect and Compton. The total mass attenuation coefficient can be attributed to the contribution of these two interactions with each having different energy dependencies on the photon energy. Specifically, the contribution of photoelectric effect reduces dramatically as the X-ray photon energy increases, while the Compton scatter has weak energy dependence with respect to the photon energy. Therefore, the contribution of Compton scatter increases quickly as the X-ray energy increases and it becomes the dominating interaction for the mass attenuation coefficient. With energy dependency, it is able to provide additional information by using two or more energy spectra [1, 2]. MECT, therefore, allows the differentiation of materials and quantitative characterization of material compositions by taking advantage of the energy-dependent nature of attenuation coefficients of materials.

12.2 Implementation of Multi-energy CT Imaging

Instead of decomposing a material onto the photoelectric and Compton basis functions (i.e., representing the mass attenuation coefficient of the material with the contribution of the photoelectric effect and Compton scattering), it is more convenient to represent the material as a linear combination of two basis materials (such as water and iodine) [3]. This is because photoelectric and Compton basis functions do not provide clinically relevant information, and it is hard to interpret the resulting images for clinical usage. In this context, other materials are able to be decomposed as a weighted summation of the basis materials, and the weights of each basis material correspond to the mass densities or relative mass densities of basis materials. Since there are usually more than two basis materials (in case of a material with K-edge in the energy range of interest, the material should be included as a third basis material [4, 5]), two or more projection measurements using different energy spectra should be performed to solve the weights. Due to the noise amplification nature of the material decomposition process, the projection should be acquired using spectra with distinct energy distributions to provide informative measurements and consequently high-quality energy-specific and material-specific images after material decomposition. Meanwhile, material decomposition requires consistent spectral measurements to be acquired in both spatial and temporal space. To achieve this goal, various techniques are developed to acquire distinct and consistent dual and multi-energy data [2, 6].

12.2.1 Sequential Scanning

A straightforward solution to consistent MECT data acquisition is to perform temporally sequential scans with different tube potentials. This solution is likely to suffer from patient motion because the projection measurements of dual or multi-energy scans are not acquired simultaneously. The motion results in inconsistent measurements and thus severe degradation of the decomposed material-specific images. To mitigate the patient motion between the sequential scans, the scan can be performed in a step-and-shot mode with alternative low- and high-kV potential, followed by table incrementation along the axial direction. In addition, a partial scan (180° plus the fan angle) can be further performed to improve the temporal resolution.

12.2.2 Fast Tube Potential Switching

Fast tube potential switching allows dual- or multi-energy projection measurements to be acquired in a single gantry rotation. In this scenario, the tube potential of consecutive views rapidly switches back and forth between the different energy settings, in less than a millisecond. Projection acquired in this manner results in an interleaved sinogram [7–9]. Although fast tube potential switching technology is able to provide near-simultaneous data acquisition of different energies, there are still temporal and spatial offsets between the dual- or multi-energy projections. Because material decomposition requires highly consistent projections. To achieve multi-energy data consistency, data interpolation is needed. Meanwhile, the projection sampling rate is much higher than conventional single kV CT scan to ensure that the data inconsistency is minimized, to maintain spatial resolution and to reduce aliasing artifacts. Since it is able to provide both near-simultaneous projections and images, the fast tube potential switching technology can perform material decomposition in either projection domain or image domain. Of note, beam hardening artifacts can be significantly reduced when performing material decomposition in projection domain. There are also some inherent limitations to this technology. The tube potential switches back and forth between different energies in a very short time interval, making it is impossible to change beam filtration and therefore optimize the energy spectrum to yield the best dose efficiency and material decomposition. It is also challenging to change tube current during scanning to provide material-specific images with optimal contrast-to-noise ratio. An example of spectra generated using the fast tube potential switching technology is shown in Figure 12.1a.

12.2.3 Multilayer Detector

Another commercially implemented multi-energy CT imaging solution is to use a layered detector. In this scenario, X-ray photons emitted from a single source are detected using two- or multi-layers of energy-integrating detectors [10, 11]. To enhance energy separation which is desirable for noise suppression during material decomposition while maintaining dose efficiency, the thickness and material of each layer are well-designed and optimized. For the dual-energy case, the upper layer and the lower layer acquire low-energy and high-energy projection measurements, respectively (as shown in Figure 12.1b). Different from fast tube potential switching technology where there exists slight inconsistency between

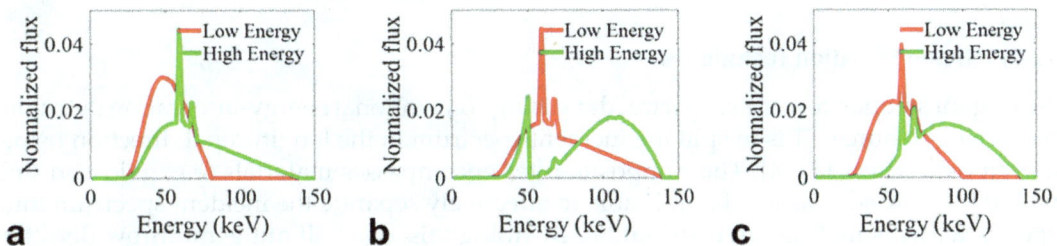

FIGURE 12.1
Energy spectra generated using different MECT implementations. (a) 80 kV and 140 kV spectra generated by fast tube potential switching dual-energy CT (DECT) scanner; (b) 140 kV low-energy and high-energy spectra detected by the DECT with layered detectors; (c) 100 kV and 140 kV dual-energy spectra generated by dual-source DECT scanner.

the low- and high-energy data, projections measured using the multilayered detector are spatially and temporally aligned, which allows perfect material decomposition in the projection domain. As the same as the fast tube potential switching which is also able to perform projection space decomposition, a layered detector yields virtual monochromatic images with significantly reduced beam hardening artifacts. This is highly desirable for imaging applications that involve bony structures or dense materials, such as hip prostheses imaging, head imaging, and coronary stent imaging. In addition, the perfect match of spectral data offers anti-correlated noise removal during basis material decomposition and yields noise-reduced material-specific images and virtual monochromatic images. It is also desirable for imaging moving anatomy and contrast enhancement. It places no constraints on field-of-view, tube current modulation, or other data acquisition protocols, but cross-talk between the layers may reduce the accuracy of the spectral data. Since a layered detector can only provide spectral data with the same tube potential, spectral separation is limited compared to other multi-energy CT imaging solutions.

12.2.4 Dual-Source Acquisitions

Dual-source CT (DSCT) is another straightforward solution to dual-energy CT imaging. It uses two X-ray measurement systems (i.e., two pairs of source-detector), both of which are installed on the rotating gantry [12]. By using different energy spectra for these two X-ray measurement systems. DSCT can acquire spectral data in a single gantry rotation. An example of spectra used in DSCT is shown in Figure 12.1c. Since each measurement system operates individually, its scanning parameters can be optimized to balance the radiation dose between dual-energy data acquisition to yield the material-specific images with the optimal noise-to-contrast ratio. Different from DECT using fast tube potential switching and layered detector technologies, the two X-ray tubes are able to use different filtrations (such as a tin filter) and tube potentials to yield the best spectral separation which is desirable for artifact reduction and material decomposition.

Since DSCT acquires raw spectral projection data with an angle offset of about 90^0, it cannot provide ray-consistent projection data for low- and high-energy measurements. Therefore, DSCT can only perform material decomposition in the image domain in principle, which is undesirable for beam hardening artifacts reduction. Meanwhile, the two X-ray imaging systems do not provide the same field-of-view due to limited gantry space, and cross-scatter radiation (scattered photons emitted from one X-ray imaging system and detected by the other system) can be induced, resulting in cupping or shading artifacts, which have the potential to distort CT numbers.

12.2.5 Beam Filtration Techniques

An elegant solution to acquire spectral data using conventional energy-integrating detector-based single source CT is to split the incident spectrum in the longitudinal direction using a composite filter [13, 14]. The composite filter encompasses materials (e.g. gold and tin) with distinct K-edges such that it is able to effectively separate the incident spectrum into spectra with distinct energy distribution. By doing this, one half of the multirow detector receives photons (e.g., filtered by tin) that have higher mean energy, while the other half of the detector receives photons (e.g., filtered by gold) that have lower mean energy. Beam filtration techniques enable spectral data to be measured with conventional single-energy CT systems, by only updating the beam collimator. Hence, multi-energy CT scanners based on beam filtration techniques have relatively low costs. A practice implementation of the beam filtration technique for MECT imaging is the TwinBeam CT scanner (Siemens SOMATOM Definition Edge).

As layered detector techniques, beam filtration techniques acquire spectral data with the same tube potential and therefore the spectral separation is limited, compared to DSCT. Since K-edge materials are employed for filtration to generate spectra with distinct energy distribution, a significant portion of photons have been absorbed to reshape the spectra. To meet the clinical requirement of rapid imaging, the source of the CT system must be powerful.

12.2.6 Energy-Resolved Detector

Different from conventional energy-integrating detectors which first convert X-ray photons into visible light and then electric signals, an energy-resolved detector is able to directly convert the X-ray photons into electrical signals. The direct convert feature enables the energy of each incident photon to be individually measured [5, 15–20]. In practice, with a few predefined energy thresholds, the incident photons can be classified into corresponding energy ranges. Each energy range is treated as an independent X-ray measurement, and therefore multi-energy projections are able to be acquired simultaneously using conventional CT scanners without system modifications other than the X-ray detector. By doing this, spectral measurements are spatially and temporally aligned, suggesting beam hardening artifacts can be well removed in principle with material decomposition in the projection domain. Scanning parameters (e.g., tube potential) of MECT based on an energy-resolved detector are also the same as in standard CT scans.

By using more than two energy thresholds, the energy-resolved detector classifies incident photons into more than three energy ranges. This is practically useful when materials with K-edges within the diagnostic energy range are involved in CT imaging. In this case, a basic material other than water, bone, or soft tissues must be included in material decomposition. Usually, the K-edge material itself needs to be the basic material as other materials are difficult to characterize its attenuation properties. MECT using the energy-resolved detector is extremely suitable for multiple contrast-enhanced CT imaging, which is able to differentiate venous and arterial blood vessels, or to evaluate inflammatory and oncologic lesions [18, 21]. At the current stage, energy-resolved detectors still suffer from various limitations, including pile-up effect at high incident flux, count-rate drift induced by detector defects, and spectral distortion caused by charge sharing and K-edge escape.

12.3 Motivation of Artificial Intelligence-based Multi-energy CT Imaging

Artificial intelligence (AI) is a general concept which denotes intelligence, such as decision-making and inferencing, demonstrated by computer techniques. As a subset of AI, machine learning is much more specific and it uses algorithms to learn from data via either human-engineered features or implicit features extracted by neural networks. The algorithms then apply the learned prior knowledge to perform tasks like classification and regression. While classical machine learning algorithms perform learning via human-extracted features and include algorithms such as linear regression, principal component analysis, support vector machine, and k-mean clustering which are mathematically explainable, recent advances in machine learning allow high-level features extracted using deep neural networks to represent complex relationship within training data (termed deep learning) [22]. Deep learning is a specific field of machine learning, and it is very suitable for mapping the relationship between datasets where an explicit formulation does not exist or the explicit formulation has been corrupted by noise or incorrect signals. With the rapid development in computing capability and data explosion in the past decade, deep learning has been extensively investigated and successfully applied to widespread applications, including biomedicine, computer vision, and natural language processing. There are numerous deep learning architectures, and they can be targeted to various types of data and tasks. For medical image-related tasks, popular architectures include convolutional neural networks (CNN) [23, 24], generative adversarial networks (GANs) [25, 26], vision transformer networks [27], and so on.

By comprehensively processing and analyzing the patient morphological information in different imaging modalities in spatial, temporal, or spectral domains, deep learning is able to perform tasks (such as medical image segmentation) that are routinely done by radiologists and oncologists. Additionally, it can be envisioned to perform tasks that are beyond human efforts or classical imaging frameworks by synergistically learning from prior knowledge [28, 29]. For example, the learning from the planning CT images and pretreatment of daily cone-beam CT images, deep learning models can provide real-time volumetric CT imaging by using a single-view X-ray projection during beam delivery [30, 31]. This highly ill-posed inverse problem is impossible under the classical image reconstruction theorem.

With their powerful reasoning logic and inference, deep learning approaches are of particular benefit to the field of medical image synthesis. With regard to MECT imaging, the previous various elegant solutions to MECT imaging are all based on hardware implementation, and the corresponding spectral CT scanners are mainly high end, which are usually unaffordable for undeveloped regions. It is therefore of interest to perform image transformation between different energy levels or between energy and material spaces. With these image transformation technologies, routine MECT applications can be performed using software-based solutions. To this end, more and more studies have been performed to provide synthetic spectral images and material-specific images for MECT imaging. In the following sections, we will present the application of deep learning on medical image synthesis, with a focus on MECT image transformation and material-specific and energy-specific image synthesis.

12.4 Energy Domain MECT Image Synthesis

To remove the barrier of MECT imaging via hardware implementation and make it possible for undeveloped regions, image synthesis has been extensively investigated for MECT imaging using deep learning approaches. By utilizing the intrinsic characteristics (such as global correlation and high similarity) of the routine DECT images, Li *et al* proposed a deep learning-based framework to yield high-energy CT images from measured low-energy images [32]. To achieve this, a cascade deep ConvNet (CD-ConvNet) is designed to learn a non-linear relationship between dual-energy CT images, with the low- and high-energy images as input and output, respectively. A total of 2121 CT slices and 334 slices were used to train and test the CD-ConvNet, respectively. The predicted high-energy CT images show close visual profiles with respect to the reference high-energy images. Qualitative evaluations using line profiles show the pseudo and reference high-energy CT images agree closely. Subsequent material decomposition using the synthetic DECT images shows water and bone equivalent fraction images are visually similar to that decomposed from reference DECT images [32].

Synthetic DECT images with diagnostic accuracy have been generated using a predenoising and difference learning approach [33, 34]. Specifically, to learn the inherent dual-energy relationship between the routine DECT images, a fully convolutional network (FCN) is first employed to reduce the noise of the training DECT data. Difference images between the denoised low- and high-energy CT images are then calculated. In the training phase, a U-Net-type deep learning model is trained using the denoised low-energy CT images and the difference images, with the former as input and the latter as output. In the testing phase, independent low-energy CT images which are unseen during the training phase are first denoised using the FCN and then sent into the predictive model to predict the corresponding difference images, and these difference images are finally added to the raw low-energy image in turn to yield the high-energy image counterpart. Contrast-enhanced abdominal CT scans are performed to evaluate the approach. The synthetic high-energy CT images are assessed using clinically relevant metrics including HU accuracy and noise power spectrum (NPS). The results show the synthetic CT images are highly consistent with the real 140 kV high-energy images, as shown in Figure 12.2. Quantitative assessments show the HU accuracy of the predicted images is within 3 HU for key organs including the aorta, liver, kidney, spine, and stomach [33]. More importantly, the synthetic images preserve the original noise texture in the raw CT images directly acquired from the CT scanners which is particularly desirable for clinical diagnosis [35].

There are two salient features for the predenoising and difference learning mechanisms. First, by predenoising before sending the images to learn the spectral relationship, the adverse influence of the image noise is mitigated when constructing the dual-energy mapping predictive model. Therefore, the intrinsic attenuation properties of the materials at low- and high-energy levels are learned with high fidelity, improving the robustness of the model. For comparison, a predictive model was also built by using images without denoising, and it was found that the resultant model yields many inferior images. Second, the model predicts the difference images between the low- and high-energy images instead of directly predicting the high-energy CT images from the input low-energy images. Since CT images are always calibrated using water images acquired using the corresponding scanning settings, the difference images between DECT images are able to show the prominent spectral information, mainly contributed by dense materials (such as bone and contrast). With the different images, the mapping network is enabled to be focused on these regions where there have distinct HU differences during the training phase. By doing so, the model not only yields accurate synthetic images but also correlates the noise of the synthetic images with the noise of the raw low-energy images.

The noise correlation between the synthetic DECT images is favorable for material decomposition. In popular DECT applications, virtual noncontrast (VNC) imaging shows a 7-fold noise reduction by using synthetic DECT images, compared to the standard DECT images [33]. This can be attributed to that conventional material decomposition based on matrix inversion yields amplified image noise, which can be significantly mitigated by the noise correlation.

The synthetic DECT images have also been subjected to virtual monochromatic (VM) imaging. Excellent agreements in terms of HU accuracy and spatial resolution were found

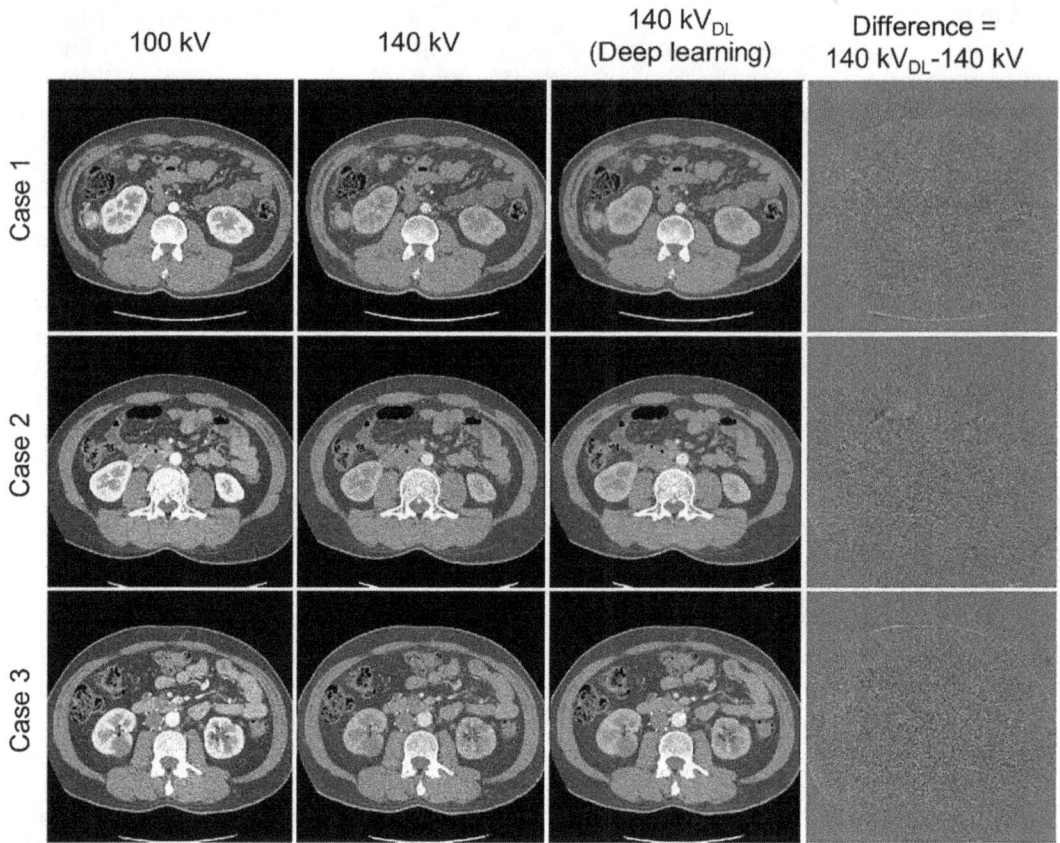

FIGURE 12.2
Raw 100 kV (1st column) and 140 kV (2nd column) images acquired from dual-source DECT scanners. The deep learning predicted 140 kV images show good consistent with the raw 140 kV images. Difference images between the predicted and raw 140 kV images provide marginal anatomical information of the patients, suggesting the HU accuracy and spatial resolution are well preserved in the predicted 140 kV images.

between the VM images reconstructed using the synthetic DECT images and those using the raw DECT images. Attributed to the noise correlation between the synthetic high-energy images and the raw low-energy images, a noise reduction of up to 68% and iodine contrast-to-noise improvement of up to 55% have been achieved as compared with that reconstructed using the raw DECT images [36].

More recently, by using the U-net model for conversions between different kV CT images, similar results have been reported by using brain DECT scans [37]. In radiotherapy, deep learning model predicted synthetic DECT images from SECT images have shown to be feasible for stopping power ratio (SPR) estimation [38]. Specifically, a residual attention generative adversarial network is proposed to synthesize twin-beam DECT images from images reconstructed from the twin-beam raw projection dataset by disregarding spectral differences. Residual blocks with attention gates enable the model to focus on the differences between the DECT images. The synthetic DECT images are then used to generate SPR maps with a dual-energy-based stoichiometric method. Comparison studies with SPR maps generated from routine twin-beam DECT images show reduced noise levels and artifacts by using synthetic DECT images. Dosimetric comparison using original DECT and synthetic DECT calculated dose demonstrates dose-volume histograms for the clinical target volume agree within 1%, suggesting the synthetic DECT images are clinically useful for proton radiotherapy [38].

Deep learning approaches are also used to directly generate VM images from conventional SECT (i.e., polychromatic photons from the single spectrum) images [39]. A modified residual neural network (ResNet) model is developed to map polychromatic images to VM images. A total of 3182 slices are used to evaluate the model and the results show the predicted VM images are very closely approximated that reconstructed using the routine DECT data (shown in Figure 12.3), with a relative error of less than 2% [39].

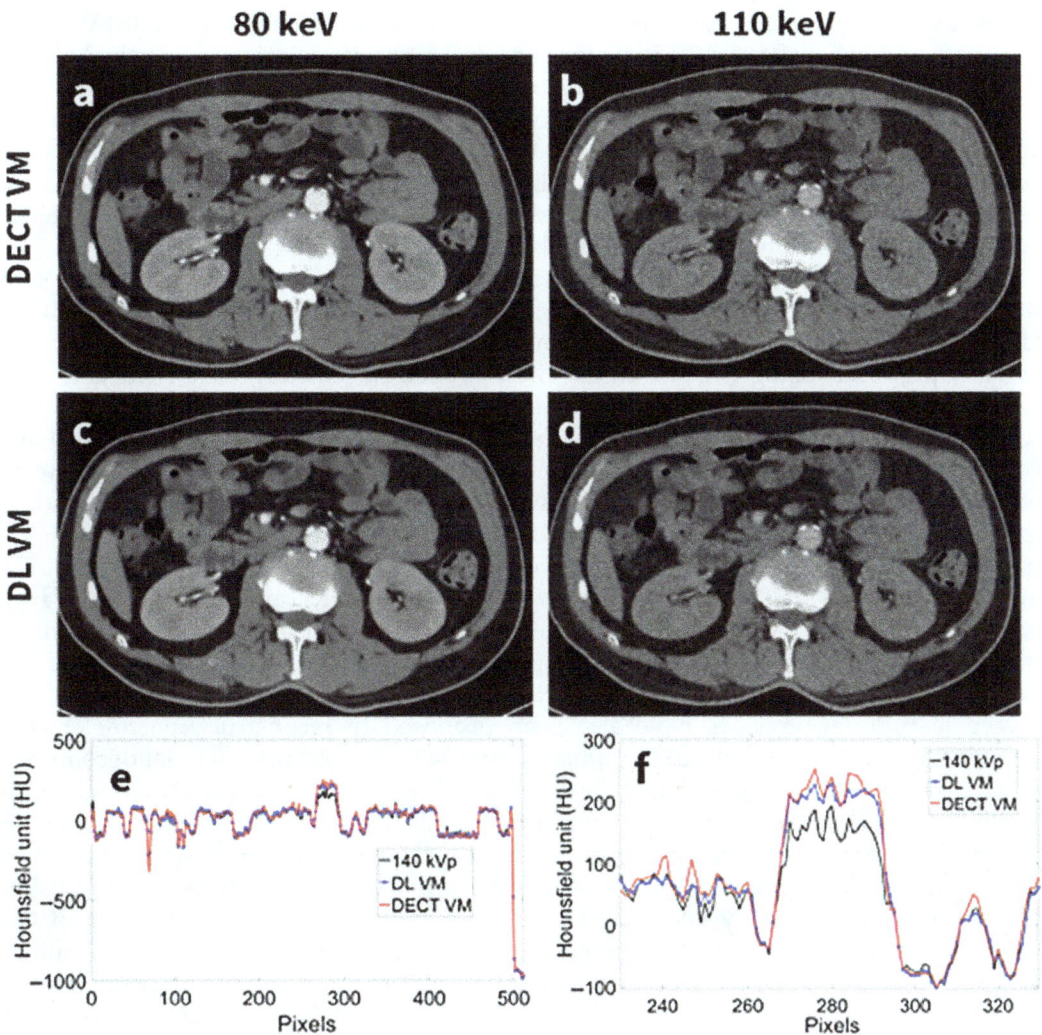

FIGURE 12.3
(a) and (b) VM images reconstructed using real 80/140 kV DECT images with monochromatic energy levels at 80 keV and 110 keV, respectively. (c) and (d) the corresponding VM images reconstructed at the same energy levels using SECT images via a deep learning approach. (e) Line profiles through the abdominal aorta in the 140 kV, and 80 keV VM images, (f) zoomed in the central region of the line profiles in (e), including the aorta. As can be seen, the deep learning predicted profiles match with the reference profiles quite well. (Reprint with permission from [39].)

12.5 Cross-Domain MECT Image Synthesis

Instead of synthesizing images in energy domain, deep learning approaches have also been proposed to synthesize material-specific images directly from the SECT images. Liao *et al* used the CD-ConvNet to obtain water and bone equivalent fractions from the input SECT images. Quantitative evaluations using feature similarity and root mean square deviation shows the predicted images are comparable to that reconstructed using the gold standard [32]. A recent study synthesizes bone and water images from a single kV CT image using a GAN-based CNN architecture. In addition, EAN images are directly predicted from 120 kV images. The HU difference between the synthetic material-specific images and that obtained from routine fast tube potential switching 80/140 kV DECT images are within 5.3 HU and 20.3 HU for bone and fat basis materials, respectively [40].

In the above studies, material decomposition is performed via the deep neural networks and conventional material decomposition procedures are not needed anymore. Conventional material decomposition usually yields amplified image noise because the spectral signals are subtracted while the noise is their summation during material decomposition in either the projection domain or the image domain. However, deep learning-based material-specific image synthesis is able to yield noise-mitigated images.

EAN is of interest to calculation SPR which is of utmost importance for treatment planning in particle therapy. Previous EAN calculation methods rely on the DECT or MECT scanners. A deep learning-based image synthetic framework for obtaining EAN images has been proposed using a CNN-based GAN model. Specifically, DECT images were acquired using the fast tube potential switching technology with tube potentials of 80 and 140 kV. The EAN and routinely used equivalent 120 kV SECT images for diagnosis have been reconstructed using the DECT images. The SECT images and the EAN images were then used together to train a GAN model, with the SECT and the EAN images as model input and output, respectively. The predicted EAN images were compared to the real EAN images in terms of mean absolute error (MAE), relative root mean square error (RMSE), structural similarity (SSIM), peak signal-to-noise ratio (PSNR), and so on. A difference between the predicted and real EAN images was reported within 9.7% in all regions of interest. This study demonstrates EAN images can be effectively synthesized from SECT, without extra DECT scans [41].

Deep learning approaches have also been proposed to reconstruct macro-elements (including H, C, N, O, P, and Ca) in human tissues [42]. Specifically, material decomposition was performed in image domain using the U-Net and VGG-based CNN models, with the SECT or DECT as input and density images of the six macro-elements as output. To generate density images of the macro-elements with ground truth for model training, virtual phantoms with predefined compositions of the macro-elements were scanned using polychromatic X-ray energy spectra. The results demonstrate the mass fraction of the macro-elements can be precisely predicted, even using SECT images [42]. A deep learning-based synthetic framework is proposed to map polychromatic projection measurement to monochromatic projection at a predefined energy level, which consequently yields VM images [43]. To take advantage of the information redundancy (such as anatomical structures) in the MECT images, a comprehensive domain network (CD-Net) has been developed to restore information for sparse-view sampling DECT data in both the projection domain and image domain. The CD-Net has been demonstrated to be able to reduce the sampling views to as low as 72 views, while providing promising images for VNC imaging and iodine contrast agent quantification [44].

12.6 Remaining Challenges and Future Work

While a few studies have shown both energy-specific and material-specific images can be generated via deep-learning approaches, few studies have used clinically relevant metrics to assess the images. On the contrary, metrics originating from computer vision (such as SSIM and PSNR) are extensively employed for image assessment. Since the medical images are usually not piece-wise constant and are with distinct noise textures, they are quite different from natural images. Metrics that are well-suited for natural images may not be suitable for medical image assessment [45]. For example, when using the SSIM or PSNR, we have no idea to what extent the resultant images are acceptable for clinical use or not. Because both the metrics cannot provide clinical meaning which is helpful for clinical diagnosis and treatment. Therefore, introducing clinically relevant metrics for synthetic image assessment is an urgent need, and clinicians should be involved in the assessment procedure, interpreting the results in a clinically meaningful way and finally approving the results.

Different from real DECT scanners which can acquire spectral data with various protocols. The spectral information of the synthetic images predicted using the deep learning models should only be consistent with the training data (i.e., the synthetic images are task-specifically generated and can only fulfill the predetermined imaging task). For example, if a model is trained using DECT images acquired with the 100 kV/140 kV energy setting, then it can only generate 140 kV images with an input of 100 kV images. Besides, data acquisition, image reconstruction parameters (such as kernel, and slice thickness), and patient positioning (such as off-centering) have a strong impact on the image spatial resolution and noise texture. Therefore, the testing data are expected to be acquired and reconstructed using the parameters used for training data.

The synthetic approaches still lack physical explanations. Since these approaches are mainly based on deep learning strategies which are usually considered to be "black boxes," it is challenging to provide physical interpretation for the synthetic approaches [46–49]. Therefore, it is rather hard to establish trust from physicists and clinicians. Meanwhile, these algorithms also lack transparency, which ensures routine quality assurance of effectiveness and accuracy (within the acceptance of clinical use). In clinical practice, CT images usually suffer from various artifacts and the deep learning-based approaches are vulnerable to these cases, especially severe metal artifacts present.

12.7 Summary

MECT imaging provides valuable information to enable new clinically relevant applications or to enhance the performance or to reduce radiation dose for routine SECT applications. Various MECT solutions have been developed and implemented. However, these high-end premium scanners are usually not affordable for undeveloped regions. With the dramatic advances in the past decade, deep learning approaches have been extensively investigated to synthesize spectral images from conventional SECT images. Although thorough evaluations still needed to be performed, the synthetic energy-specific and material-specific images show promising quality for predefined MECT imaging tasks. Compared to the MECT techniques via hardware implementation, the task-specific synthetic frameworks significantly alleviate the barrier for MECT applications in terms of cost and radiation dose. The deep learning-based synthetic strategies enable us to obtain energy- and material-specific images without paying the overhead cost of high-end premium MECT scanners and therefore offer a new paradigm of MECT imaging. Despite the challenges, looking forward, medical image synthesis via deep learning approaches has the potential to significantly change the field of MECT imaging.

Acknowledgment

This work was supported in part by the National Natural Science Foundation of China (No. 12175012).

References

1. McCollough, C.H., et al., Principles and applications of multienergy CT: report of AAPM Task Group 291. *Medical Physics*, 2020. **47**(7): pp. e881–e912.
2. McCollough, C.H., et al., Dual-and multi-energy CT: principles, technical approaches, and clinical applications. *Radiology*, 2015. **276**(3): p. 637.
3. Alvarez, R.E. and A. Macovski, Energy-selective reconstructions in x-ray computerised tomography. *Physics in Medicine & Biology*, 1976. **21**(5): p. 733.
4. Schlomka, J., et al., Experimental feasibility of multi-energy photon-counting K-edge imaging in pre-clinical computed tomography. *Physics in Medicine & Biology*, 2008. **53**(15): p. 4031.
5. Roessl, E. and R. Proksa, K-edge imaging in x-ray computed tomography using multi-bin photon counting detectors. *Physics in Medicine & Biology*, 2007. **52**(15): p. 4679.
6. Willemink, M.J., et al., Photon-counting CT: technical principles and clinical prospects. *Radiology*, 2018. **289**(2): pp. 293–312.
7. Kalender, W.A., et al., Evaluation of a prototype dual-energy computed tomographic apparatus. I. Phantom studies. *Medical Physics*, 1986. **13**(3): pp. 334–339.
8. Matsumoto, K., et al., Virtual monochromatic spectral imaging with fast kilovoltage switching: improved image quality as compared with that obtained with conventional 120-kVp CT. *Radiology*, 2011. **259**(1): pp. 257–262.
9. Silva, A.C., et al., Dual-energy (spectral) CT: applications in abdominal imaging. *Radiographics*, 2011. **31**(4): pp. 1031–1046.
10. Carmi, R., G. Naveh, and A. Altman, Material separation with dual-layer CT. in *IEEE Nuclear Science Symposium Conference Record, 2005*. 2005. IEEE.
11. Hao, J., et al., A novel image optimization method for dual-energy computed tomography. *Nuclear Instruments and Methods in Physics Research Section A: Accelerators, Spectrometers, Detectors and Associated Equipment*, 2013. **722**: pp. 34–42.
12. Johnson, T.R., et al., Material differentiation by dual energy CT: initial experience. *European Radiology*, 2007. **17**(6): pp. 1510–1517.
13. Rutt, B. and A. Fenster, Split-filter computed tomography: a simple technique for dual energy scanning. *Journal of Computer Assisted Tomography*, 1980. **4**(4): pp. 501–509.
14. Euler, A., et al., Initial results of a single-source dual-energy computed tomography technique using a split-filter: assessment of image quality, radiation dose, and accuracy of dual-energy applications in an in vitro and in vivo study. *Investigative Radiology*, 2016. **51**(8): pp. 491–498.

15. Gutjahr, R., et al., Human imaging with photon counting-based computed tomography at clinical dose levels. *Investigative Radiology*, 2016. **51**(1): pp. 421–429.

16. Yu, Z., et al., Noise performance of low-dose CT: comparison between an energy integrating detector and a photon counting detector using a whole-body research photon counting CT scanner. *Journal of Medical Imaging*, 2016. **3**(4): p. 043503.

17. Si-Mohamed, S., et al., Review of an initial experience with an experimental spectral photon-counting computed tomography system. *Nuclear Instruments and Methods in Physics Research Section A: Accelerators, Spectrometers, Detectors and Associated Equipment*, 2017. **873**: pp. 27–35.

18. Muenzel, D., et al., Spectral photon-counting CT: initial experience with dual–contrast agent K-edge colonography. *Radiology*, 2017. **283**(3): pp. 723–728.

19. Symons, R., et al., Photon-counting CT for simultaneous imaging of multiple contrast agents in the abdomen: an in vivo study. *Medical Physics*, 2017. **44**(10): pp. 5120–5127.

20. Bornefalk, H. and M. Danielsson, Photon-counting spectral computed tomography using silicon strip detectors: a feasibility study. *Physics in Medicine & Biology*, 2010. **55**(7): p. 1999.

21. Cormode, D.P., et al., Atherosclerotic plaque composition: analysis with multicolor CT and targeted gold nanoparticles. *Radiology*, 2010. **256**(3): p. 774.

22. Du, M., N. Liu, and X. Hu, Techniques for interpretable machine learning. *Communications of the ACM*, 2019. **63**(1): pp. 68–77.

23. Gu, J., et al., Recent advances in convolutional neural networks. *Pattern Recognition*, 2018. **77**: pp. 354–377.

24. Li, Z., et al., A survey of convolutional neural networks: analysis, applications, and prospects. *IEEE Transactions on Neural Networks and Learning Systems*, 2021.

25. Goodfellow, I., et al., Generative adversarial networks. *Communications of the ACM*, 2020. **63**(11): pp. 139–144.

26. Creswell, A., et al., Generative adversarial networks: an overview. *IEEE Signal Processing Magazine*, 2018. **35**(1): pp. 53–65.

27. Han, K., et al., A survey on vision transformer. *IEEE Transactions on Pattern Analysis and Machine Intelligence*, 2022.

28. Zhao, Wei, et al., "Markerless pancreatic tumor target localization enabled by deep learning." *International Journal of Radiation Oncology* Biology* Physics*, 2019a. **105**(2): pp. 432–439.

29. Zhao, Wei, et al., "Incorporating imaging information from deep neural network layers into image guided radiation therapy (IGRT)." *Radiotherapy and Oncology*, 2019b. **140**: pp. 167–174.

30. Shen, L., W. Zhao, and L. Xing, Patient-specific reconstruction of volumetric computed tomography images from a single projection view via deep learning. *Nature Biomedical Engineering*, 2019. **3**(11): pp. 880–888.

31. Lei, Y., et al., Deep learning-based real-time volumetric imaging for lung stereotactic body radiation therapy: a proof of concept study. *Physics in Medicine & Biology*, 2020. **65**(23): p. 235003.

32. Liao, Y., et al., Pseudo dual energy CT imaging using deep learning-based framework: basic material estimation. in *Medical Imaging 2018: Physics of Medical Imaging*. 2018. SPIE.

33. Zhao, W., et al., Obtaining dual-energy computed tomography (CT) information from a single-energy CT image for quantitative imaging analysis of living subjects by using deep learning. in *Pacific Symposium on Biocomputing 2020*. 2019. World Scientific.

34. Zhao, W., et al., A deep learning approach for dual-energy CT imaging using a single-energy CT data. in *15th International Meeting on Fully Three-Dimensional Image Reconstruction in Radiology and Nuclear Medicine*. 2019. SPIE.

35. Zhao, W., et al., Dual-energy Computed Tomography Imaging from Contrast-enhanced Single-energy Computed Tomography. *arXiv preprint*, arXiv:2010.13253, 2020.

36. Zhao, W., et al., A deep learning approach for virtual monochromatic spectral CT imaging with a standard single energy CT scanner. *arXiv preprint*, arXiv:2005.09859, 2020.

37. Cong, W., et al., Virtual monoenergetic CT imaging via deep learning. *Patterns*, 2020. **1**(8): p. 100128.

38. Liu, C.-K., et al., Generation of brain dual-energy CT from single-energy CT using deep learning. *Journal of Digital Imaging*, 2021. **34**(1): p. 149–161.

39. Charyyev, S., et al., Learning-based synthetic dual energy CT imaging from single energy CT for stopping power ratio calculation in proton radiation therapy. *The British Journal of Radiology*, 2022. **95**(1129): p. 20210644.

40. Kawahara, D., et al., Image synthesis with deep convolutional generative adversarial networks for material decomposition in dual-energy CT from a kilovoltage CT. *Computers in Biology and Medicine*, 2021. **128**: p. 104111.

41. Kawahara, D., et al., Image synthesis of effective atomic number images using a deep convolutional neural network-based generative adversarial network. *Reports of Practical Oncology and Radiotherapy*, 2022. 27: pp. 848–855.

42. Fujiwara, D., et al., Virtual computed-tomography system for deep-learning-based material decomposition. *Physics in Medicine & Biology*, 2022. **67**(15): p. 155008.

43. Cong, W., et al., Monochromatic image reconstruction via machine learning. *Machine Learning: Science and Technology*, 2021. **2**(2): p. 025032.

44. Zhang, Y., et al., CD-Net: comprehensive domain network with spectral complementary for DECT sparse-view reconstruction. *IEEE Transactions on Computational Imaging*, 2021. **7**: pp. 436–447.

45. Pambrun, J.-F. and R. Noumeir. Limitations of the SSIM quality metric in the context of diagnostic imaging. in *2015 IEEE International Conference on Image Processing (ICIP)*. 2015. IEEE.

46. Murdoch, W.J., et al., Definitions, methods, and applications in interpretable machine learning. *Proceedings of the National Academy of Sciences*, 2019. **116**(44): pp. 22071–22080.

47. Linardatos, P., V. Papastefanopoulos, and S. Kotsiantis, Explainable ai: a review of machine learning interpretability methods. *Entropy*, 2020. **23**(1): p. 18.

48. Lundberg, S.M., et al., From local explanations to global understanding with explainable AI for trees. *Nature Machine Intelligence*, 2020. **2**(1): p. 56–67.

49. Jiménez-Luna, J., F. Grisoni, and G. Schneider, Drug discovery with explainable artificial intelligence. *Nature Machine Intelligence*, 2020. **2**(10): pp. 573–584.

13

Metal Artifact Reduction

Zhicheng Zhang
JancsiTech, Hangzhou, China

Lingting Zhu
The University of Hong Kong, Hong Kong, China

Lei Xing
Stanford University, Stanford, CA, USA

Lequan Yu
The University of Hong Kong, Hong Kong, China

CONTENTS

DOI: 10.1201/9781003243458-16

13.1 Introduction

Medical imaging is widely used for disease diagnosis and treatment in modern medicine. High-quality medical images help clinicians to make a more accurate diagnosis. However, mechanical accuracy, reconstruction error, partial implants, etc., will degrade the image quality in clinical applications and affect the diagnosis and treatment accuracy. For example, during computed tomography (CT) and magnetic resonance imaging (MRI), some patients will carry metal implants, which are highly attenuated metal implants, and these metal implants can lead to defective X-ray projections or K-space raw data. Strong metal artifacts will appear after the reconstruction of medical images, affecting the visual diagnosis of medical images and the radiotherapy dose calculation [1, 2]. In the last decades, a number of methods for metal artifact removal (MAR) have been developed. However, the big gap between in-house research and actual clinical practice is unavoidable [3]. Therefore, how to restore the information in the metal regions is still an important yet challenging problem in the field of CT and MRI [4]. Recently, deep learning techniques, especially deep generative models, have played an important role in MAR. In this chapter, we will first make a brief overview of current MAR problems in CT and MRI, and then we will introduce our recent efforts in applying deep learning for metal artifact reduction in CT imaging, including the supervised and self-supervised deep learning methods.

13.2 Review of MAR in Different Modalities

In this section, we will go through some basic methods of MAR in different modalities, including MRI and CT.

13.2.1 MAR in Magnetic Resonance Imaging (MRI)

In general, metals can cause various artifacts in MRI, including signal loss, signal accumulation, image distortion, fat suppression failure, and ineffective signal nullification, which are derived from differences in susceptibility between the metals and adjacent soft tissue [5]. For ferromagnetic substances (such as cobalt, iron, and nickel), compared with paramagnetic material, the variation of local magnetic field around metals is more obvious.

To alleviate metal artifacts, there are several strategies for MR imaging: 1) utilizing lower magnetic fields; 2) choosing specific phases and frequency encoding directions; 3) employing high radio frequency pulse bandwidth; and 4) using special MAR sequences [5, 6]. At the same time, these strategies have their limitation of lower signal-to-noise and long scanning time [7, 8]. For advanced sequences, some techniques use bipolar spin echo images to obtain two distorted images and then estimate the inhomogeneity map of the field. Based on these maps, we can numerically reduce the distortion of the metal object image [9]. In addition, [10] proposed to remove the residual pile-up and ripple artifacts in a short scanning time via estimating off-resonance frequency maps from spectral-bin images with two distorted images.

With the assistance of bipolar readout gradients, U-Net is used to decrease the off-resonance frequencies related artifacts [11]. Furthermore, Kim et al. [12] adopted the attention mechanism to reduce the residual ripple and pile-up artifacts around metal implants. In addition to removing metal artifacts directly, we can accelerate the MAR sequence indirectly to achieve the effect of MAR. Seo et al. [13] used a neural network to decrease the scanning time of the SEMAC sequence instead of directly applying deep learning to MRI artifact correction. Due to the difficulty of data collection, unsupervised learning was employed for MAR [14]. This method used a deep neural network with the assistance of an MRI physics-based image generation module to estimate the frequency shift maps between two distorted images in the training stage. In the testing stage, the MRI physical image generation module was employed to obtain the final high-quality images.

13.2.2 MAR in Computed Tomography (CT)

Considering the structural and non-local nature of metal artifacts in CT images, the previous methods have carried out MAR from the raw projection. In these works, the metal-corrupted projections are either corrected by modeling the physical effects of CT imaging [15–17] or replaced by estimated values [18–21]. Linear interpolation (LI) [1] considers the sinogram area affected by the metal as missing data and directly replaces the missing region with the linear interpolation of the adjacent unaffected projection. However, the error from the LI often leads to strong secondary artifacts in the reconstructed CT image. To this end, some prior-image-based interpolation methods [22–24] were proposed. The key idea is to use the prior sinogram from the metal artifact image to guide the sinogram interpolation. For example, [21] proposed using the multi-threshold segmentation method to obtain prior images, normalize the projection data, and then interpolate.

Deep convolutional neural networks (DCNNs) have achieved promising results in medical image reconstruction and analysis [25–31]. Recently, CNN has been taken into account to solve the MAR problem [32–37]. Park et al. [4] corrected sinogram inconsistency with U-Net [38], and Gjesteby et al. [39] improved the normalized metal artifact reduction (NMAR) [21] approach in the sinogram domain. In addition to solving MAR in the sinogram domain using deep learning, there are some works on reducing metal artifacts through image post-processing [39–41]. For example, Zhang et al. [42] generate a prior image with reduced artifacts from raw and other methods corrected images to help correct areas of metal damage in sinograms. Wang et al. [43] employed a conditional generative adversarial network (cGAN) [44] for MAR in CT images of the ear. Huang et al. [45] used a novel CNN-based residual learning strategy in cervical CT images. Additionally, perceptual loss is used [41] to improve the quality of deep-learning-based MAR, where residual learning is also adopted in the framework. Due to the success of deep learning-based MAR in the image domain, combining deep learning techniques with traditional MAR techniques attracted more attention to fusing the merits of different approaches. Zhang et al. [42] used raw data and initial corrected CT images to generate a high-quality prior image, with which we can correct metal artifacts.

Considering the lack of a high-quality paired training dataset, most of these methods employed the simulated training dataset from metal-free CT images. Consequently, the performance of these supervised methods would decrease in different site data or real clinical practice, as the simulated metal artifacts may have difference with that in the real clinical scenarios [36]. Recent work has focused on developing MAR methods that do not utilize synthetic metal artifacts. Ghani et al. [33] considered the metal-related projection pixels as missing data, which can be completed using a self-supervised sinogram inpainting framework. Furthermore, an unsupervised disentanglement network is proposed to decompose the metal artifacts from CT images and all operators are in the latent feature space [36].

For medical imaging devices, high-performance hardware will result in high-quality medical images and derive more accurate diagnoses. However, any defect in the hardware will induce the corresponding artifact, as described in this work for metal artifacts. We should analyze each case on a case-by-case basis according to our technical background and dataset.

13.3 Supervised Dual-domain Learning for MAR

In this section, we will introduce novel supervised deep-learning-based MAR methods for CT imaging.

13.3.1 Overview of Dual-domain Learning Framework

For complex metal artifacts, in the era of big data, the most intuitive way is to prepare a high-quality paired training dataset and train a well-designed neural network. To this end, we develop a generalizable framework for MAR in CT images. In this framework, we jointly utilize sinogram and image domain information by combining the data-driven advantage of deep learning and the generality of conventional MAR approaches. Following the previous works of image restoration, we treat the MAR as the deep-learning-based sinogram completion. Therefore, we use a deep neural network, i.e., SinoNet, to complete the missing parts, which are within the metal trace region in the projection domain.

As we know, the pixel-wise error will be reflected in the global area of the filtered back-prorogation (FBP)-reconstructed CT image. To ensure the quality of the restored sinogram, we design an image-based neural network, i.e., PriorNet, which introduces the prior information into the sinogram completion task and produces a well-estimated prior image with less metal artifact. With the guidance from the forward projection of the prior image (i.e., prior sinogram), the SinoNet can achieve good completion results. In this work, we adopt an end-to-end strategy to train both two neural networks. In this case, the prior image generation and sinogram completion tasks can benefit from each other. The following Figure 13.1 shows the architecture of the proposed framework. To better utilize the prior

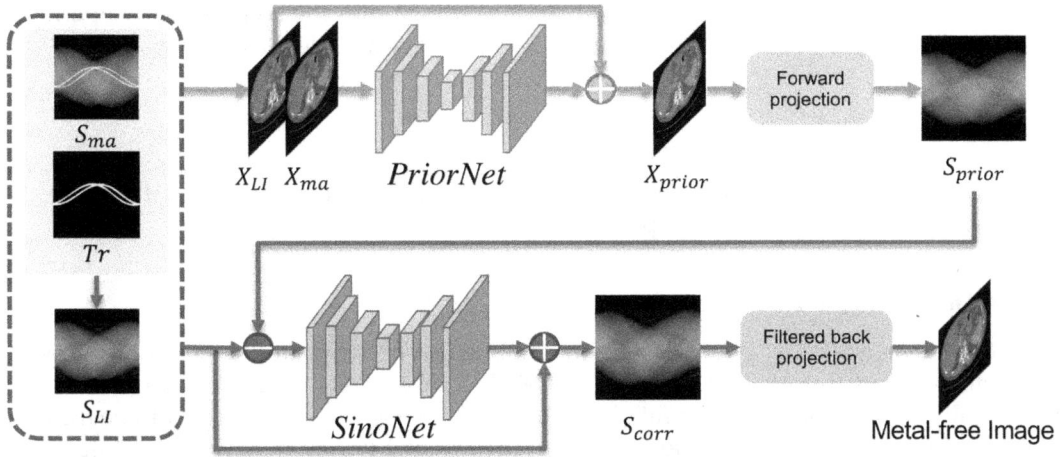

FIGURE 13.1
Schematic diagram of our proposed framework.

sinogram guidance, we further adopted a residual-sinogram-learning strategy by refining the generated prior sinogram, not directly regressing the missing projections. Finally, we adopt the conventional FBP approaches to reconstruct the CT images from the completed sinogram.

Before introducing the whole framework in detail, we should depict all the involved notations in this work. We assume that the original metal-corrupted sinogram is represented as $S_{ma} \in R^{H \times W}$ and the metal trace mask is represented as $Tr \in \{0, 1\}^{H \times W}$. To obtain a better initial estimation in the metal trace region, we first apply the LI [1] to predict the missing area and acquire the LI-corrected sinogram S_{LI} for the following procedures. After the PriorNet, a prior image X_{prior} with less metal artifact and a prior sinogram S_{prior} from the generated prior image X_{prior} can be obtained. After the SinoNet, the corrected sinogram S_{corr} is obtained. $\odot$ denotes element-wise multiplication. $\mathcal{P}^{-1}$ is the FBP operator and M represents the metal mask.

13.3.2 Generation of the Prior Image

In the proposed framework, we believe that the closer the prior image is to the ground truth, the better the sinogram completion is. To this end, we design a U-Net-based PriorNet with half-channel numbers to generate a prior image for the sinogram completion procedure. We treat the original metal-corrupted CT image as input and train the PriorNet to obtain the prior image with fewer metal artifacts. Note that the strength of metal artifacts in the image is closely related to the size of the metal. Therefore, besides the original CT image, we also take the LI-corrected image as the input for the PriorNet with the residual learning strategy. Specifically, we first use S_{ma} to get the original metal-corrupted image X_{ma} and then LI-corrected CT image X_{LI} is also reconstructed from S_{LI}. After that, the artifact-reduced prior image can be represented as

$$X_{\text{prior}} = X_{LI} + f_P\left([X_{ma}, X_{LI}]\right),$$

where f_p denotes the function of PriorNet, and $[a_1, a_2]$ denotes the concatenation operation of image a_1 and a_2. To optimize the PriorNet, we employ the L_1 loss as follow:

$$\mathcal{L}_{\text{prior}} = \left\| X_{\text{prior}} - X_{gt} \right\|_1.$$

The prior sinogram S_{prior} from the prior image X_{prior} is further employed to guide the sinogram-domain network to complete the missing projections.

13.3.3 Deep Sinogram Completion

With the assistance of S_{prior}, we design another neural network in the sinogram domain, i.e., SinoNet, and train it to restore the unreliable projections in the metal trace region. Specifically, S_{LI}, S_{prior}, and Tr are treated as the input of SinoNet, and then we utilize the contextual information of the sinogram to output the missing part in the metal trace region Tr. To improve the continuity of the predicted projections at the boundary of the metal trace region, we employ a residual learning strategy to refine the residual-sinogram between S_{LI} and S_{prior}. Particularly, to obtain a smooth transition between S_{prior} and S_{LI}, we acquire the residual-sinogram map S_{res} and then use the SinoNet to further refine the residual projections within Tr. The corrected sinogram S'_{corr} can be written as:

$$S'_{\text{corr}} = f_S \left(\left[S_{\text{prior}} - S_{LI}, Tr \right] \right) + S_{LI},$$

where f_S denotes the sinogram completion network, i.e., SinoNet. Instead of the direct estimation of the projection pixel, the SinoNet estimates the residual values, which can alleviate the discontinuity at the boundary of Tr [46]. After that, we composite the output of SinoNet and S_{LI} w.r.t. Tr to obtain the final corrected sinogram result:

$$S_{\text{corr}} = S'_{\text{corr}} \odot Tr + S_{LI} \odot (1 - Tr) = f_S \left(\left[S_{\text{prior}} - S_{LI}, Tr \right] \right) \odot Tr + S_{LI}.$$

Due to the small metal trace regions within the sinogram, we use the mask pyramid U-Net [35] to explicitly retain the metal trace information in each layer, avoiding the direct concatenation of the remaining residual-sinogram map and Tr as network input. Therefore, the SinoNet can learn more discriminative features to recover the missing projections in the metal trace area.

In this part, we use the L_1 loss to optimize the SinoNet. We only minimize the metal trace region difference between S_{corr} and the ground truth S_{gt}. Due to the small metal trace regions, we also use L_1 loss to make S'_{corr} close to S_{gt} to accelerate the training speed. In this way, the loss function can also supervise those network outputs, which are outside the metal trace region. The objective of training the SinoNet is

$$\mathcal{L}_{\text{sino}} = \left\| S_{gt} - S_{\text{corr}} \right\|_1 + \beta \left\| S_{gt} - S'_{\text{corr}} \right\|_1,$$

where β is a hyper-parameter to balance the two different items. In our preliminary experiments, we found that it is not sensitive to our framework performance, and we set it as 0.1 in our experiments.

13.3.4 Objective Function

Generally, with the assistance of prior sinogram S_{prior}, we can reduce most metal artifacts in the image domain and the SinoNet can produce a better corrected sinogram. However, the L_1 loss function only works in the sinogram domain, ignoring the geometry-consistency of CT. In our framework, the final CT image is FBP-reconstructed from the S_{corr}. Therefore, we further use an image loss with FBP to avoid the new artifacts in the image domain,

$$\mathcal{L}_{FBP} = \left\| \left(\mathcal{P}^{-1}\left(S_{\text{corr}}\right) - X_{gt} \right) \odot \left(1 - M\right) \right\|_1.$$

Here, we use the masked L_1 loss to limit the pixel-wise difference of the non-metal regions in CT images. With the assistance of the differentiable FBP operation $\mathcal{P}^{-1}$, the gradient of $\mathcal{L}_{FBP}$ can back-propagate to SinoNet, generating geometry-consistent completion results. We train both the PriorNet and SinoNet at the same time in an end-to-end manner. The whole objective function can be represented as

$$\mathcal{L}_{\text{total}} = \mathcal{L}_{\text{prior}} + \alpha_1 \mathcal{L}_{\text{sino}} + \alpha_2 \mathcal{L}_{FBP},$$

where α_1 and α_2 are the weighting hyperparameters. We empirically set them as 1.0 in our experiments.

13.3.5 Dataset and Implementation

Considering the lack of a real supervised training dataset, we used simulated metal artifacts to train and evaluate our method. For the simulation data, we synthesized metal artifacts based on the random subset of the recently released DeepLesion dataset [47]. We followed the previous work in [6] for the simulated metal masks, containing 100 manually segmented metal implants of different sizes and shapes. To be specific, 90 metal masks and 1000 CT images were chosen to synthesize the training data. 2000 combinations of the remaining ten metal masks and 200 CT images from 12 patients were used for network evaluation. To evaluate the generalization of the proposed network over different site data, we also evaluated the trained network on head CT images with the 1860 combinations of 186 head CT slices and paired them with ten metal shapes, since the CT slices of DeepLesion dataset are mainly in the abdomen and thorax.

In this work, we first insert metallic implants into clean CT images and then simulate the metal-corrupted sinograms and CT images with the integration of simulated Poisson noise and beam hardening. We use a polychromatic X-ray source and assume the incident X-ray has 2×10^7 photons with the consideration of the partial volume effect. We use a fan-beam geometry that sampled 640 views uniformly between 0 and 360 degrees. We resize the CT images to 416×416 before the simulation and obtain the final simulated of size 641×640.

We take the original metal-corrupted sinogram and metal trace as the network input. We use the simulated data to train the whole framework with the assistance of the metal trace Tr. In the testing phase, given the metal-affected sinogram S_{ma}, we first obtain the metal mask M from X_{ma} with a thresholding method or other advanced segmentation algorithms, and then apply the forward projection to get the metal trace Tr. We implement the whole framework in Python with PyTorch [48] deep learning library. We adopt the Operator Discretization Library (ODL) to simulate the sinogram from CT images and utilize the

PyTorch wrapper of the ODL library to implement the differential FBP operation in the network. The PriorNet and SinoNet are trained in an end-to-end scheme. The Adam optimizer [49] is adopted to optimize the whole framework, where $(\beta_1, \beta_2) = (0.5, 0.999)$. The training epoch is set at 400 with batch size of 8 and the learning rate is set as $1e^{-4}$. The whole framework is trained on one NVIDIA 1080Ti GPU.

13.3.6 Experimental Results on DeepLesion Data

13.3.6.1 Quantitative Comparisons

We compare our approach with conventional interpolation-based methods, including LI [1] and NMAR [21]. Also, we compare our approach with deep learning-based methods, including CNNMAR [42], cGANMAR [43], and DuDoNet [34]. Note that we re-implemented cGANMAR and DuDoNet due to the lack of public implementations. Table 13.1 shows the quantitative results of the DeepLesion dataset. Regarding both the root mean square error (RMSE) and structured similarity index (SSIM) metrics, we can see that NMAR outperforms LI approach with the prior image information. On the other hand, compared with traditional MAR methods, CNNMAR and cGANMAR achieve lower RMSE and higher SSIM, demonstrating the advantages of deep learning in MAR. DuDoNet has better RMSE and SSIM performance than cGANMAR due to the integration of artifacts removal in the sinogram domain before image refinement. Compared with DuDoNet, our method further decreases RMSE by 6.85 HU and achieves slightly better SSIM. Since we use deep networks for sinogram completion and train the entire framework in an end-to-end manner, our approach outperforms the CNNMAR approach. Overall, in terms of RMSE and SSIM, our framework outperforms among different methods, demonstrating the effectiveness of our approach in MAR. Note that we should compare the relative values of SSIM since the absolute values of SSIM vary according to different window sizes.

13.3.6.2 Qualitative Analysis

Figure 13.2 and Figure 13.3 show the visual inspections on the DeepLesion simulation data. For better visualization, we color the simulated metal masks in red. Severe striped products have a serious dark zone between two metal implants in the original metal image (see Figure 13.2(B1)). For the small metal implants (Figure 13.2), compared with CNNMAR, DuDoNet and cGANMAR can achieve better visual effects. However, some slight artifacts still exist in DuDoNet and cGANMAR (marked by the dashed blue ovals in Figure 13.2 (F3, G3)). Compared with all the other methods, our method effectively obtains the best performance of MAR and the fine structure retention. Figure 13.3 shows the results from

TABLE 13.1

Quantitative Comparison of Different Methods on DeepLesion Dataset

Method	RMSE(HU)	SSIM
LI	50.31 ± 19.41	0.9455 ± 0.0315
NMAR	47.03 ± 20.67	0.9594 ± 0.0299
CNNMAR	43.27 ± 14.44	0.9706 ± 0.0159
cGANMAR	39.01 ± 12.66	0.9754 ± 0.0055
DuDoNet	38.00 ± 13.31	0.9766 ± 0.0072
Ours	**31.15 ± 5.81**	**0.9784 ± 0.0048**

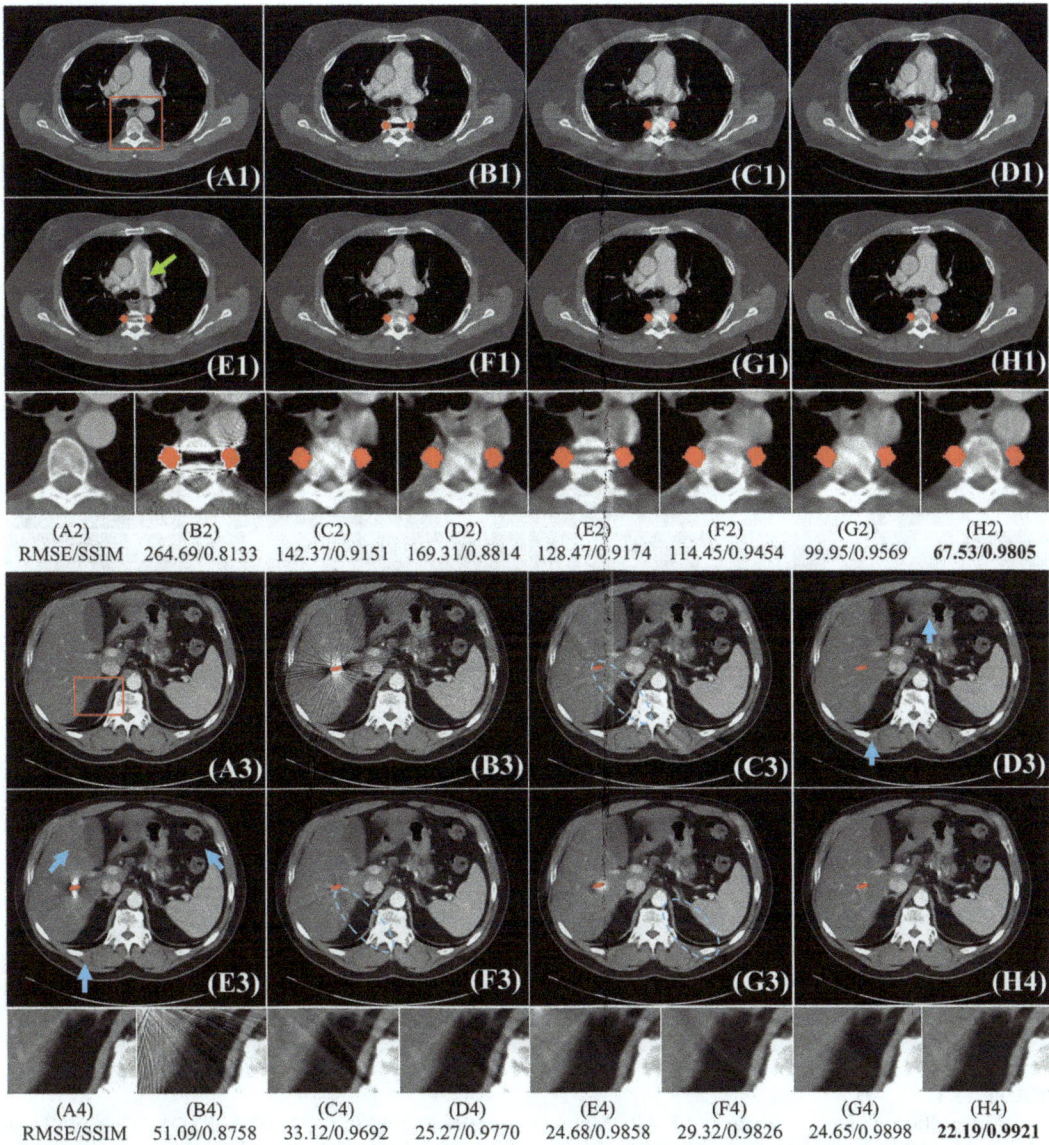

FIGURE 13.2

Visual inspection. Reference images (A1–A4). Metal images (B1–B4). LI (C1–C4), NMAR (D1–D4), CNNMAR (E1–E4), cGANMAR (F1–F4), DuDoNet (G1–G4), and our method (H1–H4). The first and second sample display Windows are [−480, 560] and [−175, 275] HU, respectively. Quantitative results were calculated using ROI marked by the red rectangle for better comparison. The simulated metal masks are painted in red for better visualization.

the large metal implants. The interpolation-based methods LI and NMAR and the image domain-based methods cGANMAR and DuDoNet lose the details of the original structure. Also, there are some additional secondary artifacts for the cGANMAR results (marked by the green arrows in Figure 13.3 (F1, F2)). Although the CNNMAR method preserves the structure to a certain extent, our method can retain more structural details due to the deep sinogram completion. Besides the qualitative comparison, we calculate the RMSE and SSIM metrics for the region of interest region (ROI) marked by the red box in Figure

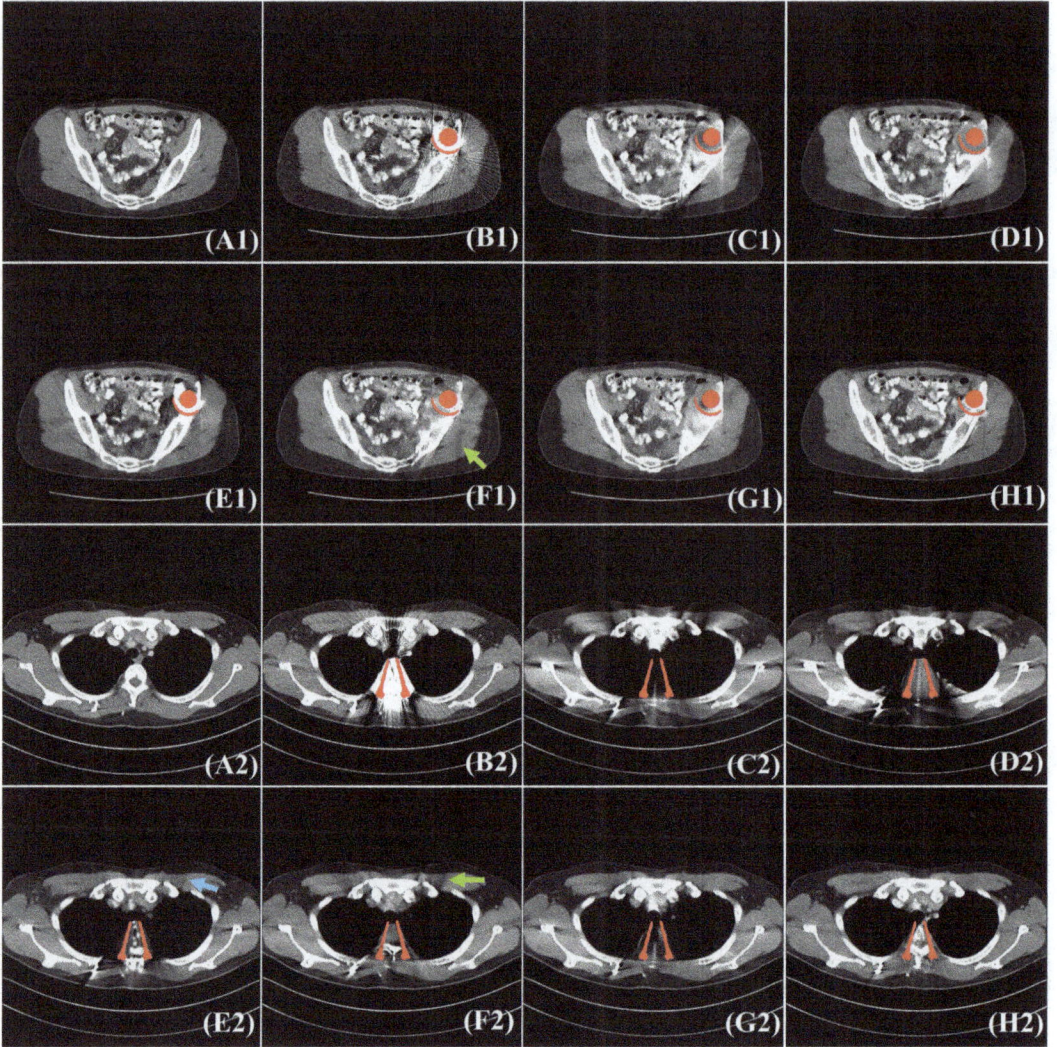

FIGURE 13.3
Visual inspection. Reference images (A1–A2), Metal images (B1–B2), LI (C1–C2), NMAR (D1–D2), CNNMAR (E1–E2), cGANMAR (F1–F2), DuDoNet (G1–G2), and our method (H1–H2). The display window is [−175, 275] HU.

13.2. As shown in Figure 13.2 (A2-H2, A4-H4), our method procedures the lowest RMSE values and highest SSIM values compare with all the other methods.

13.3.7 Generalization to Different Site Data

In the above experiments, the CT images selected from the DeepLesion dataset are abdominal and chest samples. To show the feasibility of applying our method to different organs, we directly evaluate the well-trained model with DeepLesion data on head CT images with simulated metal artifacts, where the head CT images are collected from the website. Figure 13.4 shows the visual results of our method. We observed that both LI and NMAR

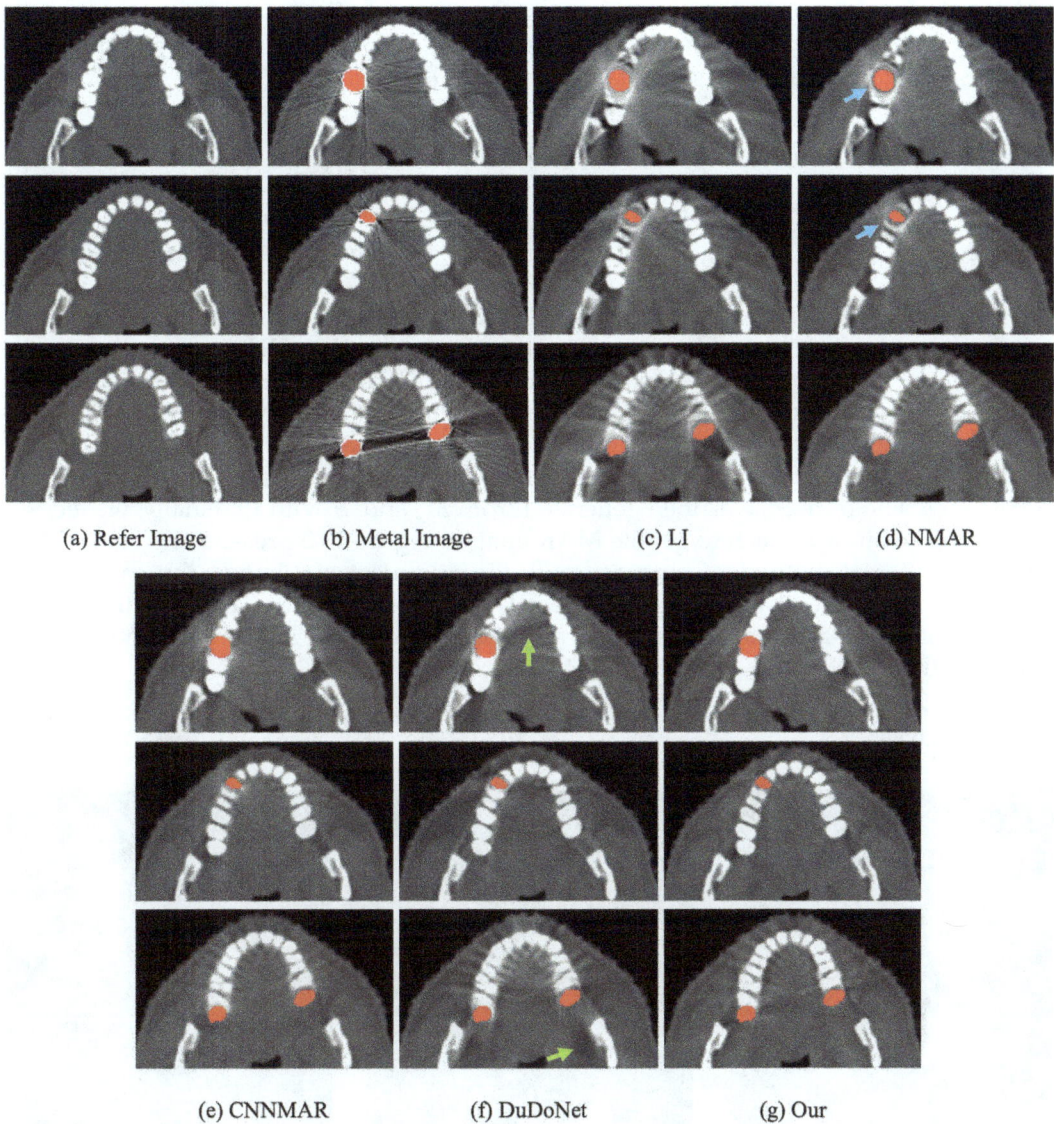

FIGURE 13.4
Visual results on head CT images with different numbers of simulated dental fillings. The simulated dental fillings are shown in red for better visualization. The display window is [−1000, 1600] HU.

will introduce some new secondary artifacts and can alter the anatomical structure of the teeth (marked by blue arrows in Figure 13.4(d)). The DuDoNet can reduce the artifacts, while several shading artifacts are still existed (marked by green arrows in Figure 13.4(f)). Although head CT images are not used for training, the excellent MAR performance indicates the potential of the proposed method to process data from different site data. Note that the MAR results of our method are comparable to those of CNNMAR trained with head CT images. LI and NMAR could not remove strong artifacts because of the high attenuation values of the teeth. Since the network is mainly trained with abdominal and chest CT images, image domain learning (i.e., previous image results) can only reduce some artifacts on the unseen head CT images. Our method effectively reduces most of the artifacts on the new head CT images, demonstrating the potential of the proposed method to process data from different site data.

13.3.8 Experiments on CT Images with Real Metal Artifacts

13.3.8.1 Results on Real Metal-Corrupted CT Images

Because of the lack of the original metal-corrupted sinogram data in the real clinical scenario, we evaluate our approach to clinical metal-corrupted CT images, following the previous work [34]. To be specific, some clinical metal-corrupted CT images were prepared, and a simple threshold segmentation method (i.e., 2000 HU in our experiments) was employed to segment metal masks from clinical CT images. We further employ the forward projection of the metal masks to generate the metal projection. We treat the projection pixels with a value greater than zero as the metal trace region Tr. We also acquire the metal-corrupted sinogram S_{ma} from the clinical CT image with the same geometry. The LI-corrected sinogram S_{LI} was then generated from S_{ma} and Tr with LI. Finally, we fed S_{ma}, S_{LI} and Tr into our approach to get the MAR images. Figure 13.5 presents the visual MAR results of different methods. Compared with the original metal image, this method can effectively reduce metal artifacts. From the enlarged yellow patches, other methods have changed some fine-grained anatomical structures in the original image, while our method can retain them well.

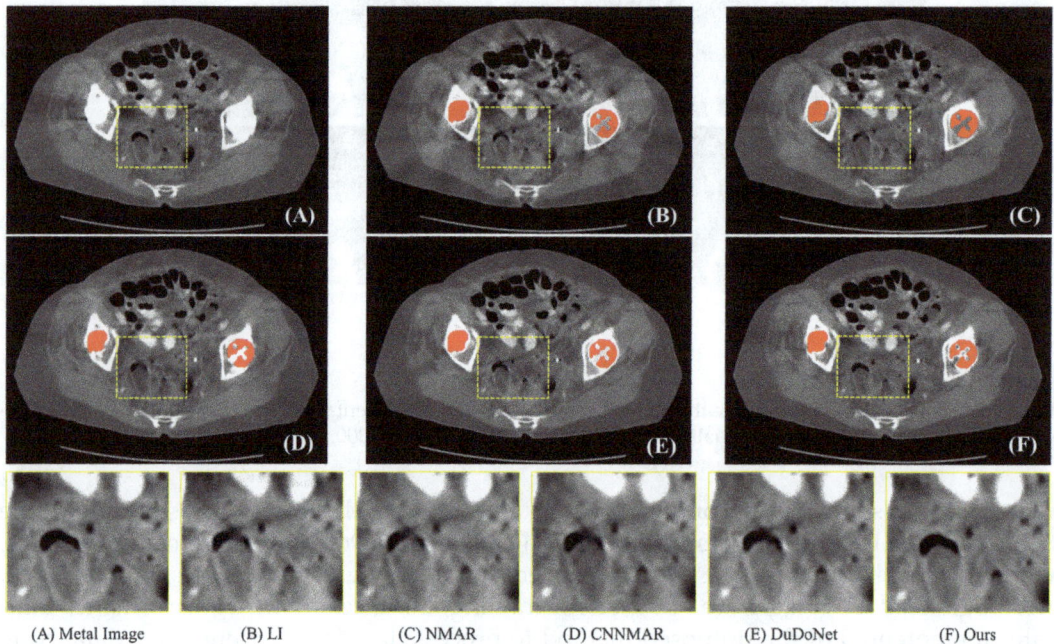

(A) Metal Image   (B) LI   (C) NMAR   (D) CNNMAR   (E) DuDoNet   (F) Ours

FIGURE 13.5
Visual inspection on real metal-corrupted CT images with real metal artifacts. The segmented metals are colored in red for better visualization. The display window of the whole image is [−480, 560] HU, and the display window of cropped patches is [−400, 300] HU.

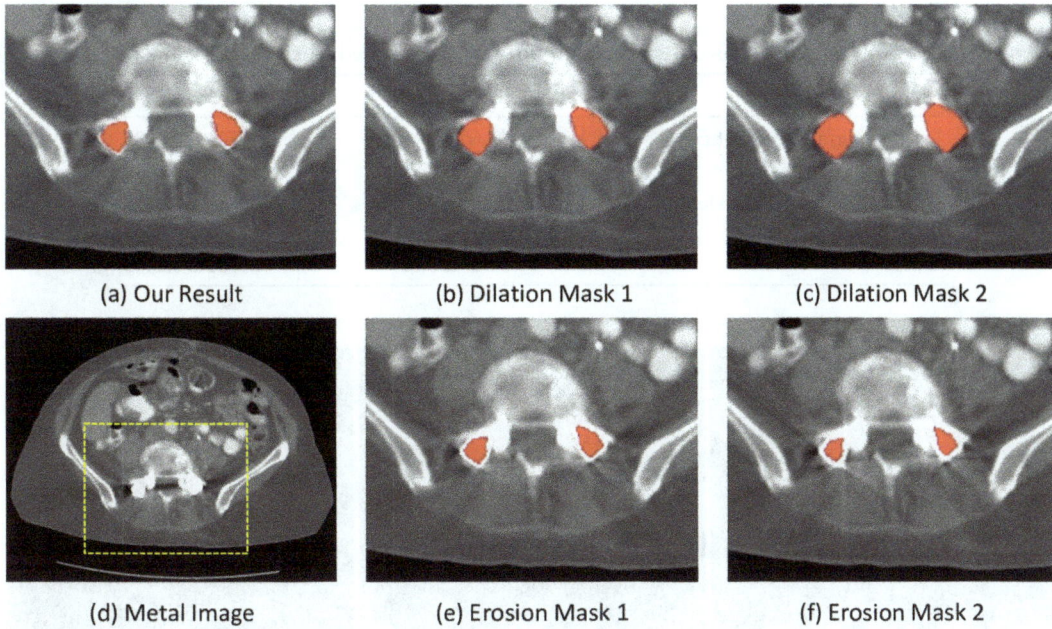

(a) Our Result (b) Dilation Mask 1 (c) Dilation Mask 2

(d) Metal Image (e) Erosion Mask 1 (f) Erosion Mask 2

FIGURE 13.6

Ablation study on different segmented metal masks. Thresholding-based metal segmentation (a); Over-segmented masks (b, c) Undersegmented masks (e, f). The display window is [−480 560] HU.

13.3.8.2 The Influence of Metal Segmentation

In our framework, an important factor in ensuring good performance is to obtain an accurate metal trace mask. In practice, we can use manual segmentation to obtain the metal. To study the influence of different metal segmentation masks, we adapt our approach to MAR with different metal traces generated from different metal masks as input. Figure 13.6 shows the MAR results of our approach by feeding the original thresholding-based metal mask (a), the dilation metal masks (b, c), and erosion metal masks (e, f). Overall, our method can obtain acceptable MAR results with the small metal segmentation error. With the consideration of different metal traces in the training stage, our method can predict missing projection pixels and obtain better reconstructed CT images, despite additional shading artifacts in the CT images. Given that the undersegmented metal trace is usually narrower than the original metal trace, the SinoNet will only complete some parts of the projection values of the original metal trace domain and keep some unreliable projection data, leading to some residual streaking artifacts.

13.3.9 Analysis of Our Approach

13.3.9.1 Effectiveness of Prior Image Generation

To verify the validity of the PriorNet, we directly train another neural network to conduct sinogram completion without the prior sinogram as guidance. Table 13.2 shows the results of a DeepLesion simulated dataset. It is observed that the performance of this strategy (deep sinogram completion) is inferior to our method, showing the effectiveness of the previous image generation process. In Figure 13.7, we can see that the original metal

TABLE 13.2

Quantitative Analysis of Analytical Studies on Simulated DeepLesion Dataset

Method	RMSE(HU)	SSIM
Deep sinogram completion	43.65 ± 17.61	0.9720 ± 0.0082
With tissue processing	35.64 ± 7.91	0.9768 ± 0.0047
w/o residual-sinogram-learning	31.86 ± 4.69	0.9781 ± 0.0048
Only metal image	31.60 ± 4.95	0.9778 ± 0.0048
Ours	**31.15 ± 5.81**	**0.9784 ± 0.0048**

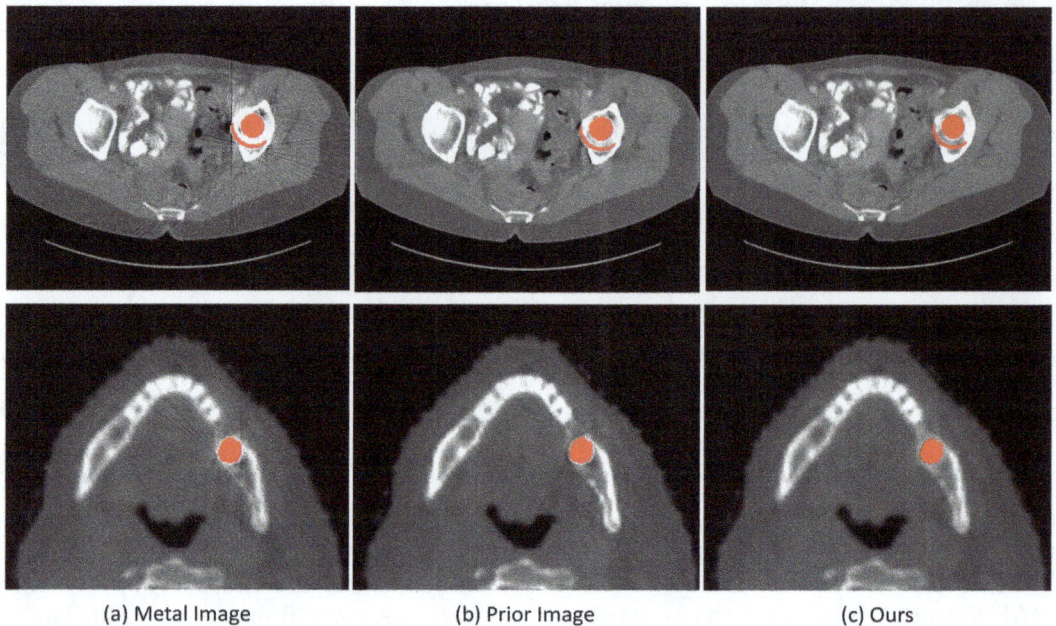

(a) Metal Image   (b) Prior Image   (c) Ours

FIGURE 13.7
Illustration of the generated prior image. The display window of the first row and second row are [−480, 560] HU and [−1000, 1600] HU, respectively.

image (Figure 13.7(a)) has more artifacts than the prior images (Figure 13.7(b)). Our results (Figure 13.7(c)) outperform others. We also train our approach with only the original metal image as the input of PriorNet. From Table 13.2, it is observed that this method (only metal images) produces slightly worse performances on simulated datasets than our method, demonstrating that using the LI-corrected image as input makes it more convenient to generate the previous image.

13.3.9.2 Effectiveness of Residual-Sinogram-Learning

To verify the validity of the residual-sinogram-learning strategy in Figure 13.8, we conduct another experiment without the proposed residual-sinogram-learning strategy. The first and second rows show the simulated metal artifact image and the real clinical CT image, respectively. Without residual-sinogram learning, SinoNet directly treats the prior sinogram and metal trace mask as the network input and generates the refined projection.

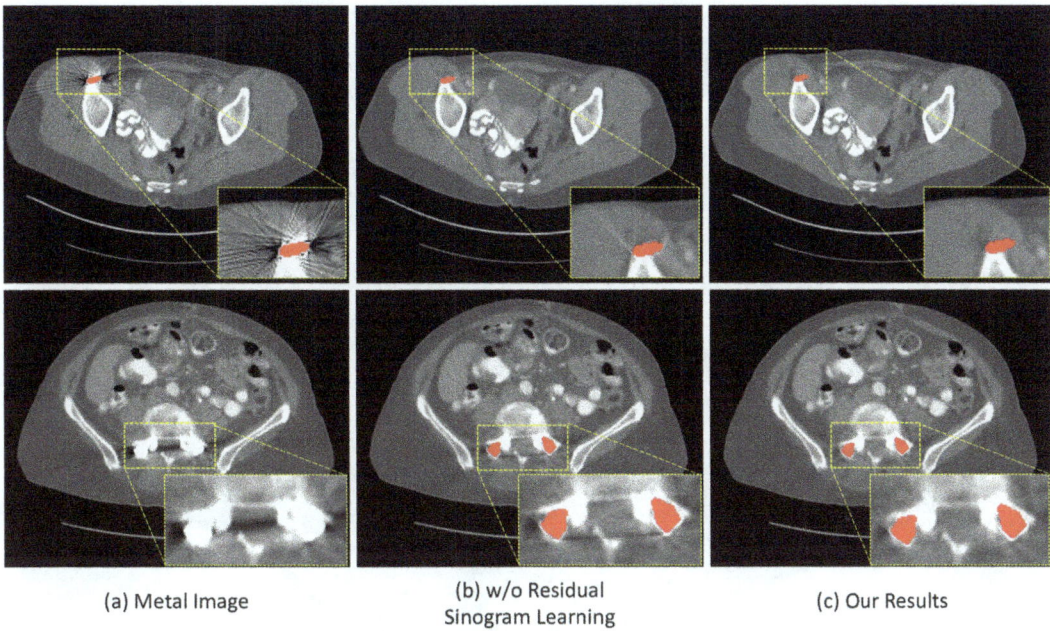

(a) Metal Image

(b) w/o Residual
Sinogram Learning

(c) Our Results

FIGURE 13.8
The results of our method w/o residual-sinogram-learning. The first row is the simulated metal image, and the
second image is the real clinical CT image. The display window is [−480, 560] HU.

From the visual inspection in Figure 13.8, with the residual-sinogram-learning strategy, our
approach further reduces artifacts on both simulated and real CT samples. The quantitative results of "w/o residual-sinogram-learning" are shown in Table 13.2. We observe that
higher RMSE and lower SSIM values can be obtained using residual-sinogram-learning.

13.3.9.3 Compared with Tissue Processing

In previous MAR methods, tissue processing is usually adopted to obtain prior images. To
this end, another experiment was conducted to verify the effect of this strategy. We used
tissue processing and metal trace replacement [30, 46] to process the prior image X_{prior},
and then obtained the final FBP-reconstructed metal-free image. The quantitative results
of this method (after tissue processing) are shown in Table 13.2. The results show satisfactory values on simulated DeepLesion datasets with tissue processing but still worse than
our end-to-end deep sinogram completion strategy, demonstrating that the deep sinogram
network can automatically learn how to reduce slight artifacts in prior images.

13.4 Self-Supervised Dual-Domain MAR

In this section, we will introduce a novel self-supervised deep-learning-based MAR
method for CT imaging, which can alleviate the requirement for anatomically identical
metal artifact CT images and metal artifact-free images during the network training.

(a) The training procedure of joint network

(b) The testing procedure

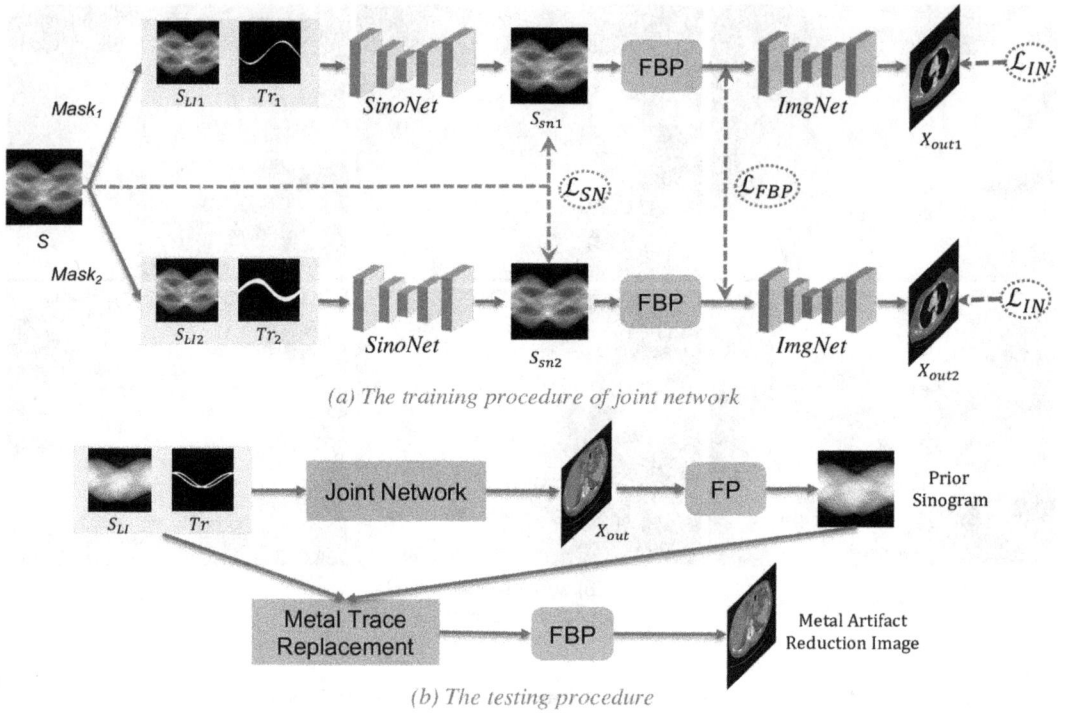

FIGURE 13.9
Illustration of the proposed self-supervised learning framework for MAR.

13.4.1 Overview

Figure 13.9 presents the overview of the proposed self-supervised learning framework. We trained SinoNet to complete the missing regions of the sinogram (i.e., metal tracks). We used an FBP consistency loss for SinoNet to generate geometric consistency, which can encourage the consistency of FBP reconstruction results. At the same time, the image refinement module, ImgNet, is used to avoid the secondary artifacts in the reconstructed image. In the testing stage, we used the metal trace replacement for the final completed sinogram. With the completed sinogram, we reconstructed the MAR image to retain the fine structural details and fidelity of the final MAR image.

13.4.2 Self-Supervised Cross-Domain Learning

13.4.2.1 Sinogram Completion

In this part, the SinoNet is used for the missing data completion within the metal trace in the projection domain. We used $S_{ma} \in R^{N \times D}$ and $Tr \in \{0, 1\}^{N \times D}$ as inputs (N: the number of projection views. D: the number of detectors) to estimate the missing pixels in Tr of S_{ma}. Since collecting the high-quality training dataset is impractical, a self-supervised learning manner is used to learn the sinogram completion network. To be specific, only the metal-free sinogram is used in this work. Given a metal-free sinogram S, we first simulate Tr from a pre-collected metal mask M as

$$Tr = \mathcal{P}(M) > 0,$$

where $\mathcal{P}$ denotes the forward projection. To simulate the metal-corrupted sinogram, we set the pixel in the metal trace at zero in S to get the incomplete sinogram S_{ma}. To ease the network learning, we use LI to provide an initial sinogram S_{LI} and then take S_{LI} and Tr as network input. Due to the unreliable projection data in the metal trace, we composite the results of SinoNet and S_{LI} w.r.t. Tr to acquire the completion result:

$$S_{sn} = f_S(S_{LI}, Tr) \odot Tr + S_{LI} \odot (1 - Tr),$$

where f_S is the procedure of SinoNet. In this work, the mask pyramid U-Net was employed as the network architecture backbone. This network structure explicitly integrates metal trace information into each layer, thus guiding the sinogram completion network to extract more discriminative features to conduct sinogram completion.

13.4.2.2 FBP Reconstruction Loss

Similarly, we adopt L_1 loss for SinoNet optimization

$$\mathcal{L}_{SN} = \left\| S_{sn} - S \right\|_1.$$

Moreover, we used FBP reconstruction loss to encourage the SinoNet to generate better completion to alleviate the secondary artifacts. With a S, we can generate a pair of metal-corrupted sinograms S_{ma1} and S_{ma2} w.r.t. Tr_1 and Tr_2. Then we feed S_{ma1} and S_{ma2} into the SinoNet and acquire two completed sinograms S_{sn1} and S_{sn2}. The corresponding FBP reconstructions from S_{sn1} and S_{sn2} should be consistent due to the same S,

$$\mathcal{L}_{FBP} = \left\| \left(\mathcal{P}^{-1}(S_{sn1}) - \mathcal{P}^{-1}(S_{sn2}) \right) \odot (1 - M_1 \mid M_2) \right\|_1.$$

M_1 and M_2 denote the metal masks. Since accurate CT images in metal regions are hard to obtain, we use a masked L_1 loss to only consider the differences of non-metal regions in CT images.

13.4.2.3 Image Refinement

The SinoNet can produce an anatomically plausible result, while the reconstructed images directly from the completed sinogram still have some secondary artifacts. To this line, we design a U-Net-based image refinement module, where we reduce the channel number of each layer to half. We first reconstruct the metal-free sinogram S into CT image as the "reference image" and then use a L_1 loss

$$\mathcal{L}_{IN} = \left\| X_{out} - \mathcal{P}^{-1}(S) \right\|_1.$$

We train our framework in an end-to-end fashion. We train both the SinoNet and ImgNet at the same time according to the full objective function

$$\mathcal{L}_{total} = \mathcal{L}_{SN} + \alpha_1 \mathcal{L}_{FBP} + \alpha_2 \mathcal{L}_{IN},$$

where α_1 and α_2 are hyperparameters (set at 1.0).

13.4.3 Prior-Image-Based Metal Trace Replacement

In our preliminary experiment [50], the images refined by the CNN network are often vague. Following the prior-image-based MAR methods, we employ a metal trace replacement to produce the final CT image. We utilize S_{prior} to only replace the metal-affected projections in the original sinogram and then we can acquire the corrected sinogram S_{corr}, which can be reconstructed into the final metal-free image, as shown in Figure 13.9 (b). Specifically, we conduct metal trace replacement to eliminate the discontinuity at the boundary of Tr. We carried out linear interpolation LI to obtain the residual sinogram S_{res} and S_{corr} can be represented as:

$$S_{\text{res}} = LI\left(S_{LI} - S_{\text{prior}}, Tr\right),$$

$$S_{\text{corr}} = S_{\text{prior}} + S_{\text{res}}.$$

Furthermore, $S_{i,k}$ denotes the kth projection pixel of ith view in the sinogram data. We performed the LI interpolation in each projection view S_i. As to S_i, if $\{S\_\{i, k\} \mid k \in [j + 1, j + \Delta]\}$ are affected by metal objects (i.e., $Tr_{i,k} = 1$) and its neighboring projection pixels $S_{i,j}$, $S_{i,j+\Delta+1}$ are not affected by metal objects, the projection pixels $S_{i,k}$ are obtained by

$$S_{i,k} = S_{i,j} + \frac{S_{i,j+\Delta+1} - S_{i,j}}{\Delta + 1}(k - j).$$

Finally, we FBP-reconstruct the MAR image from the corrected sinogram.

13.4.4 Training and Testing Strategies

For network training, only raw metal-free sinograms are essential with the assistance metal of masks, generating metal traces and lowering the burden of simulating related CT image pairs. In the inference stage, we first segment the metal masks from the metal-corrupted CT images using a threshold-based method. We then acquire Tr by forward-projecting the segment metals. We treat the metal-affected projection as missing data and get S_{LI} with LI. After that, the final MAR CT images are acquired with our proposed method.

13.4.5 Datasets and Implementation

We trained our framework with a simulated training dataset based on DeepLesion [47] CT images, due to the lack of clinical databases. All the simulation strategies are the same as those used in the above framework. We trained the SinoNet and ImgNet in an end-to-end manner. In total, we trained the framework with 180 epochs on one NVIDIA 1080Ti GPU card, where the batch size is 4. We used the Adam optimizer, $(\beta_1, \beta_2) = (0.5, 0.999)$, to train our framework. The learning rate was set at $1e^{-4}$.

13.4.6 Ablation Study

We conducted some ablation studies to verify the effectiveness of each part of our framework: (1) SinoNet: sinogram completion without FBP loss; (ii) SinoNet+ FBPcons: sinogram completion with FBP loss; (iii) JointLearning: both sinogram completion and image refinement, and (iv) JointLearning+MTR: our proposed method with the assistance of the metal trace replacement. For methods (i) and (ii), the final CT image is FBP-reconstructed from the completed sinogram. Note that SinoNet can be treated as a sinogram inpainting method. Table 13.3 shows the results of different ablation studies. With FBP loss, on the DeepLesion simulation dataset, the sinogram completion method outperforms SinoNet. Joint learning is superior to the sinogram completion methods by additionally combining with image domain refinement, which shows the effectiveness of our proposed joint dual-domain learning. Our method further reduces the RMSE metric by 3.74 HU compared to JointLearning. We also provide visual inspections. As shown in Figure 13.10, compared with SinoNet, the network with FBP loss reduces severe artifacts near the spine (marked by the blue arrows in Figure 13.10(B and C)). Compared to JointLearning, our method further reduces artifacts (see green and blue arrows (D and E) in Figure 13.10), visually demonstrating the usefulness of metal trace replacement components.

TABLE 13.3

Ablation Study on Different Parts of Our Framework

Method	RMSE(HU)	SSIM
SinoNet	42.60	0.9727
SinoNet+FBPcons	41.97	0.9731
JointLearning	36.28	0.9755
JointLearning+MTR **(Ours)**	**34.20**	**0.9773**

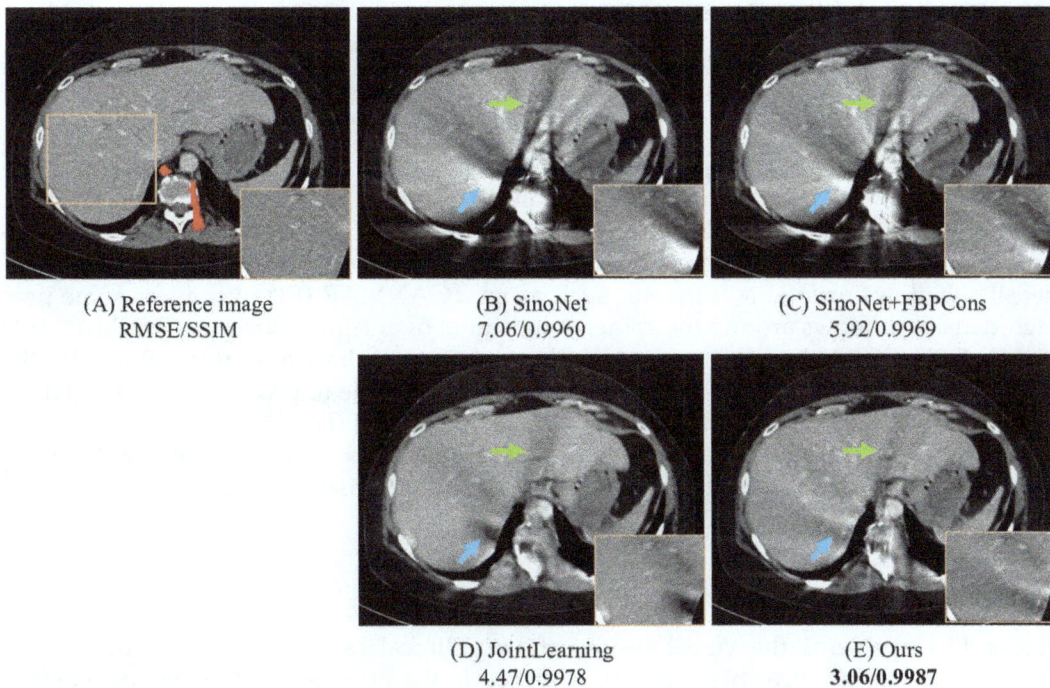

(A) Reference image
RMSE/SSIM

(B) SinoNet
7.06/0.9960

(C) SinoNet+FBPCons
5.92/0.9969

(D) JointLearning
4.47/0.9978

(E) Ours
3.06/0.9987

FIGURE 13.10

Visual inspection. The metal mask is marked red in (A). The numbers are the ROI metrics for quantitative comparison marked by the orange patches. The display window is [−140, 240] HU.

TABLE 13.4

Comparison of Different Methods on DeepLesion Simulation Dataset

Method	RMSE(HU)	SSIM
LI	50.31	0.9455
NMAR	47.03	0.9594
CNNMAR	43.27	0.9706
cGANMAR	39.01	0.9754
DuDoNet	38.00	0.9766
Ours	**34.20**	**0.9773**

13.4.7 Comparison with Other Methods

In this part, we also compare our method with baseline methods LI, NMAR, CNNMAR, cGANMAR and DuDoNet. As shown in Table 13.4, the CNNMAR and cGANMAR methods based on deep learning are superior to the traditional MAR methods in terms of RMSE and SSIM, which shows the advantages of deep neural networks. DuDoNet achieves lower RMSE and comparable SSIM values compared to CNNMAR and cGANMAR due to its integration of sinogram completion and image refinement to reduce metal artifacts. Compared to DuDoNet, our method further gets the best RMSE and SSIM. This may be due to the fact that the final CT image of our method is reconstructed directly from the completed sinogram, which better preserves the original intensity values. Note that we should compare the relative values of SSIM measures since the absolute values of SSIM vary according to different window sizes.

13.4.8 Qualitative Analysis

Figure 13.11 shows the visual inspection of all the involved methods on two DeepLesion samples from two organs (chest and abdomen). From the chest (A1 - H1) and the abdomen (A2 - H2), severe streaking artifacts can be observed in (B1, B2). After correction, all the methods can reduce streaking artifacts to some extent. However, they have different degrees of image detail restoration. In the LI and NMAR (C1, D1), numerous low-density shadows exist in soft tissues, such as fat. cGANMAR (F1) introduces some new high-density shadows around the spine, as indicated by the green arrows. CNNMAR (E1) and DuDoNet (G1) change the gray distribution of the pulmonary artery shown by the red arrow. cGANMAR, DuDoNet and our method were able to preserve some capillaries well. Compared with other correction methods, our method (H1) is closer to the reference CT image (A1), and the same trend can be observed in the experimental results of the abdomen. Here, we can see lots of macroscopic streaking artifacts (marked by blue dashed ellipses in (C2, F2, G2), blue arrows in (D2, E2)).

13.4.9 Experiments on CT Images with Real Artifacts

Figure 13.12 presents the visual results of two clinical samples. A severe dark band between the two hip prostheses can be observed in the original metal image (See Figure 13.12(A1, A2)). Compared with the original metal image, our method can optimize the metal artifacts of the two organs (see Figure 13.12(G1, G2)). Although other methods can

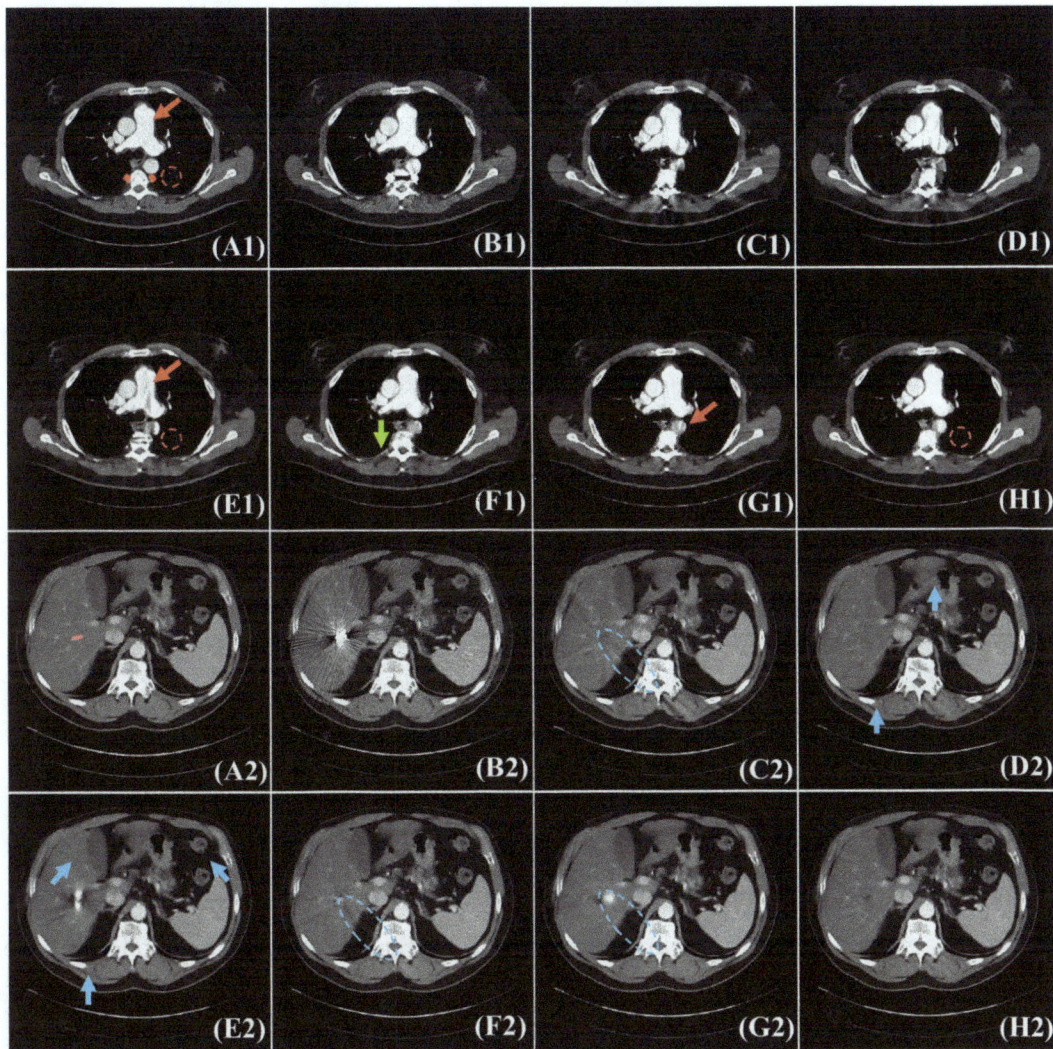

FIGURE 13.11
Qualitative results. Reference images (A1, A2) and metal-corrupted CT images (B1, B2) are showed. LI (C1, C2), NMAR (D1, D2), CNNMAR (E1, E2), cGANMAR (F1, F2), and DuDoNet (G1, G2) are compared with our method (H1, H2). The figures are shown in [−200, 200] HU.

also mitigate dark bands and artifacts to some extent, they do not remove strong artifacts near the metal as effectively as the green arrow in Figure 13.12(D1), nor do they introduce new artifacts as the blue arrow in Figure 13.12(F1). In addition, cGANMAR changes the tiny anatomical structure of the original image (Figure 13.12(E)). Also, in terms of the dark strip artifacts, our method outperforms DuDoNet (Figure 13.12(F2)). Since the abdomen sample has already been corrected by some MAR approaches, the metal artifacts are a little mild. Therefore, the differences among Figure 13.12(E2, F2, G2) are not more obvious than that in the chest sample.

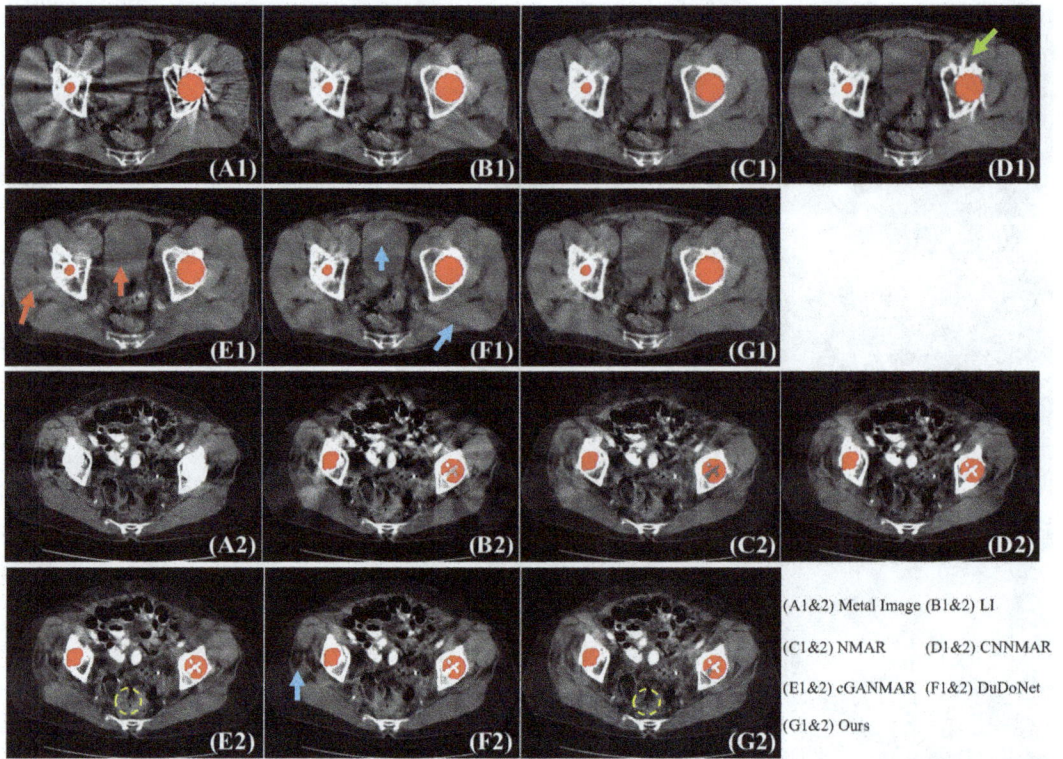

FIGURE 13.12
Visual result. The figures are shown in [−160, 310] HU.

13.4.10 Experiments on Different Site Data

To show the generalization capability of our approach over different site data, following the above work of the supervised MAR, we also evaluated our well-trained model on the head CT images collected from MICCAI 2015 Head and Neck Auto Segmentation Challenge dataset [51]. The original size of the new head CT image is 512 × 512. Specifically, we randomly chose some CT images from QIN-HEADNECK dataset [52] and simulated metal artifacts with the same simulation strategy. We then directly adopt the trained network on DeepLesion to correct those images. The results are shown in Figure 13.13. As we can see from the result, the original metal image is affected by the severe artifacts because of the metal implants. Traditional interpolated LI and NMAR methods can reduce artifacts to a certain extent, but there are still serious artifacts, especially near the teeth. Although trained with images from other organs, our approach shows excellent ability to reduce the artifacts of the new head CT image and produces better results than other methods based on deep learning, suggesting that the proposed approach has the potential to handle different organ data.

13.4.11 Comparison with More Recent Methods

We use the more recent method ADN [36], DuDoNet++ [37], and the above deep sinogram completion work [50] as comparisons. Table 13.5 shows the comparison results. ADN has got a good SSIM metric, while RMSE metrics are far less than our methods and other

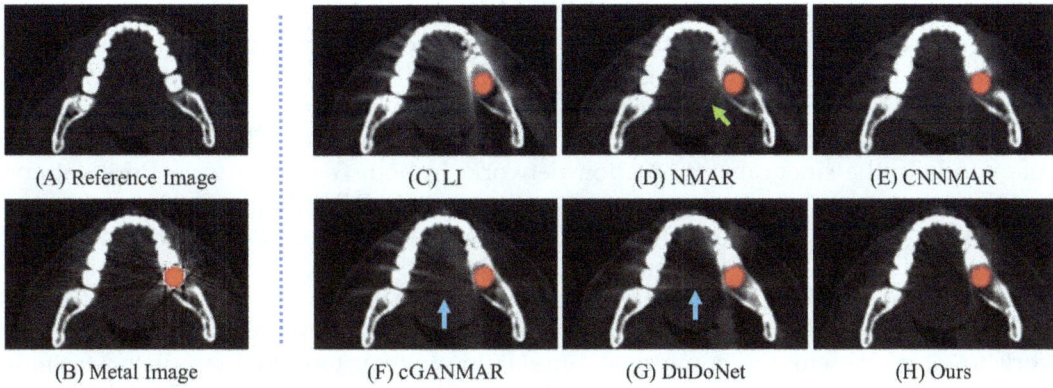

FIGURE 13.13

The visual comparison of different methods on head CT images. The gray window is set at [−200, 1400] HU.

TABLE 13.5

Comparison of Different Methods on DeepLesion Simulation Dataset

Method	RMSE(HU)	SSIM
ADN	89.34	0.9764
DuDoNet++*	38.29	0.9770
Deep sinogram completion*	31.15	0.9784
Ours	**34.20**	**0.9773**

methods. This is because the adversarial loss is more concerned with the distribution similarity between the generated metal reduction images and other non-metallic images, and there are no hard constraints on the strength difference of the target function. DuDoNet++ has better performance than ADN. But it is still inferior to our method. It is worth noting that the above deep sinogram completion is supervised learning, which has a huge burden of collecting training datasets, while this work follows self-supervised learning strategies without supervised signals in network training. Regarding this, the absolute performance of our methods is inferior to the above supervised deep sinogram completion work.

13.5 Discussion and Conclusion

MAR is a long-standing problem in CT imaging. Previous deep learning-based methods usually define MAR as an image restoration problem. In both works, we aim to address this problem by designing a data-driven framework that leverages large amounts of training data. To improve the generalization and robustness of the framework, we draw on the spirit of the traditional multi-objective fitting method and formulate the multi-objective fitting problem as a deep sinogram completion problem. Since it is difficult to directly predict accurate missing data in the projection domain, we integrate a deep prior image generation process with a residual-sinogram completion strategy. This method can improve the continuity of the projected pixel at the boundary of the metal traces, which cannot be better addressed by other MAR methods based on sinogram completion. In this way, our framework can not only better utilize the advantages of deep learning but also mitigates the risk of overfitting some training data.

Previous MAR methods based on prior images all use tissue processing to post-process the generated prior images. In both methods, we directly use the CNN output as a prior image to help the sinogram completion network. In both cases, we train PriorNet and SinoNet together, and the prior image generation and SinoNet complement processes can benefit each other. Although some slight artifacts are preserved in the generated prior images, the sinogram completion network can complete the sinogram from the prior sinogram so that our final output can eliminate these slight artifacts.

Clinically, it is impractical to prepare high-quality supervised training datasets for network training. We employ a common simulation strategy to obtain simulated training pairs. In both cases, the quality of simulated data greatly affects the network performance. At present, we simulate metal artifacts without carefully designing simulated metal masks. In the future, we will further investigate how to create a well-simulated dataset for higher network performance on real clinical CT images.

A limitation of our work is that we only evaluate our approaches in the CT image domain. We will work with our clinical collaborators to collect raw prediction data and evaluate the performance of our approach in real clinical applications in the future. When our framework is deployed into clinical practice, an important issue is how to obtain an accurate metal mask. In our work, we use a threshold segmentation approach to segment metal masks in CT images. Since deep learning has achieved promising performance in different medical image segmentation tasks, we believe that we can benefit from the integration of these deep-learning-based methods. In addition, how to perform metal mask recognition and MAR simultaneously is still to be investigated. In the future, we will investigate how binary reconstruction can be incorporated into our framework for better MAR. Another limitation of our approach is that it only works in the 2D mode. Extending our framework to 3D scanning is an interesting and important topic.

References

[1] W. A. Kalender, R. Hebel, and J. Ebersberger, "Reduction of CT artifacts caused by metallic implants," *Radiology*, vol. 164, no. 2, pp. 576–577, 1987.

[2] B. Meng, J. Wang, and L. Xing, "Sinogram preprocessing and binary reconstruction for determination of the shape and location of metal objects in computed tomography (CT)," *Med. Phys.*, vol. 37, no. 11, pp. 5867–5875, 2010.

[3] L. Gjesteby et al., "Metal artifact reduction in CT: Where are we after four decades?," *IEEE Access*, vol. 4, pp. 5826–5849, 2016.

[4] H. S. Park, S. M. Lee, H. P. Kim, J. K. Seo, and Y. E. Chung, "CT sinogram-consistency learning for metal-induced beam hardening correction," *Med. Phys.*, vol. 45, no. 12, pp. 5376–5384, 2018.

[5] E. M. Lee et al., "Improving mr image quality in patients with metallic implants," *Radiographics*, vol. 41, no. 4, pp. E126–E137, 2021.

[6] C. Huang et al., "Metal artifact reduction sequences MRI: A useful reference for preoperative diagnosis and debridement planning of periprosthetic joint infection," *J. Clin. Med.*, vol. 11, no. 15, p. 4371, 2022.

[7] I. Khodarahmi, A. Isaac, E. K. Fishman, D. Dalili, and J. Fritz, "Metal about the hip and artifact reduction techniques: From basic concepts to advanced imaging," *Semin. Musculoskelet. Radiol.*, vol. 23, no. 3, pp. 68–81, 2019.

[8] D. Tamada, "Review: Noise and artifact reduction for MRI using deep learning," *arXiv*, arXiv:2002.12889, 2020.

[9] S. Skare and J. L. R. Andersson, "Correction of MR image distortions induced by metallic objects using a 3D cubic B-spline basis set: Application to stereotactic surgical planning," *Magn. Reson. Med.*, vol. 54, no. 1, pp. 169–181, 2005.

[10] X. Shi, B. Quist, and B. Hargreaves, "No TitlePile-up and ripple artifact correction near metallic implants by alternating gradients," in *Presented at the 25th Annual Meeting International Society for Magnetic Resonance in Medicine*, 2017, p. 574.

[11] K. Kwon, D. Kim, and H. W. Park, "A learning-based metal artifacts correction method for MRI using dual-polarity readout gradients and simulated data," in *Lecture Notes in Computer Science (including subseries Lecture Notes in Artificial Intelligence and Lecture Notes in Bioinformatics)*, 2018.

[12] J. W. Kim, K. Kwon, B. Kim, and H. W. Park, "Attention guided metal artifact correction in mri using deep neural networks," *arXiv*, 2019.

[13] S. Seo, W. J. Do, H. M. Luu, K. H. Kim, S. H. Choi, and S. H. Park, "Artificial neural network for slice encoding for metal artifact correction (SEMAC) MRI," *Magn. Reson. Med.*, vol. 84, no. 1, pp. 263–276, 2020.

[14] K. Kwon, D. Kim, B. Kim, and H. W. Park, "Unsupervised learning of a deep neural network for metal artifact correction using dual-polarity readout gradients," *Magn. Reson. Med.*, vol. 83, no. 1, pp. 124–138, 2020.

[15] H. S. Park, D. Hwang, and J. K. Seo, "Metal artifact reduction for polychromatic X-ray CT based on a beam-hardening corrector," *IEEE Trans. Med. Imaging*, vol. 35, no. 2, pp. 480–487, 2016.

[16] J. Hsieh, R. C. Molthen, C. A. Dawson, and R. H. Johnson, "An iterative approach to the beam hardening correction in cone beam CT," *Med. Phys.*, vol. 27, no. 1, pp. 23–29, 2000.

[17] M. Kachelrieß, O. Watzke, and W. A. Kalender, "Generalized multi-dimensional adaptive filtering for conventional and spiral single-slice, multi-slice, and cone-beam CT," *Med. Phys.*, vol. 28, no. 4, pp. 475–490, 2001.

[18] Y. Zhang, Y. F. Pu, J. R. Hu, Y. Liu, Q. L. Chen, and J. L. Zhou, "Efficient CT metal artifact reduction based on fractional-order curvature diffusion," *Comput. Math. Methods Med.*, vol. 2011, p. 173748, 2011.

[19] Y. Zhang, Y. F. Pu, J. R. Hu, Y. Liu, and J. L. Zhou, "A new CT metal artifacts reduction algorithm based on fractional-order sinogram inpainting," *J. Xray. Sci. Technol.*, vol. 19, no. 3, pp. 373–384, 2011.

[20] A. Mehranian, M. R. Ay, A. Rahmim, and H. Zaidi, "X-ray CT metal artifact reduction using wavelet domain L0 sparse regularization," *IEEE Trans. Med. Imaging*, vol. 32, no. 9, pp. 1707–1722, 2013.

[21] E. Meyer, R. Raupach, M. Lell, B. Schmidt, and M. Kachelrieß, "Normalized metal artifact reduction (NMAR) in computed tomography," *Med. Phys.*, vol. 37, no. 10, pp. 5482–5493, 2010.

[22] M. Bal and L. Spies, "Metal artifact reduction in CT using tissue-class modeling and adaptive prefiltering," *Med. Phys.*, vol. 33, no. 8, pp. 2852–2859, 2006.

[23] D. Prell, Y. Kyriakou, M. Beister, and W. A. Kalender, "A novel forward projection-based metal artifact reduction method for flat-detector computed tomography," *Phys. Med. Biol.*, vol. 54, no. 21, p. 6575, 2009.

[24] J. Wang, S. Wang, Y. Chen, J. Wu, J. L. Coatrieux, and L. Luo, "Metal artifact reduction in CT using fusion based prior image," *Med. Phys.*, vol. 40, no. 8, p. 081903, 2013.

[25] K. H. Jin, M. T. McCann, E. Froustey, and M. Unser, "Deep convolutional neural network for inverse problems in imaging," *IEEE Trans. Image Process.*, vol. 26, no. 9, pp. 4509–4522, 2016.

[26] H. Chen et al., "Low-dose CT with a residual encoder-decoder convolutional neural network," *IEEE Trans. Med. Imaging*, vol. 36, no. 12, pp. 2524–2535, 2017.

[27] H. U. C. Hen et al., "Low-dose CT via convolutional neural network," *Biomed. Opt. Express*, vol. 8, no. 2, pp. 679–694, 2017.

[28] J. M. Wolterink, T. Leiner, M. A. Viergever, and I. Isgum, "Generative adversarial networks for noise reduction in low-dose CT," *IEEE Trans. Med. Imaging*, vol. 36, no. 12, pp. 2536–2545, 2017.

[29] G. Wang, J. C. Ye, K. Mueller, and J. A. Fessler, "Image reconstruction is a new frontier of machine learning," *IEEE Trans. Med. Imaging*, vol. 37, no. 6, pp. 1289–1296, 2018.

[30] Z. Zhang, X. Liang, X. Dong, Y. Xie, and G. Cao, "A sparse-view CT reconstruction method based on combination of densenet and deconvolution," *IEEE Trans. Med. Imaging*, vol. 37, no. 6, pp. 1407–1417, 2018.

[31] X. Xie, J. Niu, X. Liu, Z. Chen, S. Tang, and S. Yu, "A survey on domain knowledge powered deep learning for medical image analysis," *Med. Image Anal. Imaging*, vol. 69, p. 101985, 2021.

[32] S. Xu and H. Dang, "Deep residual learning enabled metal artifact reduction in CT," in *Medical Imaging 2018: Physics of Medical Imaging*, 2018.

[33] M. U. Ghani and W. C. Karl, "Fast enhanced CT metal artifact reduction using data domain deep learning," *IEEE Trans. Comput. Imaging*, vol. 6, pp. 181–193, 2019.

[34] W. A. Lin et al., "DuDoNet: Dual domain network for CT metal artifact reduction," in *Proceedings of the IEEE Computer Society Conference on Computer Vision and Pattern Recognition*, 2019.

[35] H. Liao et al., "Generative mask pyramid network for CT/CBCT metal artifact reduction with joint projection-sinogram correction," in *Lecture Notes in Computer Science (including subseries Lecture Notes in Artificial Intelligence and Lecture Notes in Bioinformatics)*, 2019.

[36] H. Liao, S. Member, W. Lin, and S. Member, "ADN: Artifact disentanglement network for unsupervised metal artifact reduction," *IEEE Trans. Med. Imaging*, vol. 39, no. 3, pp. 634–643, 2019.

[37] Y. Lyu, W.-A. Lin, J. Lu, and S. K. Zhou, "DuDoNet++: Encoding mask projection to reduce CT metal artifacts," *arXiv*, 2020.

[38] O. Ronneberger, P. Fischer, and T. Brox, "U-Net: Convolutional networks for biomedical image segmentation," in *Medical Image Computing and Computer-Assisted Intervention–MICCAI 2015: 18th International Conference, Munich, Germany, October 5–9, 2015, Proceedings, Part III 18*, 2016, pp. 432–440.

[39] L. Gjesteby, Q. Yang, Y. Xi, Y. Zhou, J. Zhang, and G. Wang, "Deep learning methods to guide CT image reconstruction and reduce metal artifacts," in *Medical Imaging 2017: Physics of Medical Imaging*, 2017, vol. 10132, pp. 1–7.

[40] L. Gjesteby et al., "Reducing metal streak artifacts in CT images via deep learning: Pilot results," *The 14th International Meeting on Fully Three-Dimensional Image Reconstruction in Radiology and Nuclear Medicine*, 2017.

[41] L. Gjesteby et al., "Deep neural network for CT metal artifact reduction with a perceptual loss function," in *Proceedings of The Fifth International Conference on Image Formation in X-ray Computed Tomography*, 2019.

[42] Y. Zhang and H. Yu, "Convolutional neural network based metal artifact reduction in X-ray computed tomography," *IEEE Trans. Med. Imaging*, vol. 37, pp. 1370–1381, 2018.

[43] J. Wang, Y. Zhao, J. H. Noble, and B. M. Dawant, "Conditional generative adversarial networks for metal artifact reduction in CT images of the ear," in *Lecture Notes in Computer Science (Including Subseries Lecture Notes in Artificial Intelligence and Lecture Notes in Bioinformatics)*, 2018.

[44] P. Isola, J.-Y. Zhu, T. Zhou, and A. A. Efros, "Image-to-image translation with conditional adversarial networks," in *Proceedings of the IEEE Conference on Computer Vision and Pattern Recognition*, 2016.

[45] X. Huang, J. Wang, F. Tang, T. Zhong, and Y. Zhang, "Metal artifact reduction on cervical CT images by deep residual learning," *Biomed. Eng. Online*, vol. 17, pp. 1–15, 2018.

[46] Y. Zhang, H. Yan, X. Jia, J. Yang, S. B. Jiang, and X. Mou, "A hybrid metal artifact reduction algorithm for x-ray CT," *Med. Phys.*, vol. 40, no. 4, p. 041910, 2013.

[47] K. Yan et al., "Deep lesion graph in the wild: Relationship learning and organization of significant radiology image findings in a diverse large-scale lesion database," in *Proceedings of the IEEE Conference on Computer Vision and Pattern Recognition*, pp. 9261–9270, 2018.

[48] A. Paszke et al., "PyTorch: An imperative style, high-performance deep learning library," in *Advances in Neural Information Processing Systems, 32*, 2019.

[49] D. P. Kingma and J. Ba, "Adam: A Method for Stochastic Optimization," *arXiv preprint*, arXiv1412.6980, 2014.

[50] L. Yu, Z. Zhang, X. Li, and L. Xing, "Deep sinogram completion with image prior for metal artifact reduction in CT images," *IEEE Trans. Med. Imaging*, vol. 40, no. 1, pp. 228–238, 2020.

[51] P. F. Raudaschl et al., "Evaluation of segmentation methods on head and neck CT: Auto-segmentation challenge 2015," *Med. Phys.*, vol. 44, no. 5, pp. 2020–2036, 2017.

[52] A. Fedorov et al., "DICOM for quantitative imaging biomarker development: A standards based approach to sharing clinical data and structured PET/CT analysis results in head and neck cancer research," *PeerJ*, vol. 4, p. e2057, 2016.

Section IV

Other Applications of Medical Image Synthesis

14

Synthetic Image-Aided Segmentation

Yang Lei, Richard L.J. Qiu, and Xiaofeng Yang
Emory University, Atlanta, GA, USA

CONTENTS

14.1 Introduction

Segmentation is one of the most important tasks in medical image analysis, which are routinely done on microscopy images, X-ray, ultrasound, computed tomography (CT), magnetic resonance imaging (MRI), positron emission tomography (PET) and so on. In radiation therapy cancer treatment [1], accurate radiation target and organs at risk (OARs) delineation is crucial for success [2, 3], especially for highly conformal techniques such as intensity modulated radiation therapy (IMRT), proton therapy, and stereotactic body radiation therapy (SBRT). The sharp dose fall-off from target to OARs is the reason that those highly conformal treatments can greatly spare OARs while maintaining target coverage. To harness its benefits, accurate structure delineation and precise treatment are required. Contrarily, errors in those processes could cause severe misadministration of radiation doses to the target and OARs.

In current clinical practice, the process of structure delineations is manually done by physicians, which is time-consuming, labor-intensive, and prone to errors and inter/intra-observer variabilities. For instance, manual delineation of soft tissues in planning CT images is challenging due to low soft-tissue contrast [4–9]. Therefore, researchers have spent tremendous effort in developing automatic segmentation methods to achieve accurate and consistent organ delineation over the past decade.

Recently, machine learning (ML) has gained significant interest in the field of medicine [10–12]. Artificial Neural Networks (ANNs), a subfield of ML, use multiple layers of connected neurons with learnable weights and biases to simulate the function of the human brain and perform high-level tasks [13–18]. Deep Learning (DL) is a term for ANNs that arose from advances in ANNs architectures and algorithms since 2006. It refers to ANNs

DOI: 10.1201/9781003243458-18

with many hidden layers, although there is no consensus on the specific number of layers required to be considered 'deep'. As a result, the distinction between ANNs and DL is not clearly defined [19]. DL has been particularly successful in computer vision [20], using a data-driven approach to explore vast amounts of image features for tasks such as image classification [21], object detection [22] and segmentation [23]. Inspired by the success of DL in computer vision, researchers have proposed various methods to apply DL techniques to medical imaging. It has been extensively studied in medical image segmentation [24–38], image synthesis [39–50], image enhancement and correction [51–56], and registration [57–65]. DL-based medical image segmentation technique represents a significant innovation in daily practices of radiation therapy, as they can expedite the contouring process, improve contour accuracy and consistency, and promote compliance with delineation guidelines [66–72]. Additionally, rapid DL-based multi-organ segmentation could facilitate online adaptive radiotherapy, potentially improving clinical outcomes.

DL-based methods have achieved state-of-the-art performance in medical image segmentation, particularly in multi-organ segmentation. In contrast to traditional methods that use handcrafted features, DL-based methods adaptively learn representative features from medical images [73–76]. Here in this chapter, we review DL-based approaches for medical image segmentation, with an emphasis on multi-organ segmentation. We classify the reviewed approaches into two categories: pixel-wise classification and end-to-end segmentation. We provide a detailed review of each category, examining its latest developments, contributions, and challenges. We also present benchmark evaluations of recently published multi-organ segmentation methods for thoracic CT and Head and Neck (HN) CT segmentations.

Image synthesis is the process of generating synthetic or pseudo images in the target image modality or domain (called the "target domain") from input images in a different image modality or domain (called the "source domain") [77, 78]. The goal of image synthesis is to use synthetic images to replace the physical patient imaging procedure. This is motivated by a variety of factors, such as the infeasibility of specific image acquisition, the added cost and labor of the imaging procedure, the ionizing radiation exposure to patients by some of those imaging procedures, or the introduction of uncertainties from the image registration between different modalities. In recent years, research on image synthesis has gained great interest in fields such as radiation oncology, radiology, and biology [79]. Its potential benefit has led to numerous investigations into its potential clinical applications, including MRI-only radiation therapy treatment planning [80–82], PET/MRI scanning [83, 84], proton-stopping power estimation [85–87], synthetic image-aided auto-segmentation [88–93], low dose CT denoising [52, 94, 95], image quality enhancement [96–99], reconstruction [100, 101], high-resolution visualization [102], and etc.

14.2 Modality Enhancement-Based Segmentation

Modality enhancement-based segmentation methods involve synthesizing an additional modality image from the primary modality image, in order to provide more information for the purpose of segmentation. Based on the properties of the imaging modality, those methods can be divided into two groups: 1) those that generate a very different modality image from the primary modality image; 2) and those that generate a similar-modality image from the primary modality image. Both two groups use the generated or synthetic image in combination with a trained segmentation model to derive the final segmentation.

14.2.1 Multi-Modality Image Synthesis

One important application of this group is synthetic MRI-aided CT organ segmentation [103]. The therapeutic benefit of external beam radiation therapy (EBRT) relies on the ability to deliver adequate prescription doses covering the target volumes while sparing OARs by optimizing beam parameters. Accurate target and OARs delineation are crucial for the quality of radiation therapy plans and subsequent clinical outcomes. Traditional machine learning-based and deep learning-based segmentation methods use planning CT images to contour OARs [104]. CT provides anatomic information with a high contrast of bones and the necessary electron density information for radiation dose calculation [105], but the low soft-tissue contrast can lead to inter- and intra-observer variability in manually delineated contours [7, 106]. This can be particularly problematic in challenging cases where target and OARs are abutting, such as the case for the prostate, bladder, and rectum. Using imaging modalities with better soft-tissue contrast, such as MRI, may reduce inter-observer variability and improve organ delineation accuracy and reproducibility [107]. However, MRIs are not always routinely acquired in clinical practice due to availability and cost [108].

DL has been shown to be effective for medical image segmentation because it can extract representative image features from large datasets and can be applied to a wide range of segmentation scenarios. Multiple studies have used different DL architectures to segment prostate and/or OARs for CT-based radiation therapy planning [69, 109, 110]. However, the low soft-tissue contrast issue in CT introduces inherent ambiguity to the segmentation process. To address this issue, a synthetic MRI (sMRI)-based deep attention network was developed to segment the prostate [89]. This method used a cycle-consistent generative adversarial network (CycleGAN) to generate sMRI from CT, which has superior soft-tissue contrast for prostate segmentation. The sMRI images were then segmented using a deep attention fully convolution network (DAFCN). In a separate study, this method was extended to demonstrate its feasibility for multi-organ segmentation for the prostate, bladder, and rectum (Figure 14.1) [72].

In a recent study, Xu et al. proposed using dual pyramid networks (DPNs) to combine the complementary soft-tissue information from sMRIs with the superior bony structure information from CTs [112] in order to automatically segment five key organs required for prostate EBRT treatment planning. This approach differs from earlier studies that used both CT and sMRI as inputs for multi-OAR segmentation [72, 89]. while CTs provide information for bone segmentation [24], sMRIs are more useful for soft-tissue segmentation [113, 114]. By combining both modalities, it is possible to simultaneously segment soft-tissue organs and bone with much better accuracy [111, 115–118].

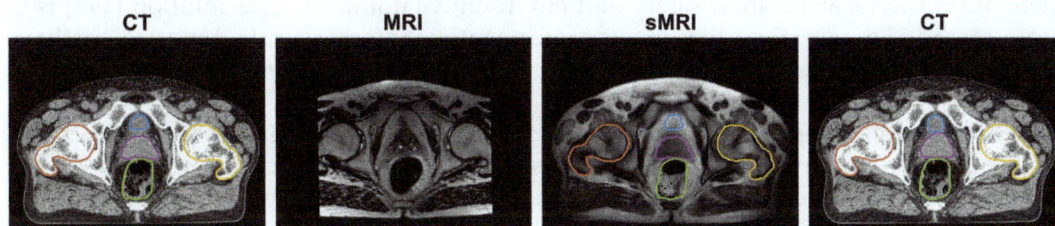

FIGURE 14.1
From left to right: CT with manual contours overlaid, MRI, sMRI with manual contours overlaid, and CT image with segmented contours overlaid. (From [111].)

Similar to the use of sMRIs to improve CT segmentation, synthetic images can also be used to improve the accuracy of segmentation in cone-beam CT (CBCT) and PET. For CBCT image segmentation, the image quality can be poor due to artifacts [119, 120]. One way to improve segmentation is to generate sMRIs from CBCTs [121, 122]. For PET image segmentation [123–126], using DL to perform PET attenuation correction or to estimate low-contrast PETs as full-contrast PETs can increase the image quality before segmentation [127–130].

14.2.2 Similar-Modality Image Synthesis

Due to its capability in providing meaningful anatomical and functional information, MRI is increasingly being used in radiation therapy to assist with target and OAR delineation [131, 132]. For instance, the complexity of HN anatomy can make it difficult to accurately contour targets and OARs using only CT, while the use of MRI can improve the accuracy of both [133–135]. By using different MRI pulse sequences, multi-contrast images can be obtained for the same anatomy, providing physicians with complementary information for the assessment, diagnosis, and treatment planning of various diseases. For example, T1-weighted (T1) MRI images show distinguishable white and gray matters, T1-weighted and contrast-enhanced (T1c) images can be used for assessment of tumor shape changes with its enhanced demarcation around tumor, T2-weighted (T2) images clearly separate fluid from cortical tissue, while lesion can be clearly delineated on fluid-attenuated inversion recovery (Flair) images [136, 137]. Therefore, combining the strengths of each modality, it is more likely to uncover rich underlying information that can facilitate diagnosis and treatment management [104, 138, 139]. However, it is difficult to use consistent scanning protocols for every patient due to a variety of factors such as limited scan time, human operating errors, scan artifacts and corruption, and situations in which some patients are allergic to contrast agents., This lack of consistency in imaging protocols across patients and institutions presents challenges for clinical practice and longitudinal research. To tackle this challenge, cross-modal image synthesis has been proposed as a way to generate the missing modalities using available modalities as input [140]. In a study by Dai et al., a multimodal MRI synthesis method based on a unified generative adversarial network was proposed [141]. The network takes an arbitrary modality image as input and synthesizes multimodal images in a single forward pass. By using multi-modality MRI, the segmentation performance of single-modality MRI can be improved.

Similar to the multi-modality MRI task, the contrast of MRI can be enhanced to make the segmentation task easier. generative adversarial networks (GANs) have recently been used for this purpose [142–144]. While gadolinium-based contrast agents (GBCAs) are commonly administered for MRI imaging for diagnostic studies and treatment planning, there are concerns about their safety and environmental impact. One solution [144] is to use a cascade DL workflow that incorporates contour information to generate synthetic contrast-enhance MRIs from unenhanced counterpart images, eliminating the need for GBCAs.

In deep learning-based segmentation methods, high-quality CT images with good contrast could help improve performance because the neural network can more easily extract representative image features from large datasets to cover various segmentation scenarios. However, conventional CT images are typically acquired with a single-energy spectrum of X-ray, meaning that different materials may have the same Hounsfield Unit (HU) value in these single-energy CT (SECT) images, which makes it difficult for the network to differentiate them. Recently, dual-energy CT (DECT) has been introduced, which uses two

CBCT	Planning CT	Synthetic CT

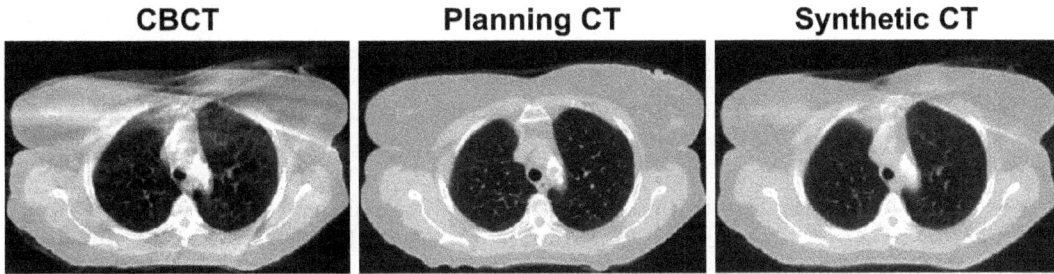

FIGURE 14.2
Synthetic CT generated from lung CBCT. (From [120])

different energy spectra to differentiate materials based on their energy dependence on photoelectric and Compton interactions [145–148]. DECT provides the CT images reconstructed with low- and high-energy X-rays, and derivative images can be generated for advanced clinical uses such as bone maps for bone removal in angiography [149–152] and relative stopping power maps for proton radiation therapy treatment planning [153, 154]. Studies have shown that DECT and its derivatives may outperform SECT for manual contouring due to superior contrast-noise-ratio [155–160]. However, the potential benefits of DECT for DL networks have not yet been studied. One potential approach is to generate synthetic DECT from SECT, as has been done in recent works [161–164], and use DL-based methods such as mask scoring R-CNN to segment the DECT by learning deep features from two-channel CTs input [165].

As we discussed previously, the CBCT image quality can affect the performance of segmentation due to the image artifact. In addition to synthetic MRI-aided segmentation, another potential approach is to produce enhanced quality CBCT. This can be done by training the machine learning-based or deep learning-based model using CTs as the learning target and CBCTs as input. After training the model, the synthetic CT can be generated from CBCT with enhanced image quality. These synthetic CT images can then be used input for the segmentation model to increase the performance (Figure 14.2) [166–170].

Image resolution can also be a factor that affects the quality of segmentation. One potential solution is to use deep learning-based method to increase the image resolution [171, 172] and then perform segmentation.

14.3 Training Data Enlargement-Based Segmentation

The success of segmentation models is highly dependent on the size and quality of the data used to train the model [173]. However, the costly and time-consuming process of medical data labeling and annotation process can be an obstacle to producing large datasets for artificial intelligence (AI) algorithms training and evaluation [174]. Pixelwise data annotation is particularly demanding in terms of time [175], and if expert annotations are not possible, at least a review process by experts should be done to ensure accuracy before using the data for training AI algorithms [176]. The importance of having accurate annotations from experts for medical data is, for example, discussed by [177] using a pelvic OAR segmentation dataset of CBCT images. In this regard, it is important to research methods for producing synthetic segmentation datasets (synthetic images and the corresponding accurate ground truth masks) to extend the training datasets [178–181].

In order to overcome these issues, fully synthetic data is proposed as a possible solution that reduces the cost and time needed for annotations, and bypass patient data privacy issues. For instance, some medical datasets inevitably need to include patient-identifying data such as images from the faces of stroke patients [182, 183], where synthetic data would be perfect. Synthetic data generated by GANs is used in almost all domains of medical imaging [184]. Rather than generating synthetic data only [185–187], some other studies generate synthetic data and the corresponding ground truth [188]. The authors [188] generated synthetic glioblastoma multiforme (GBM) in 2D magnetic resonance images with a segmentation mask. For this method, they manually placed artificially generated GBM in real images using MeVisLab (https://www.mevislab.de), calling it semi-supervised data generation. However, placing the generated synthetic GBM is a time-consuming task for generating a big synthetic dataset using this approach. Moreover, the background image is still real while the segmented GBM is synthetic. The real sections of synthetic images may raise privacy concerns.

Another study [189] used the conditional GAN to generate synthetic polyp data conditioned on real edge-filtering images and a randomly generated mask. This approach can generate diverse synthetic data, but a large dataset is required to train the model because it uses the deep convolution GAN, which needs tons of data to train. Moreover, extracted edge-filtering images and random polyp masks should be merged to generate the corresponding synthetic data. Therefore, input data preparation is a time-consuming process.

Recently, a group of methods, called the few-shot learning-based methods, aim to solve this inflexibility by learning to segment an arbitrary unseen semantically meaningful class by referring to only a few labeled examples without involving fine-tuning [190]. DL-base model is used to generate synthetic data to enlarge the training data variation and thus improve the segmentation performance without the need of large data [191].

14.4 Summary and Discussion

We provide a comprehensive review of recently published synthetic-aided medical image segmentation. Using synthetic data, we can either improve the image quality, provide sufficient information and better contrast to reduce the difficulty of medical image segmentation, and synthesize image data with annotated labels to enlarge the training data variation, introducing the robustness for small dataset tasks, which is often appeared in clinic tasks. There is a clear trend in using DL-based methods to perform end-to-end data synthesis and medical image segmentation. We expect to see steady growth in the combination of these two approaches in medical image segmentation.

Data collected directly from clinical databases are usually not suitable for network training. It is necessary to perform data preprocessing, such as image resizing, cropping, normalization, and augmentation, before training a network. Other preprocessing techniques include registration [57], bias/scatter/attenuation correction [83, 119], voxel intensity normalization [192, 193], and cropping [36]. Data augmentation is used to increase the number of training samples and reduce network overfitting [194]. Common data augmentation techniques include image rotation, translation, scaling, flipping, distortion, linear warping, elastic deformation, and noise contamination [195]. During training, the ground truth contours are obtained through manual delineation by physicians [196]. However, this method is subject to bias toward the contouring style of the physicians and contouring uncertainty [197, 198], which is a limitation common to all supervised learning-based methods [199].

Depending on the network design and graphics processing unit (GPU) availability, some methods use the whole image volume as input to train the network [200], while others uses 2D image slices [201]. The 2D-based approaches can reduce the memory requirement and computational cost, but they fails to utilize the spatial information in the third dimension. Another way is to use 2D kernels on multidirectional 2D images to exploit the 3D feature information while avoiding computer memory overflow. Segmentation results from different image planes such as axial, coronal, and sagittal planes, can be combined using a surface-based contour refinement method [28]. 3D image patches are also widely used as network input [202, 203] to retain 3D spatial information and reduce computational cost.

To refine the DL segmentation results, post-processing is often applied to generate realistic structures with smooth boundaries. Morphological operations are widely used to remove small erroneous labels. Conditional random field is also widely adopted as a post-processing step to refine segmentation results [200]. CRFs models the pixelwise spatial relationships in order to preserve the plausibility of the segmented structures. As GAN-based methods are increasingly used to penalize implausible structures and preserve the spatial integrity of the segmentation results, CRF post-processing is expected to be used less in the future.

Another challenge in multi-organ segmentation is the class imbalance problem. For example, the esophagus and spinal cord are often much smaller than the lung in thorax CT images. Training a network with class imbalanced datasets would bias the result toward the classes of large organs. Therefore, the choice of loss functions is crucial for these tasks [204, 205]. Johnson et al. published a survey on deep learning with class imbalance [206].

Disclosures

The authors declare no conflicts of interest.

References

[1] Z. X. Liao et al., "Bayesian adaptive randomization trial of passive scattering proton therapy and intensity-modulated photon radiotherapy for locally advanced non-small-cell lung cancer (vol 36, pg 1813, 2018)," (in eng), *Journal of Clinical Oncology*, vol. 36, no. 24, p. 2570, Aug 20 2018, doi: 10.1200/Jco.2018.18.00817.

[2] J. K. Molitoris et al., "Advances in the use of motion management and image guidance in radiation therapy treatment for lung cancer," (in eng), *Journal of Thoracic Disease*, vol. 10, pp. S2437–S2450, y2ercm 2018, doi: 10.21037/jtd.2018.01.155.

[3] M. A. L. Vyfhuis et al., "Advances in proton therapy in lung cancer," (in eng), *Therapeutic Advances in Respiratory Disease*, vol. 12, Jul 17 2018, doi: 10.1177/1753466618783878.

[4] C. W. Hurkmans, J. H. Borger, B. R. Pieters, N. S. Russell, E. P. Jansen, and B. J. Mijnheer, "Variability in target volume delineation on CT scans of the breast," (in eng), *International Journal of Radiation Oncology, Biology, Physics*, vol. 50, no. 5, pp. 1366–1372, Aug 1 2001, doi: 10.1016/s0360-3016(01)01635-2.

[5] C. Rasch, R. Steenbakkers, and M. van Herk, "Target definition in prostate, head, and neck," (in eng), *Seminars in Radiation Oncology*, vol. 15, no. 3, pp. 136–145, Jul 2005, doi: 10.1016/j.semradonc.2005.01.005.

[6] J. Van de Steene et al., "Definition of gross tumor volume in lung cancer: Inter-observer variability," (in eng), *Radiotherapy and Oncology*, vol. 62, no. 1, pp. 37–49, Jan 2002, doi: 10.1016/S0167-8140(01)00453-4.

[7] S. K. Vinod, M. G. Jameson, M. Min, and L. C. Holloway, "Uncertainties in volume delineation in radiation oncology: A systematic review and recommendations for future studies," (in eng), *Radiotherapy and Oncology*, vol. 121, no. 2, pp. 169–179, Nov 2016, doi: 10.1016/j.radonc.2016.09.009.

[8] J. Breunig et al., "A system for continual quality improvement of normal tissue delineation for radiation therapy treatment planning," (in eng), *International Journal of Radiation Oncology Biology Physics*, vol. 83, no. 5, pp. E703–E708, Aug 1 2012, doi: 10.1016/j.ijrobp.2012.02.003.

[9] B. E. Nelms, W. A. Tome, G. Robinson, and J. Wheeler, "Variations in the contouring of organs at risk: Test case from a patient with oropharyngeal cancer," (in eng), *International Journal of Radiation Oncology Biology Physics*, vol. 82, no. 1, pp. 368–378, Jan 2012, doi: 10.1016/j.ijrobp.2010.10.019.

[10] A. L. Beam and I. S. Kohane, "Translating artificial intelligence into clinical care," (in eng), *Jama-Journal of the American Medical Association*, vol. 316, no. 22, pp. 2368–2369, Dec 13 2016, doi: 10.1001/jama.2016.17217.

[11] A. Pella et al., "Use of machine learning methods for prediction of acute toxicity in organs at risk following prostate radiotherapy," (in eng), *Medical Physics*, vol. 38, no. 6, pp. 2859–2867, Jun 2011, doi: 10.1118/1.3582947.

[12] X. Yang et al., "Automated segmentation of the parotid gland based on atlas registration and machine learning: A longitudinal MRI study in head-and-neck radiation therapy," *International Journal of Radiation Oncology, Biology, Physics*, vol. 90, no. 5, pp. 1225–1233, 2014.

[13] T. J. Bryce, M. W. Dewhirst, C. E. Floyd, V. Hars, and D. M. Brizel, "Artificial neural network model of survival in patients treated with irradiation with and without concurrent chemotherapy for advanced carcinoma of the head and neck," (in eng), *International Journal of Radiation Oncology Biology Physics*, vol. 41, no. 2, pp. 339–345, May 1 1998, doi: 10.1016/S0360-3016(98)00016-9.

[14] S. L. Gulliford, S. Webb, C. G. Rowbottom, D. W. Corne, and D. P. Dearnaley, "Use of artificial neural networks to predict biological outcomes for patients receiving radical radiotherapy of the prostate," (in eng), *Radiotherapy and Oncology*, vol. 71, no. 1, pp. 3–12, Apr 2004, doi: 10.1016/j.radonc.2003.03.001.

[15] S. Tomatis et al., "Late rectal bleeding after 3D-CRT for prostate cancer: Development of a neural-network-based predictive model," (in eng), *Physics in Medicine and Biology*, vol. 57, no. 5, pp. 1399–1412, Mar 7 2012, doi: 10.1088/0031-9155/57/5/1399.

[16] S. F. Chen, S. M. Zhou, J. N. Zhang, F. F. Yin, L. B. Marks, and S. K. Das, "A neural network model to predict lung radiation-induced pneumonitis," (in eng), *Medical Physics*, vol. 34, no. 9, pp. 3420–3427, Sep 2007, doi: 10.1118/1.2759601.

[17] M. Su, M. Miften, C. Whiddon, X. J. Sun, K. Light, and L. Marks, "An artificial neural network for predicting the incidence of radiation pneumonitis," (in eng), *Medical Physics*, vol. 32, no. 2, pp. 318–325, Feb 2005, doi: 10.1118/1.1835611.

[18] T. Ochi, K. Murase, T. Fujii, M. Kawamura, and J. Ikezoe, "Survival prediction using artificial neural networks in patients with uterine cervical cancer treated by radiation therapy alone," (in eng), *Radiology*, vol. 217, pp. 142, Nov 2000. [Online]. Available: <Go to ISI>://WOS:000090071300178.

[19] L. Boldrini, J. E. Bibault, C. Masciocchi, Y. T. Shen, and M. I. Bittner, "Deep learning: A review for the radiation oncologist," (in eng), *Frontiers in Oncology*, vol. 9, Oct 1 2019, doi: ARTN 977. 10.3389/fonc.2019.00977.

[20] Y. LeCun, Y. Bengio, and G. Hinton, "Deep learning," *Nature*, vol. 521, no. 7553, pp. 436–444, May 1 2015, doi: 10.1038/nature14539.

[21] A. Krizhevsky, I. Sutskever, and G. E. Hinton, "ImageNet classification with deep convolutional neural networks," (in eng), *Communications of the ACM*, vol. 60, no. 6, pp. 84–90, Jun 2012, doi: 10.1145/3065386.

[22] P. Sermanet, D. Eigen, X. Zhang, M. Mathieu, R. Fergus, and Y. LeCun, "OverFeat: Integrated recognition, localization and detection using convolutional networks," *CoRR*, arXiv:1312.6229, 2013.

[23] E. Shelhamer, J. Long, and T. Darrell, "Fully convolutional networks for semantic segmentation," (in eng), *IEEE Transactions on Pattern Analysis and Machine Intelligence*, vol. 39, no. 4, pp. 640–651, Apr 2017, doi: 10.1109/Tpami.2016.2572683.

[24] M. H. Hesamian, W. Jia, X. J. He, and P. Kennedy, "Deep learning techniques for medical image segmentation: Achievements and challenges," (in eng), *Journal of Digital Imaging*, vol. 32, no. 4, pp. 582–596, Aug 2019, doi: 10.1007/s10278-019-00227-x.

[25] T. Zhou, S. Ruan, and S. Canu, "A review: Deep learning for medical image segmentation using multi-modality fusion," *Array*, vol. 3–4, p. 100004, Sept 1 2019, doi: https://doi.org/10.1016/j.array.2019.100004

[26] Y. Lei et al., "Whole-body PET estimation from low count statistics using cycle-consistent generative adversarial networks," (in eng), *Physics in Medicine and Biology*, vol. 64, no. 21, p. 215017, Nov 4 2019, doi: 10.1088/1361-6560/ab4891.

[27] B. van der Heyden et al., "Dual-energy CT for automatic organs-at-risk segmentation in brain-tumor patients using a multi-atlas and deep-learning approach," (in eng), *Scientific Reports*, vol. 9, Mar 11 2019, doi: ARTN 4126 10.1038/s41598-019-40584-9.

[28] Y. Lei et al., "Ultrasound prostate segmentation based on multidirectional deeply supervised V-Net," (in eng), *Medical Physics*, vol. 46, no. 7, pp. 3194–3206, Jul 2019, doi: 10.1002/mp.13577.

[29] T. Wang et al., "Learning-based automatic segmentation of arteriovenous malformations on contrast CT images in brain stereotactic radiosurgery," (in eng), *Medical Physics*, vol. 46, no. 7, pp. 3133–3141, Jul 2019, doi: 10.1002/mp.13560.

[30] B. Wang et al., "Deeply supervised 3D fully convolutional networks with group dilated convolution for automatic MRI prostate segmentation," (in eng), *Medical Physics*, vol. 46, no. 4, pp. 1707–1718, Apr 2019, doi: 10.1002/mp.13416.

[31] Y. Lei et al., "Ultrasound prostate segmentation based on 3D V-Net with deep supervision," in *Proceeding of SPIE Medical Imaging 2019: Ultrasonic Imaging and Tomography*, 2019, vol. 10955, no. 109550V.

[32] B. Wang et al., "Automated prostate segmentation of volumetric CT images using 3D deeply supervised dilated FCN," in *Proceeding of SPIE Medical Imaging 2019: Image Processing*, 2019, vol. 10949, no. 109492S.

[33] T. Wang et al., "A learning-based automatic segmentation method on left ventricle in SPECT imaging," in *Proceeding of SPIE Medical Imaging 2019: Image Processing*, 2019, vol. 10953, no. 109531M.

[34] B. Wang et al., "Automatic MRI prostate segmentation using 3D deeply supervised FCN with concatenated atrous convolution," in *Proceeding of SPIE Medical Imaging 2019: Image Processing*, 2019, vol. 10950, no. 109503X.

[35] Y. Lei et al., "Automatic multi-organ segmentation in thorax CT images using U-net-GAN," in *Proceeding of SPIE Medical Imaging 2019: Image Processing*, 2019, vol. 10950, no. 1095010.

[36] T. Wang et al., "Learning-based automatic segmentation on arteriovenous malformations from contract-enhanced CT images," in *Proceeding of SPIE Medical Imaging 2019: Image Processing*, 2019, vol. 10950, no. 109504D.

[37] T. Wang et al., "A learning-based automatic segmentation and quantification method on left ventricle in gated myocardial perfusion SPECT imaging: A feasibility study," (in eng), *Journal of Nuclear Cardiology*, vol. 27, no. 3, pp. 976–987, Jun 2020, doi: 10.1007/s12350-019-01594-2.

[38] J. Wu et al., "Deep morphology aided diagnosis network for segmentation of carotid artery vessel wall and diagnosis of carotid atherosclerosis on black-blood vessel wall MRI," (in eng), *Medical Physics*, vol. 46, no. 12, pp. 5544–5561, Dec 2019, doi: 10.1002/mp.13739.

[39] X. Dong et al., "Synthetic CT generation from non-attenuation corrected PET images for whole-body PET imaging," (in eng), *Physics in Medicine and Biology*, vol. 64, no. 21, p. 215016, Nov 4 2019, doi: 10.1088/1361-6560/ab4eb7.

[40] Y. Liu et al., "Evaluation of a deep learning-based pelvic synthetic CT generation technique for MRI-based prostate proton treatment planning," (in eng), *Physics in Medicine and Biology*, vol. 64, no. 20, p. 205022, Oct 21 2019, doi: 10.1088/1361-6560/ab41af.

[41] G. Shafai-Erfani et al., "MRI-based proton treatment planning for base of skull tumors," *International Journal of Particle Therapy*, vol. 6, no. 2, pp. 12–25, Fall 2019, doi: 10.14338/IJPT-19-00062.1.

[42] X. Yang et al., "CBCT-guided prostate adaptive radiotherapy with CBCT-based synthetic MRI and CT," *International Journal of Radiation Oncology, Biology, Physics*, vol. 105, no. 1, p. S250, 2019, doi: 10.1016/j.ijrobp.2019.06.372.

[43] X. Yang et al., "MRI-based proton radiotherapy for prostate cancer using deep convolutional neural networks," *International Journal of Radiation Oncology, Biology, Physics*, vol. 105, no. 1, p. S200, 2019, doi: 10.1016/j.ijrobp.2019.06.263.

[44] T. Wang et al., "MRI-based treatment planning for brain stereotactic radiosurgery: Dosimetric validation of a learning-based pseudo-CT generation method," (in eng), *Medical Dosimetry*, vol. 44, no. 3, pp. 199–204, Autumn 2019, doi: 10.1016/j.meddos.2018.06.008.

[45] Y. Liu et al., "MRI-based treatment planning for proton radiotherapy: Dosimetric validation of a deep learning-based liver synthetic CT generation method," (in eng), *Physics in Medicine and Biology*, vol. 64, no. 14, p. 145015, Jul 16 2019, doi: 10.1088/1361-6560/ab25bc.

[46] Y. Liu et al., "MRI-based treatment planning for liver stereotactic body radiotherapy: Validation of a deep learning-based synthetic CT generation method," (in eng), *The British Journal of Radiology*, vol. 92, no. 1100, p. 20190067, Aug 2019, doi: 10.1259/bjr.20190067.

[47] Y. Lei et al., "MRI-only based synthetic CT generation using dense cycle consistent generative adversarial networks," (in eng), *Medical Physics*, vol. 46, no. 8, pp. 3565–3581, Aug 2019, doi: 10.1002/mp.13617.

[48] Y. Lei et al., "MRI-based synthetic CT generation using deep convolutional neural network," in *Proceeding of SPIE Medical Imaging 2019: Image Processing*, 2019, vol. 10949, no. 109492T.

[49] G. Shafai-Erfani et al., "Dose evaluation of MRI-based synthetic CT generated using a machine learning method for prostate cancer radiotherapy," (in eng), *Medical Dosimetry*, vol. 44, no. 4, pp. e64–e70, Winter 2019, doi: 10.1016/j.meddos.2019.01.002.

[50] X. Yang et al., "MRI-based synthetic CT for radiation treatment of prostate cancer," *International Journal of Radiation Oncology, Biology, Physics*, 2018, vol. 102, no. 3, pp. S193–S194, doi: 10.1016/j.ijrobp.2018.07.086.

[51] X. Dong et al., "Deep learning-based attenuation correction in the absence of structural information for whole-body positron emission tomography imaging," (in eng), *Physics in Medicine and Biology*, vol. 65, no. 5, p. 055011, Mar 2 2020, doi: 10.1088/1361-6560/ab652c.

[52] T. Wang et al., "Deep learning-based image quality improvement for low-dose computed tomography simulation in radiation therapy," (in eng), *Journal of Medical Imaging (Bellingham)*, vol. 6, no. 4, p. 043504, Oct 2019, doi: 10.1117/1.JMI.6.4.043504.

[53] J. Harms et al., "Paired cycle-GAN-based image correction for quantitative cone-beam computed tomography," (in eng), *Medical Physics*, vol. 46, no. 9, pp. 3998–4009, Sep 2019, doi: 10.1002/mp.13656.

[54] Y. Lei et al., "Image quality improvement in cone-beam CT using deep learning," in *Proceeding of SPIE Medical Imaging 2019: Image Processing*, 2019, vol. 10948, no. 1094827

[55] X. Yang et al., "A learning-based method to improve pelvis cone beam CT image quality for prostate cancer radiation therapy," in *International Journal of Radiation Oncology, Biology, Physics*, vol. 102, no. 3, pp. E377–E378, 2018, doi: 10.1016/j.ijrobp.2018.07.1124.

[56] X. Yang et al., "Whole-body PET estimation from ultra-short scan durations using 3D cycle-consistent generative adversarial networks," in *Journal of Nuclear Medicine*, vol. 60, no. supplement 1, p. 247, 2019.

[57] Y. Fu, Y. Lei, T. Wang, W. J. Curran, T. J. Liu, and X. J. A. Yang, "Deep learning in medical image registration: A review," *ArXiv*, arXiv:1912.12318, 2019.

[58] G. Haskins, U. Kruger, and P. Yan, "Deep learning in medical image registration: A survey," *ArXiv*, arXiv:1903.02026, 2019.

[59] Y. Lei et al., "4D-CT Deformable Image Registration Using an Unsupervised Deep Convolutional Neural Network," in *Artificial Intelligence in Radiation Therapy*, Cham, D. Nguyen, L. Xing, and S. Jiang, Eds., 2019, doi: https://doi.org/10.1007/978-3-030-32486-5_4: Springer International Publishing, pp. 26–33.

[60] X. Yang et al., "MRI-US registration using label-driven weakly-supervised learning for multiparametric MRI-guided HDR prostate brachytherapy," *International Journal of Radiation Oncology, Biology, Physics*, vol. 105, no. 1, p. E727, 2019, doi: 10.1016/j.ijrobp.2019.06.911.

[61] H. M. Li and Y. Fan, "Non-rigid image registration using self-supervised fully convolutional networks without training data," (in eng), *2018 Ieee 15th International Symposium on Biomedical Imaging (Isbi 2018)*, pp. 1075–1078, 2018. [Online]. Available: <Go to ISI>://WOS:000455045600246.

[62] Y. Fu et al., "LungRegNet: An unsupervised deformable image registration method for 4D-CT lung," *Medical Physics*, vol. 47, no. 4, pp. 1763–1774, Apr 2020, doi: 10.1002/mp.14065.

[63] Y. Fu et al., "Biomechanically constrained non-rigid MR-TRUS prostate registration using deep learning based 3D point cloud matching," (in eng), *Medical Image Analysis*, vol. 67, p. 101845, Jan 2021, doi: 10.1016/j.media.2020.101845.

[64] Y. Fu et al., "Deformable MR-CBCT prostate registration using biomechanically constrained deep learning networks," *Medical Physics*, vol. 48, no. 1, pp. 253–263, Jan 2021, doi: https://doi.org/10.1002/mp.14584

[65] Y. Lei et al., "4D-CT deformable image registration using multiscale unsupervised deep learning," (in eng), *Physics in Medicine and Biology*, vol. 65, no. 8, p. 085003, Apr 20 2020, doi: 10.1088/1361-6560/ab79c4.

[66] X. Dong et al., "Automatic multiorgan segmentation in thorax CT images using U-net-GAN," (in eng), *Medical Physics*, vol. 46, no. 5, pp. 2157–2168, May 2019, doi: 10.1002/mp.13458.

[67] N. Tong, S. Gou, S. Yang, D. Ruan, and K. Sheng, "Fully automatic multi-organ segmentation for head and neck cancer radiotherapy using shape representation model constrained fully convolutional neural networks," (in eng), *Medical Physics*, vol. 45, no. 10, pp. 4558–4567, Oct 2018, doi: 10.1002/mp.13147.

[68] K. Men et al., "Deep deconvolutional neural network for target segmentation of nasopharyngeal cancer in planning computed tomography images," (in eng), *Frontiers in Oncology*, vol. 7, Dec 20 2017, doi: ARTN 315 10.3389/fonc.2017.00315.

[69] S. Kazemifar et al., "Segmentation of the prostate and organs at risk in male pelvic CT images using deep learning," (in eng), *Biomedical Physics & Engineering Express*, vol. 4, no. 5, Sep 2018, doi: UNSP 055003 10.1088/2057-1976/aad100

[70] U. Javaid, D. Dasnoy, and J. A. Lee, "Multi-organ segmentation of Chest CT images in radiation oncology: Comparison of standard and dilated UNet," (in eng), *Advanced Concepts for Intelligent Vision Systems, Acivs 2018*, vol. 11182, pp. 188–199, 2018, doi: 10.1007/978-3-030-01449-0_16.

[71] S. Elguindi et al., "Deep learning-based auto-segmentation of targets and organs-at-risk for magnetic resonance imaging only planning of prostate radiotherapy," *Physics and Imaging in Radiation Oncology*, vol. 12, pp. 80–86, Oct 1 2019, doi: https://doi.org/10.1016/j.phro.2019.11.006

[72] X. Dong et al., "Synthetic MRI-aided multi-organ segmentation on male pelvic CT using cycle consistent deep attention network," (in eng), *Radiotherapy and Oncology*, vol. 141, pp. 192–199, Dec 2019, doi: 10.1016/j.radonc.2019.09.028.

[73] T. Wang, Y. Lei, Y. Fu, W. Curran, T. Liu, and X. Yang, "Machine Learning in Quantitative PET Imaging," *ArXiv*, arXiv:2001.06597, 2020.

[74] Y. Fu, Y. Lei, T. Wang, W. J. Curran, T. Liu, and X. Yang, "A review of deep learning based methods for medical image multi-organ segmentation," *Physica Medica*, vol. 85, pp. 107–122, May 2021, doi: 10.1016/j.ejmp.2021.05.003.

[75] M. Lin et al., "Artificial intelligence in tumor subregion analysis based on medical imaging: A review," *Journal of Applied Clinical Medical Physics*, vol. 22, no. 7, pp. 10–26, Jul 2021, doi: 10.1002/acm2.13321.

[76] J. J. Jeong, A. Ali, T. Liu, H. Mao, W. J. Curran, and X. Yang, "Radiomics in cancer radiotherapy: A review," *ArXiv*, arXiv:1910.02102, 2019.

[77] T. Wang, Y. Lei, Y. Fu, W. Curran, T. Liu, and X. Yang, "Medical imaging synthesis using deep learning and its clinical applications: A review," *arXiv: Medical Physics*, arXiv:2004.10322, 2020.

[78] T. Wang et al., "Machine learning in quantitative PET: A review of attenuation correction and low-count image reconstruction methods," *Physica Medica*, vol. 76, pp. 294–306, Aug 2020, doi: 10.1016/j.ejmp.2020.07.028.

[79] T. Wang et al., "A review on medical imaging synthesis using deep learning and its clinical applications," *Journal of Applied Clinical Medical Physics*, vol. 22, no. 1, pp. 11–36, Jan 1 2021, doi: https://doi.org/10.1002/acm2.13121

[80] Y. Lei et al., "MRI-based pseudo CT synthesis using anatomical signature and alternating random forest with iterative refinement model," (in eng), *Journal of Medical Imaging (Bellingham)*, vol. 5, no. 4, p. 043504, Oct 2018, doi: 10.1117/1.JMI.5.4.043504.

[81] Y. Lei et al., "Pseudo CT estimation using patch-based joint dictionary learning," *presented at the IEEE Eng Med Biol Soc*, Jul, 2018.

[82] X. Yang et al., "Pseudo CT estimation from MRI using patch-based random forest," in *Proceeding of SPIE Medical Imaging 2017: Image Processing*, 2017, vol. 10133, no. 101332Q.

[83] X. Yang et al., "MRI-based attenuation correction for brain PET/MRI based on anatomic signature and machine learning," (in eng), *Physics in Medicine and Biology*, vol. 64, no. 2, p. 025001, Jan 7 2019, doi: 10.1088/1361-6560/aaf5e0.

[84] X. Yang and B. Fei, "Multiscale segmentation of the skull in MR images for MRI-based attenuation correction of combined MR/PET," *Journal of the American Medical Informatics Association*, vol. 20, no. 6, pp. 1037–1045, Nov–Dec 2013, doi: 10.1136/amiajnl-2012-001544.

[85] T. Wang et al., "Learning-based stopping power mapping on dual energy CT for proton radiation therapy," *International Journal of Particle Therapy*, vol. 7, pp. 46–60, 2021.

[86] S. Charyyev et al., "Learning-based synthetic dual energy CT imaging from single energy CT for stopping power ratio calculation in proton radiation therapy," *The British Journal of Radiology*, vol. 95, pp. 20210644, 2022.

[87] S. Charyyev et al., "High quality proton portal imaging using deep learning for proton radiation therapy: A phantom study," *Biomedical Physics & Engineering Express*, vol. 6, no. 3, p. 035029, Apr 27 2020, doi: 10.1088/2057-1976/ab8a74.

[88] X. Dai et al., "Synthetic MRI-aided head-and-neck organs-at-risk auto-delineation for CBCT-guided adaptive radiotherapy," *arXiv: Medical Physics*, arXiv:2010.04275, 2020.

[89] Y. Lei et al., "CT prostate segmentation based on synthetic MRI-aided deep attention fully convolution network," *Medical Physics*, vol. 47, no. 2, pp. 530–540, Feb 2020, doi: 10.1002/mp.13933.

[90] Y. Lei et al., "CBCT-Based Synthetic MRI Generation for CBCT-Guided Adaptive Radiotherapy," in *Artificial Intelligence in Radiation Therapy*, Cham, D. Nguyen, L. Xing, and S. Jiang, Eds., 2019: Springer International Publishing, pp. 154–161.

[91] Y. Lei et al., "Male pelvic multi-organ segmentation aided by CBCT-based synthetic MRI," (in eng), *Physics in Medicine and Biology*, vol. 65, no. 3, p. 035013, Feb 4 2020, doi: 10.1088/1361-6560/ab63bb.

[92] Y. Liu et al., "Head and neck multi-organ auto-segmentation on CT images aided by synthetic MRI," *Medical Physics*, vol. 47, no. 9, pp. 4294–4302, Sep 2020, doi: 10.1002/mp.14378.

[93] Y. Fu et al., "Pelvic multi-organ segmentation on cone-beam CT for prostate adaptive radiotherapy," *Medical Physics*, vol. 47, no. 8, pp. 3415–3422, Aug 2020, doi: 10.1002/mp.14196.

[94] J. M. Wolterink, T. Leiner, M. A. Viergever, and I. Išgum, "Generative adversarial networks for noise reduction in low-dose CT," *IEEE Transactions on Medical Imaging*, vol. 36, no. 12, pp. 2536–2545, 2017, doi: 10.1109/TMI.2017.2708987.

[95] Y. Lei et al., "A denoising algorithm for CT image using low-rank sparse coding," in *Proceeding of SPIE Medical Imaging 2018: Image Processing*, 2018, vol. 10574, no. 105741P.

[96] X. Dai et al., "Intensity non-uniformity correction in MR imaging using residual cycle generative adversarial network," *Physics in Medicine and Biology*, vol. 65, no. 21, p. 215025, Nov 27 2020, doi: 10.1088/1361-6560/abb31f.

[97] T. Wang et al., "Dosimetric study on learning-based cone-beam CT correction in adaptive radiation therapy," (in eng), *Medical Dosimetry*, vol. 44, no. 4, pp. e71–e79, Winter 2019, doi: 10.1016/j.meddos.2019.03.001.

[98] Y. Lei et al., "Improving image quality of cone-beam CT using alternating regression forest," in *Proceeding of SPIE Medical Imaging 2018: Image Processing*, 2018, vol. 10573, no. 1057345.

[99] X. Yang et al., "Attenuation and scatter correction for whole-body PET using 3D generative adversarial networks," in *Journal of Nuclear Medicine*, 2019, vol. 60, no. supplement 1, p. 174.

[100] G. Yang et al., "DAGAN: Deep de-aliasing generative adversarial networks for fast compressed sensing MRI reconstruction," *IEEE Transactions on Medical Imaging*, vol. 37, no. 6, pp. 1310–1321, 2018, doi: 10.1109/TMI.2017.2785879.

[101] J. Harms et al., "Cone-beam CT-derived relative stopping power map generation via deep learning for proton radiotherapy," *Medical Physics*, vol. 47, no. 9, pp. 4416–4427, Sep 2020, doi: 10.1002/mp.14347.

[102] Y. Lei et al., "High-resolution CT image retrieval using sparse convolutional neural network," in *Proceeding of SPIE Medical Imaging 2018: Image Processing*, 2018, vol. 10573, no. 105733F

[103] G. Sharp et al., "Vision 20/20: Perspectives on automated image segmentation for radiotherapy," (in eng), *Medical Physics*, vol. 41, no. 5, p. 050902, May 2014, doi: 10.1118/1.4871620.

[104] Y. Lei et al., "MRI-based synthetic CT generation using semantic random forest with iterative refinement," (in eng), *Physics in Medicine and Biology*, vol. 64, no. 8, p. 085001, Apr 5 2019, doi: 10.1088/1361-6560/ab0b66.

[105] R. Farjam, N. Tyagi, J. O. Deasy, and M. A. Hunt, "Dosimetric evaluation of an atlas-based synthetic CT generation approach for MR-only radiotherapy of pelvis anatomy," *Journal of Applied Clinical Medical Physics*, vol. 20, no. 1, pp. 101–109, Jan 1 2019, doi: 10.1002/acm2.12501.

[106] J. Breunig et al., "A system for continual quality improvement of normal tissue delineation for radiation therapy treatment planning," (in eng), *International Journal of Radiation Oncology, Biology, Physics*, vol. 83, no. 5, pp. e703–e708, Aug 1 2012, doi: 10.1016/j.ijrobp.2012.02.003.

[107] G. M. Villeirs et al., "Interobserver delineation variation using CT versus combined CT + MRI in intensity-modulated radiotherapy for prostate cancer," (in eng), *Strahlentherapie und Onkologie: Organ der Deutschen Rontgengesellschaft ... [et al]*, vol. 181, no. 7, pp. 424–430, Jul 2005, doi: 10.1007/s00066-005-1383-x.

[108] X. Dong et al., "Air, bone and soft-tissue segmentation on 3D brain MRI using semantic classification random forest with auto-context model," *ArXiv*, arXiv:1911.09264, 2019.

[109] C. Liu et al., "Automatic segmentation of the prostate on CT images using deep neural networks (DNN)," (in eng), *International Journal of Radiation Oncology, Biology, Physics*, vol. 104, no. 4, pp. 924–932, Jul 15 2019, doi: 10.1016/j.ijrobp.2019.03.017.

[110] A. Balagopal et al., "Fully automated organ segmentation in male pelvic CT images," (in eng), *Physics in Medicine and Biology*, vol. 63, no. 24, p. 245015, Dec 14 2018, doi: 10.1088/1361-6560/aaf11c.

[111] Y. Lei et al., "Male pelvic CT multi-organ segmentation using synthetic MRI-aided dual pyramid networks," (in eng), *Physics in Medicine and Biology*, vol. 66, no. 8, Apr 16 2021, doi: 10.1088/1361-6560/abf2f9.

[112] X. Xu, J. Chen, H. Zhang, and G. Han, "Dual pyramid network for salient object detection," *Neurocomputing*, vol. 375, pp. 113–123, Jan 29 2020, doi: https://doi.org/10.1016/j.neucom.2019.09.077.

[113] Y. Lei et al., "MRI-based synthetic CT generation using cycle consistent adversarial network," *Medical Physics*, 2019, vol. 46, no. 6, pp. E516–E517.

[114] X. Yang et al., "Synthetic MRI-aided multi-organ CT segmentation for head and neck radiotherapy treatment planning," *International Journal of Radiation Oncology, Biology, Physics*, vol. 108, no. 3, p. e341, 2020.

[115] X. Dai et al., "Synthetic MRI-aided multi-organ segmentation in head-and-neck cone beam CT," in *Proceeding of SPIE Medical Imaging 2021: Image-Guided Procedures, Robotic Interventions, and Modeling*, 2021, vol. 11598, no. 115981M SPIE. [Online]. Available: https://doi.org/10.1117/12.2581128

[116] X. Dai et al., "Synthetic MRI-aided delineation of organs at risk in head-and-neck radiotherapy," *Medical Physics*, 2021, vol. 48, no. 6. arXiv:2010.04275.

[117] X. Dai et al., "Automated delineation of head and neck organs at risk using synthetic MRI-aided mask scoring regional convolutional neural network," *Medical Physics*, https://doi.org/10.1002/mp.15146 vol. 48, no. 10, pp. 5862–5873, Oct 2021, doi: 10.1002/mp.15146.

[118] Y. Lei et al., "Synthetic MRI-aided prostate segmentation in CT image," *Medical Physics*, 2019, vol. 46, no. 6, p. E505.

[119] Y. Lei et al., "Learning-based CBCT correction using alternating random forest based on auto-context model," (in eng), *Medical Physics*, vol. 46, no. 2, pp. 601–618, Feb 2019, doi: 10.1002/mp.13295.

[120] R. L. Qiu et al., "Chest CBCT-based synthetic CT using cycle-consistent adversarial network with histogram matching," in *Proceeding of SPIE Medical Imaging 2021: Image Processing*, 2021, vol. 11596, no. 115961Z SPIE. [Online]. Available: https://doi.org/10.1117/12.2581094

[121] X. Dai et al., "Synthetic CT-aided multiorgan segmentation for CBCT-guided adaptive pancreatic radiotherapy," *Medical Physics*, https://doi.org/10.1002/mp.15264 vol. 48, no. 11, pp. 7063–7073, Nov 2021, doi: 10.1002/mp.15264.

[122] Y. Lei et al., "Accurate CBCT prostate segmentation aided by CBCT-based synthetic MRI," *Medical Physics*, 2019, vol. 46, no. 6, p. E132.

[123] L. A. Matkovic et al., "Prostate and dominant intraprostatic lesion segmentation on PET/CT using cascaded regional-net," *Physics in Medicine & Biology*, vol. 66, no. 24, p. 245006, Dec 7 2021, doi: 10.1088/1361-6560/ac3c13.

[124] T. Wang et al., "Prostate and tumor segmentation on PET/CT using dual mask R-CNN," in *Proceeding of SPIE Medical Imaging 2021: Image Processing*, 2021, vol. 11600, no. 116000S SPIE. [Online]. Available: https://doi.org/10.1117/12.2580970

[125] T. Wang et al., "Deep-learning-based lesion segmentation on 18F-fluciclovine PET/CT," *Medical Physics*, 2022, vol. 49, no. 6, p. E795.

[126] T. Wang et al., "Lung tumor segmentation of PET/CT using dual pyramid mask R-CNN," in *Proceeding of SPIE Medical Imaging 2021: Image Processing*, 2021, vol. 11596, no. 1159632 SPIE. [Online]. Available: https://doi.org/10.1117/12.2580987

[127] X. Yang et al., "PET attenuation correction using MRI-aided two-stream pyramid attention network," *Journal of Nuclear Medicine*, vol. 61, no. supplement 1, p. 110, 2020. [Online]. Available: http://jnm.snmjournals.org/content/61/supplement_1/110.abstract

[128] X. Dong et al., "Jack Krohmer Junior investigator competition winner: Deep learning-based self attenuation correction for whole-body PET imaging," *Medical Physics*, 2019, vol. 46, no. 6, p. E192.

[129] T. Wang et al., "MRI-based attenuation correction for brain PET/MRI based on anatomic signature and machine learning," *Medical Physics*, 2019, vol. 46, no. 6, pp. E192–E193.

[130] X. Yang et al., "Lesion segmentation using convolutional neural network for PET/CT-guided salvage post-prostatectomy radiotherapy," *International Journal of Radiation Oncology, Biology, Physics*, vol. 114, no. 3, p. e116, 2022.

[131] VS Khoo and DL Joon, "New developments in MRI for target volume delineation in radiotherapy," *The British Journal of Radiology*, vol. 79, no. special_issue_1, pp. S2–S15, 2006.

[132] H. Chandarana, H. Wang, RHN Tijssen, I.J. Das "Emerging role of MRI in radiation therapy," vol. 48, no. 6, pp. 1468–1478, 2018.

[133] P. Metcalfe et al., "The potential for an enhanced role for MRI in radiation-therapy treatment planning," vol. 12, no. 5, pp. 429–446, 2013.

[134] R. Prestwich, J. Sykes, B. Carey, M. Sen, K. Dyker, and A. J. C. O. Scarsbrook, "Improving target definition for head and neck radiotherapy: A place for magnetic resonance imaging and 18-fluoride fluorodeoxyglucose positron emission tomography?," vol. 24, no. 8, pp. 577–589, 2012.

[135] J. T.-C. Chang et al., "Nasopharyngeal carcinoma with cranial nerve palsy: The importance of MRI for radiotherapy," vol. 63, no. 5, pp. 1354–1360, 2005.

[136] H. Lu, L. M. Nagae-Poetscher, X. Golay, D. Lin, M. Pomper, and P. C. van Zijl, "Routine clinical brain MRI sequences for use at 3.0 Tesla," *Journal of Magnetic Resonance Imaging*, vol. 22, no. 1, pp. 13–22, Jul 2005, doi: 10.1002/jmri.20356.

[137] R. Bitar et al., "MR pulse sequences: What every radiologist wants to know but is afraid to ask," *Radiographics*, vol. 26, no. 2, pp. 513–537, 2006.

[138] T. Wang et al., "Multiparametric MRI-guided dose boost to dominant intraprostatic lesions in CT-based High-dose-rate prostate brachytherapy," (in eng), *The British Journal of Radiology*, vol. 92, no. 1097, p. 20190089, May 2019, doi: 10.1259/bjr.20190089.

[139] Y. Lei et al., "Magnetic resonance imaging-based pseudo computed tomography using anatomic signature and joint dictionary learning," (in eng), *Journal of Medical Imaging (Bellingham)*, vol. 5, no. 3, p. 034001, Jul 2018, doi: 10.1117/1.JMI.5.3.034001.

[140] K. Armanious et al., "MedGAN: Medical image translation using GANs," *Computerized Medical Imaging and Graphics*, vol. 79, p. 101684, Jan 2020, doi: 10.1016/j.compmedimag.2019.101684.

[141] X. Dai et al., "Multimodal MRI synthesis using unified generative adversarial networks," *Medical Physics*, https://doi.org/10.1002/mp.14539 vol. 47, no. 12, pp. 6343–6354, Dec 2020, doi: 10.1002/mp.14539.

[142] Y. Liu et al., "Synthetic contrast MR image generation using deep learning," *Medical Physics*, 2021, vol. 48, no. 6.

[143] T. Wang, Y. Lei, W. Curran, T. Liu, and X. Yang, "Contrast-enhanced MRI synthesis from non-contrast MRI using attention CycleGAN," in *Proceeding of SPIE Medical Imaging 2021: Biomedical Applications in Molecular, Structural, and Functional Imaging*, 2021, vol. 11600, no. 116001L SPIE. [Online]. Available: https://doi.org/10.1117/12.2581064

[144] H. Xie et al., "Magnetic resonance imaging contrast enhancement synthesis using cascade networks with local supervision," *Medical Physics*, vol. 49, no. 5, pp. 3278–3287, 2022.

[145] T. Wang and L. Zhu, "Dual energy CT with one full scan and a second sparse-view scan using structure preserving iterative reconstruction (SPIR)," *Physics in Medicine & Biology*, vol. 61, no. 18, p. 6684, 2016. [Online]. Available: http://stacks.iop.org/0031-9155/61/i=18/a=6684

[146] J. Harms, T. Wang, M. Petrongolo, T. Niu, and L. Zhu, "Noise suppression for dual-energy CT via penalized weighted least-square optimization with similarity-based regularization," *Medical Physics*, vol. 43, no. 5, pp. 2676–2686, 2016, doi: http://doi.org/10.1118/1.4947485

[147] J. Harms, T. Wang, M. Petrongolo, and L. Zhu, "Noise suppression for energy-resolved CT using similarity-based non-local filtration," in *Proceeding of SPIE Medical Imaging 2016: Physics of Medical Imaging*, 2016, vol. 9783: SPIE, p. 8.

[148] W. van Elmpt, G. Landry, M. Das, and F. Verhaegen, "Dual energy CT in radiotherapy: Current applications and future outlook," (in eng), *Radiotherapy and Oncology: Journal of the European Society for Therapeutic Radiology and Oncology*, vol. 119, no. 1, pp. 137–144, Apr 2016, doi: 10.1016/j.radonc.2016.02.026.

[149] B. Ruzsics, H. Lee, P. L. Zwerner, M. Gebregziabher, P. Costello, and U. J. Schoepf, "Dual-energy CT of the heart for diagnosing coronary artery stenosis and myocardial ischemia-initial experience," (in eng), *European Radiology*, vol. 18, no. 11, pp. 2414–2424, Nov 2008, doi: 10.1007/s00330-008-1022-x.

[150] D. N. Tran, M. Straka, J. E. Roos, S. Napel, and D. Fleischmann, "Dual-energy CT discrimination of iodine and calcium: Experimental results and implications for lower extremity CT angiography," (in eng), *Academic Radiology*, vol. 16, no. 2, pp. 160–171, Feb 2009, doi: 10.1016/j.acra.2008.09.004.

[151] Y. Watanabe et al., "Dual-energy direct bone removal CT angiography for evaluation of intracranial aneurysm or stenosis: Comparison with conventional digital subtraction angiography," (in eng), *European Radiology*, vol. 19, no. 4, pp. 1019–1024, Apr 2009, doi: 10.1007/s00330-008-1213-5.

[152] T. Kau et al., "Dual-energy CT angiography in peripheral arterial occlusive disease-accuracy of maximum intensity projections in clinical routine and subgroup analysis," (in eng), *European Radiology*, vol. 21, no. 8, pp. 1677–1686, Aug 2011, doi: 10.1007/s00330-011-2099-1.

[153] J. Zhu and S. N. Penfold, "Dosimetric comparison of stopping power calibration with dual-energy CT and single-energy CT in proton therapy treatment planning," (in eng), *Medical Physics*, vol. 43, no. 6, pp. 2845–2854, Jun 2016, doi: 10.1118/1.4948683.

[154] M. Yang, G. Virshup, J. Clayton, X. R. Zhu, R. Mohan, and L. Dong, "Theoretical variance analysis of single- and dual-energy computed tomography methods for calculating proton stopping power ratios of biological tissues," (in eng), *Physics in Medicine and Biology*, vol. 55, no. 5, pp. 1343–1362, Mar 7 2010, doi: 10.1088/0031-9155/55/5/006.

[155] T. Wang et al., "Optimal virtual monoenergetic image in "TwinBeam" dual-energy CT for organs-at-risk delineation based on contrast-noise-ratio in head-and-neck radiotherapy," *Journal of Applied Clinical Medical Physics*, vol. 20, no. 2, pp. 121–128, 2019, doi: 10.1002/acm2.12539.

[156] S. R. Pomerantz et al., "Virtual monochromatic reconstruction of dual-energy unenhanced head CT at 65–75 keV maximizes image quality compared with conventional polychromatic CT," (in eng), *Radiology*, vol. 266, no. 1, pp. 318–325, Jan 2013, doi: 10.1148/radiol.12111604.

[157] S. Lam, R. Gupta, M. Levental, E. Yu, H. D. Curtin, and R. Forghani, "Optimal Virtual Monochromatic Images for Evaluation of Normal Tissues and Head and Neck Cancer Using Dual-Energy CT," (in eng), *American Journal of Neuroradiology*, vol. 36, no. 8, pp. 1518–1524, Aug 2015, doi: 10.3174/ajnr.A4314.

[158] J. L. Wichmann et al., "Virtual monoenergetic dual-energy computed tomography: Optimization of kiloelectron volt settings in head and neck cancer," (in eng), *Investigative Radiology*, vol. 49, no. 11, pp. 735–741, Nov 2014, doi: 10.1097/rli.0000000000000077.

[159] C. Frellesen et al., "Dual-energy CT of the pancreas: Improved carcinoma-to-pancreas contrast with a noise-optimized monoenergetic reconstruction algorithm," (in eng), *European Journal of Radiology*, vol. 84, no. 11, pp. 2052–2058, Nov 2015, doi: 10.1016/j.ejrad.2015.07.020.

[160] L. D. Di Maso, J. Huang, M. F. Bassetti, L. A. DeWerd, and J. R. Miller, "Investigating a novel split-filter dual-energy CT technique for improving pancreas tumor visibility for radiation therapy," *Journal of Applied Clinical Medical Physics*, vol. 19, no. 5, pp. 676–683, 2018, doi: 10.1002/acm2.12435.

[161] S. Charyyev et al., "Synthetic dual energy CT images from single energy CT image for proton radiotherapy," *Medical Physics*, 2020, vol. 47, no. 6, pp. E378–E379.

[162] S. Charyyev et al., "Learning-based synthetic dual energy CT imaging from single energy CT for stopping power ratio calculation in proton radiation therapy," *The British Journal of Radiology*, vol. 95, no. 1129, p. 20210644, Jan 1 2022, doi: 10.1259/bjr.20210644.

[163] R. Liu et al., "Synthetic dual-energy CT for MRI-based proton therapy treatment planning using label-GAN," *Medical Physics*, 2021, vol. 48, no. 6. doi: 10.1088/1361-6560/abe736.

[164] R. Liu et al., "Synthetic dual-energy CT for MRI-only based proton therapy treatment planning using label-GAN," (in eng), *Physics in Medicine and Biology*, vol. 66, no. 6, p. 065014, Mar 9 2021, doi: 10.1088/1361-6560/abe736.

[165] T. Wang et al., "Head and neck multi-organ segmentation on dual-energy CT using dual pyramid convolutional neural networks," *Physics in Medicine & Biology*, vol. 66, no. 11, p. 115008, May 20 2021, doi: 10.1088/1361-6560/abfce2.

[166] X. Dai et al., "Synthetic CT-based multi-organ segmentation in cone beam CT for adaptive pancreatic radiotherapy," in *SPIE Medical Imaging*, 2021, vol. 11596, no. 1159623 SPIE. [Online]. Available: https://doi.org/10.1117/12.2581132

[167] J. Janopaul-Naylor et al., "Synthetic CT-aided Online CBCT Multi-Organ Segmentation for CBCT-guided Adaptive Radiotherapy of Pancreatic Cancer," *International Journal of Radiation Oncology, Biology, Physics*, vol. 108, no. 3, pp. S7–S8, 2020, doi: 10.1016/j.ijrobp.2020.07.2080.

[168] Y. Lei et al., "Thoracic CBCT-based synthetic CT for lung stereotactic body radiation therapy," *Medical Physics*, 2021, vol. 48, no. 6.

[169] Y. Liu et al., "Deep-learning-based synthetic-CT generation method for CBCT-guided proton therapy," *Medical Physics*, 2019, vol. 46, no. 6, p. E475.

[170] Y. Liu et al., "CBCT-based synthetic CT generation using deep-attention cycleGAN for pancreatic adaptive radiotherapy," (in eng), *Medical Physics*, vol. 47, no. 6, pp. 2472–2483, Jun 2020, doi: 10.1002/mp.14121.

[171] T. Wang et al., "Synthesizing high-resolution CT from low-resolution CT using self-learning," in *Proceeding of SPIE Medical Imaging 2021: Physics of Medical Imaging*, 2021, vol. 11595, no. 115952N SPIE. [Online]. Available: https://doi.org/10.1117/12.2581080

[172] H. Xie et al., "Synthesizing high-resolution magnetic resonance imaging using parallel cycle-consistent generative adversarial networks for fast magnetic resonance imaging," *Medical Physics*, https://doi.org/10.1002/mp.15380 vol. 49, no. 1, pp. 357–369, Jan 1 2022, doi: https://doi.org/10.1002/mp.15380

[173] X. Dai et al., "Head-and-neck organs-at-risk auto-delineation using dual pyramid networks for CBCT-guided adaptive radiotherapy," (in eng), *Physics in Medicine and Biology*, vol. 66, no. 4, p. 045021, Feb 11 2021, doi: 10.1088/1361-6560/abd953.

[174] Y. Lei et al., *Deep Learning Architecture Design for Multi-Organ Segmentation*, 1st Edition ed. (Auto-Segmentation for Radiation Oncology: State of the Art). Boca Raton: CRC Press, 2021, p. 32.

[175] Y. Lei et al., "Deep learning in multi-organ segmentation," *ArXiv*, arXiv:2001.10619, 2020.

[176] Y. Lei et al., "Cascaded learning-based cone beam CT head-and-neck multi-organ segmentation," *Medical Physics*, 2022, vol. 49, no. 6, pp. E551–E552.

[177] Y. Fu et al., "CBCT-based prostate and organs-at-risk segmentation using deep attention network," *Medical Physics*, 2021, vol. 48, no. 6.

[178] V. Thambawita et al., "SinGAN-seg: Synthetic training data generation for medical image segmentation," (in eng), *PLoS One*, vol. 17, no. 5, p. e0267976, 2022, doi: 10.1371/journal.pone.0267976.

[179] X. Dai et al., "Deep learning-based volumetric image generation from projection imaging for prostate radiotherapy," in *SPIE Medical Imaging*, 2021, vol. 11598, no. 115981R SPIE. [Online]. Available: https://doi.org/10.1117/12.2581053

[180] C. Chang et al., "Realtime volumetric image for target localization during proton FLASH radiotherapy," in *Medical Physics*, 2022, vol. 49, no. 6, p. E489.

[181] Y. Lei et al., "Deep learning-based real-time volumetric imaging for lung stereotactic body radiation therapy: A proof of concept study," *Physics in Medicine and Biology*, vol. 65, no. 23, p. 235003, Dec 18 2020, doi: 10.1088/1361-6560/abc303.

[182] Y. Lei et al., "Real-time patient-specific volumetric imaging for lung stereotactic body radiation therapy via InferGAN with perceptual supervision," *Medical Physics*, 2020, vol. 47, no. 6, p. E361.

[183] Y. Lei et al., "Deep learning-based real-time volumetric imaging using single CBCT projections," *Medical Physics*, 2021, vol. 48, no. 6.

[184] Y. Lei et al., "Deep learning-based fast volumetric imaging for treatment setup during lung radiotherapy," *Medical Physics*, 2022, vol. 49, no. 6, pp. E604–E605.

[185] Y. Fu et al., "CT-based volumetric strain imaging via a deep learning registration framework," in *Proceeding of SPIE Medical Imaging 2022: Biomedical Applications in Molecular, Structural, and Functional Imaging*, 2022, vol. 12036: SPIE, pp. 542–547.

[186] C.-W. Chang et al., "A deep learning approach to transform two orthogonal X-ray images to volumetric images for image-guided proton therapy," in *Proceeding of SPIE Medical Imaging 2022: Image Processing*, 2022, vol. 12032: SPIE, pp. 484–490.

[187] C.-W. Chang et al., "Deep learning-based fast volumetric image generation for image-guided proton FLASH radiotherapy," *arXiv preprint*, arXiv:2210.00971, 2022.

[188] L. Lindner, D. Narnhofer, M. Weber, C. Gsaxner, M. Kolodziej, and J. Egger, "Using synthetic training data for deep learning-based GBM segmentation," (in eng), *2019 41st Annual International Conference of the IEEE Engineering in Medicine and Biology Society (EMBC)*, vol. 2019, pp. 6724–6729, Jul 2019, doi: 10.1109/embc.2019.8856297.

[189] Y. Shin, H. A. Qadir, and I. Balasingham, "Abnormal colon polyp image synthesis using conditional adversarial networks for improved detection performance," *IEEE Access*, vol. 6, pp. 56007–56017, 2018, doi: 10.1109/ACCESS.2018.2872717.

[190] C. Ouyang, C. Biffi, C. Chen, T. Kart, H. Qiu, and D. Rueckert, "Self-supervised learning for few-shot medical image segmentation," (in eng), *IEEE Transactions on Medical Imaging*, vol. 41, no. 7, pp. 1837–1848, Jul 2022, doi: 10.1109/tmi.2022.3150682.

[191] X. Yang et al., "Learning-based real-time patient-specific volumetric imaging for lung stereotactic body radiation therapy," *International Journal of Radiation Oncology, Biology, Physics*, vol. 108, no. 3, pp. e345–e346, 2020.

[192] X.-Y. Zhou and G.-Z. Yang, "Normalization in training U-net for 2-D biomedical semantic segmentation," *IEEE Robotics Automation Letters*, vol. 4, pp. 1792–1799, 2018.

[193] Y. Liu et al., "A deep-learning-based intensity inhomogeneity correction for MR imaging," *Medical Physics*, 2019, vol. 46, no. 6, p. E386.

[194] Y. Lei et al., "Male pelvic multi-organ segmentation on transrectal ultrasound using anchor-free mask CNN," *Medical Physics*, vol. 48, no. 6, pp. 3055–3064, Jun 2021, doi: 10.1002/mp.14895.

[195] Y. Lei et al., "Breast tumor segmentation in 3D automatic breast ultrasound using Mask scoring R-CNN," *Medical Physics*, vol. 48, no. 1, pp. 204–214, Jan 2021, doi: 10.1002/mp.14569.

[196] Y. Lei et al., "Echocardiographic image multi-structure segmentation using Cardiac-SegNet," *Medical Physics*, vol. 48, no. 5, pp. 2426–2437, May 2021, doi: 10.1002/mp.14818.

[197] B. Jun Guo et al., "Automated left ventricular myocardium segmentation using 3D deeply supervised attention U-net for coronary computed tomography angiography; CT myocardium segmentation," (in eng), *Medical Physics*, vol. 47, no. 4, pp. 1775–1785, Apr 2020, doi: 10.1002/mp.14066.

[198] J. Roper et al., "Evaluation of commercial AI segmentation software," *Medical Physics*, 2022, vol. 49, no. 6, p. E552.

[199] S. Momin et al., "Lung tumor segmentation in 4D CT images using motion convolutional neural networks," *Medical Physics*, vol. 48, no. 11, pp. 7141–7153, Nov 1 2021, doi: https://doi.org/10.1002/mp.15204

[200] P. F. Christ et al., "Automatic liver and lesion segmentation in CT using cascaded fully convolutional neural networks and 3D conditional random fields," in *International Conference on Medical Image Computing and Computer-Assisted Intervention*, 2016.

[201] O. Ronneberger, P. Fischer, and T. Brox, "U-net: Convolutional networks for biomedical image segmentation," (in eng), *Medical Image Computing and Computer-Assisted Intervention*, vol. 9351, pp. 234–241, 2015, doi: 10.1007/978-3-319-24574-4_28.

[202] V. Alex, K. Vaidhya, S. Thirunavukkarasu, C. Kesavadas, and G. Krishnamurthi, "Semisupervised learning using denoising autoencoders for brain lesion detection and segmentation," (in eng), *Journal of Medical Imaging (Bellingham)*, vol. 4, no. 4, p. 041311, Oct 2017, doi: 10.1117/1.Jmi.4.4.041311.

[203] K. Vaidhya, S. Thirunavukkarasu, A. Varghese, and G. Krishnamurthi, "Multi-modal brain tumor segmentation using stacked denoising autoencoders," in *Brainles@MICCAI*, 2015.

[204] S. A. Taghanaki et al., "Combo loss: Handling input and output imbalance in multi-organ segmentation," *Computerized Medical Imaging and Graphics: The Official Journal of the Computerized Medical Imaging Society*, vol. 75, pp. 24–33, 2019.

[205] S. Momin et al., "Cascaded mutual enhancing networks for brain tumor subregion segmentation in multiparametric MRI," (in eng), *Physics in Medicine and Biology*, vol. 67, no. 8, Apr 11 2022, doi: 10.1088/1361-6560/ac5ed8.

[206] J. Johnson and T. Khoshgoftaar, "Survey on deep learning with class imbalance," *Journal of Big Data*, vol. 6, pp. 1–54, 2019.

15

Synthetic Image-Aided Registration

Yabo Fu
Memorial Sloan Kettering Cancer Center, New York, NY, USA

Xiaofeng Yang
Emory University, Atlanta, GA, USA

CONTENT

Multimodal image registration is important for various medical applications such as disease diagnosis, surgical planning, interventional image guidance, radiotherapy and so on. Multimodal image registration aims to blend information acquired by different imaging modalities which capture different physical properties of the subject by aligning the multimodal images. Inter-subject MRI image registration among different imaging sequences has been used to analyze structural variations in normal and diseased populations [1–3]. MRI brain atlas registration is an important tool in mapping regions of cerebral structures, shapes, characteristics, and its functional activations [4]. The fusion of MRI anatomical images and functional images is crucial for neurosurgeons when planning surgical procedures to target malignant lesions and avoid healthy and functional regions [5]. For interventional neurosurgery, such as tumor biopsy, cyst resection, CT-based image guidance is usually necessary to account for anatomical changes such as soft tissue deformation caused by surgical instrumentation. In this process, CT-MRI registration is indispensable for surgery planning since the fused MRI provides excellent soft tissue contrast which helps the visualization of crucial structures such as gray and white matter, cerebrospinal fluid (CSF), and subcortical structures. In radiotherapy, CT and cone beam CT (CBCT) image registration is used for daily patient setup. For Gamma Knife (GK) stereotactic radiosurgery, a brain tumor is usually delineated on MRI T1 images and registered to CBCT for stereotactic definition. For prostate cancer, MRI and transrectal ultrasound image registration could be used for both MR-guided biopsy [6] and brachytherapy needle implant [7].

Multimodal image registration has been studied for years and remains a hot research topic because of its importance and technical difficulties. Images acquired by different modalities usually have vast image appearance gaps caused by different imaging physical principles. For non-rigid image registration, traditional methods solve the deformation by minimizing a global loss function which is often the image dissimilarity loss and deformation regularization loss. For monomodal image registration, the image dissimilarity loss is usually relatively easy to define. Commonly used metrics for monomodal registration include the sum squared difference (SSD) [8], cross-correlation (CC) [8] and ratio image uniformity (RIU) [9]. These image dissimilarity metrics work quite well for monomodal

DOI: 10.1201/9781003243458-19

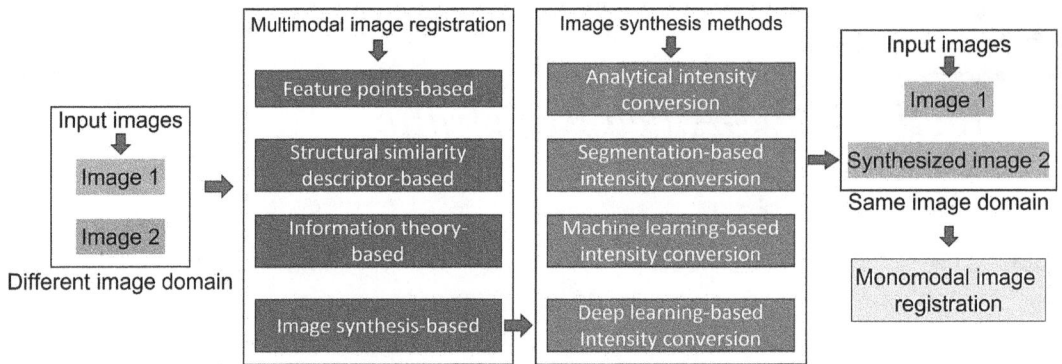

FIGURE 15.1
Image synthesis-based registration translates multimodal registration into monomodal registration.

image registration. However, these metrics do not work well for multimodal image registration due to inconsistent and uncorrelated image intensity. The lack of effective image similarity metrics for multimodal image registration has been one of the biggest challenges which prevent it from being widely used clinically. Figure 15.1 shows four major groups of multimodal image registration and focuses on the image-synthesis-based method. The four major groups are salient feature points-based registration, structural similarity-descriptor-based registration, information-theory-based registration, and image-synthesis-based registration [10].

The salient feature points-based registration first extracts a number of feature points such as the vessel bifurcations, corners, and edge points, and then utilizes point cloud registration to perform the image registration [11]. Rivest-Henault *et al.* used feature point-based registration which extracted the centerline of a coronary artery from 3D pre-interventional CT angiography and registered with live 2D fluoroscopy for percutaneous coronary interventions [12]. Closet iterative point (ICP) [13] and coherent point drifting [14] have also been used to register feature points for neuro-interventional image guidance. The limitations of salient feature points-based registration are: 1) it is difficult to extract sufficient reliable feature points that are uniformly distributed across the image domain; 2) with separately detected feature points on the fixed and moving images, it is challenging to establish robust point correspondence; and 3) the feature point itself is subject to localization errors which can propagate to the point-based registration algorithms. McLaughlin *et al.* compared the registration accuracy between intensity-based and ICP-based registration using a phantom study and found out the former to be more accurate [13].

Structural similarity descriptor-based registration utilizes various self-similarity descriptors which are independent of absolute image intensity values for multimodal image registration. Examples of such descriptors include the modality-independent neighborhood descriptor (MIND) [15] and discriminative local derivative pattern (dLDP). The MIND extracts distinctive structures in a local neighborhood and has been shown to be effective for certain MR-CT registration and outperformed conditional mutual information. Jiang *et al.* proposed to use dLDP to encode multimodal images into similar image representations. The dLDP is essentially a binary descriptor that is derived from the pattern of intensity derivatives in one voxel's neighborhood. They have shown that dLDP outperformed localized mutual information in MR-CT and MR-ultrasound registration [16]. Such structural descriptors bring multimodal image registration into a common image representation domain. However, limitations of such methods are that 1) it is difficult to handcraft a proper descriptor which might be specific to the imaging modality and region of interest; and 2) information loss during the descriptor construction, e.g., only derivative patterns are preserved whereas the original absolute intensity value which indicates a particular tissue physical property is lost.

Mutual information (MI) is a typical similarity measure for multimodal image registration. MI is an information-theoretic metric that seeks to minimize the joint histogram of the multimodal image intensities. The original MI is a global measure that lacks sufficient spatial and anatomical structural information. To alleviate some of the concerns of MI, various MI variants have been proposed such as localized mutual information [17], conditional mutual information [18], and MI incorporating spatial and geometric context [19]. Despite of these MI variants, MI-based similarity metric is generally more time-consuming to evaluate than monomodal similarity measures such as SSD and CC since MI is inherently a statistical measure. Repeated similarity evaluation using MI can be time-consuming [20].

The final group of methods for multimodal image registration is image synthesis-based registration. The basic idea is to translate one image modality to the other one to convert the multimodal image registration to monomodal image registration, which enables SSD, NCC, and RIU to be used as image similarity metrics during registration. The idea of image intensity conversion is not new. In 1995, Friston *et al.* published a method using image intensity conversion and normalization for spatial registration of MRI and PET. They partitioned the difference between the fixed and moving images as two components, one due to the spatial misalignment and the other due to the intrinsic intensity difference assuming perfect alignment. The intensity conversion was implemented as a nonlinear transformation to the voxel value followed by convolution. The MRI structural image can be transformed to approximate PET image after intensity transformation and convolution. The nonlinear intensity conversion was linearized using basis functions and Taylor series [21]. The work is one of the pioneers that used synthetic images for registration. One big limitation was the "reasonableness" and "appropriateness" of the simplified modeling of multimodal intensity mapping. As pointed out by Guimond *et al.*, this image synthesis method may not be robust to outliers. They proposed to model the intensity mapping using higher-degree polynomials, which were determined using a least trimmed square method followed by a binary reweighted least square. They claimed that the combination of the two least square methods is robust to outliers. Another concept proposed in this study is the bifunctional dependency assumption. The assumption allows two intensity mapping functions to be defined for a certain voxel. This is because the image signal from one modality could have the same response to different tissues, such as ventricles and bones have a similar response in T1 weight MRI while gray matter and white matter have a similar response in CT images. The registration was performed by iterating between correcting for intensity differences and monomodal registration. They found that image synthesis-based image registration is more accurate than MI-based registration, especially for high-dimensional deformations [22].

Another type of early image intensity conversion is to use tissue segmentation and atlas registration to assign pre-determined intensity values from one modality to the other. Roy *et al.* synthesized CT images from MRI using co-registered pair of MR and CT images as atlas [23]. Image patches from the subject MRI were registered to the atlas CT images. The synthetic CT patch cloud was combined using the Gaussian mixture model as introduced in coherent point drift. The MR-CT brain image registration was performed using the synthetic CT as a bridge and using CC as a similarity metric between synthetic CT and CT. The quality of the synthetic CT needs to be improved as holes can be observed in the skull of the synthetic CT. Despite the poor quality of synthetic CT, they showed their method outperformed the VelocityAI [24] and SyN [25] which are MI-based registration by 25%. The authors noticed the poor image quality of synthetic CT and mentioned that multi-atlas with different patch sizes and shapes could potentially improve image quality and registration performance. Patch-based image synthesis for registration has been used on not only MRI-CT brain images but also MRI brain T1-weighted, T2-weighted, proton density images. Iglesias *et al.* published a study in 2013, hoping to answer the question: is synthesizing MRI contrast useful for multimodal MRI image registration?[26] Though the synthesized images seem to be visually acceptable, it is unclear to what extent they can substitute the original images in medical image analysis such as segmentation and registration. To answer this question, they have collected around 40 MRI scans of different imaging sequences with manually labeled structures for registration evaluation. For image synthesis, they used a method similar to [27]. They have chosen to use ANTS [25], a symmetric diffeomorphic registration method, with a Gaussian regularization kernel size of 3 mm. They have found that using the synthetic MRI and NCC as similarity metrics for the registration considerably outperformed directly using MI for multimodal registration.

Chen *et al.* proposed cross-contrast multi-channel image registration using image synthesis for MR brain [28]. For image synthesis, they also used a patch-based method that was introduced by Jog *et al.* [29]. This method trains a regression forest to learn the nonlinear intensity mapping between different modalities. The registration proposed by Chen *et al.* differs from many other registrations in that they used two-channel registration rather than synthesizing only one modality to register with the other. They first synthesize the missing modality for both the fixed and moving images. These synthesized images together with the original images were taken as two pairs of correlated monomodal image registration. The synthesized images were used as a proxy in the cross-modality registration. Therefore, they call their method PROXI, which stands for proxy registration of cross-modality images. Multi-channel registration aims to register a set of moving images to a set of correlated fixed images. The image similarity measures are calculated separately for each channel. The final similarity metric is taken as a weighted sum of all the channels. Though the multi-channel registration concept is not new itself [30], it is interesting to combine both original images and synthetic images as separate registration channels. They showed that the multi-channel registration with synthetic images consistently performed better than the single-channel multimodal registration using MI. One limitation of the patch-based image synthesis is that it has only been tested on MRI brain images and may not be applicable to whole-body MR scans because small patch sizes may not be able to discern similar patches representing different soft tissues.

Cao *et al.* also noticed that many image syntheses aided cross-modality registration methods only perform the image synthesis in a single direction, usually from image modality with rich structural contrast to image modality with less image contrast, such as synthetic CT from MRI. They proposed to use bi-directional synthesis in the registration. Different from the above-mentioned multi-channel registration, they used region-adaptive key point-guided registration. For pelvic registration, they first synthesized MRI and CT in both directions and detected key points near pubic bones on CT, and key points near soft tissues including bladder, rectum, and prostate on MRI. For each key point, a local image patch was extracted as a morphological signature to search for its corresponding key point on the other image. In this study, a multi-target regression forest model was proposed for bi-directional image synthesis. Using volumetric overlap of the main pelvic organs as evaluation metrics of registration, they have found out that bi-directional registration consistently outperformed single direction synthesis-aided registration. As compared with SyN using MI as similarity metric, their proposed method has higher volumetric overlap for both intra- and inter-subject MRI-CT registration [31].

As compared to regression-based image synthesis, deep learning-based image synthesis using convolutional neural network (CNN) and generative adversarial network (GAN) has dramatically improved the image quality of synthetic images. Therefore, the synthesis-aided registration techniques were explored again based on the improved synthetic image quality. Liu *et al.* proposed to use a ten-layer full CNN to learn end-to-end intensity mapping across modalities. Due to computer memory restrictions, they trained the CNN using 2D image patches. Though the training took hours, it needs to be done only once. The multimodal MR brain image registration can be transformed into monomodal image registration which is much faster. Using SSD as similarity metrics, they nonrigidly registered both the BrainWeb phantom and IXI real patient datasets using cubic b-spline-based registration. Their method has consistently outperformed MI-based registration with the registration errors being 1.19, 2.23, and 1.57 compared to 1.53, 2.60, and 2.36 of LMI and 1.34, 2.39, and 1.76 of α-MI in T2-weighted vs PD, T1-weighted vs PD, and T1-weighted vs T2-weighted image registration, respectively, for IXI real patients' data [10]. In head-and-neck (HN) radiotherapy, MR-CT image registration can greatly improve organ contouring. However, due to the different immobilization devices used and the highly flexible neck position, it is very challenging to non-rigidly register both the bony structure and soft tissue well for MR-CT HN registration. Recognizing the challenges of HN registration, McKenzie *et al.* proposed to use a deep-learning-derived synthetic CT as a bridge for the HN registration [32]. Cycle-GAN was used to generate synthetic CT from MRI [33]. The cycle consistent loss term helps to enforce a realistic intensity mapping. As compared to regression-based or atlas-based image synthesis methods, cycleGAN-based image synthesis represents a new era and is more accurate. Multi-resolution b-spline-based registration was performed using Elastix [34]. They showed that this registration pipeline of using synthetic CT as a bridge could significantly improve the result over direct registration when the deformation is large. They also conducted inverse consistency experiments where mean square error was calculated for MR-CT-MR, CT-MR-CT, synCT-CT-synCT, and CT-synCT-CT registration. With synthetic CT, the mean square error was reduced, indicating more realistic deformation. A similar study was published by Fu *et al.* and showed improved target registration error with synthetic CT [35]. The image synthesis methods we have mentioned so far are mainly focused on anatomical sites that have small physiological motions, such as the brain, HN, and pelvis. Meanwhile, the image synthesis methods mostly require paired multimodal images to train the image synthesis network. However, these methods might not be applicable to thoracic and abdominal sites where large respiratory motion is present. Additionally, paired multimodal images are difficult to obtain for these anatomical sites, such as paired MR-CT lung images or paired MR-CT abdomen images. To investigate the challenges of both image synthesis and synthesis-aided registration for the thorax and abdomen, Tanner *et al.* used unpaired CT-MRI to train a CycleGAN to image synthesis [36]. Since unpaired images were used for training, the synthesized image sometimes has a clear different lung diaphragm from the source modality. For comparison, they performed the multimodal image registration using MIND, NMI, and MIND+NMI as similarity measures. Local NCC was used for the monomodal registration between the synthetic images and the source images. Total variation regularization was used in the deformable image registration. Grid search was used to find optimal registration parameters. The results showed that the registration performance deteriorated when a synthetic image was used for the thorax. The lung volume overlaps reduced from 76% when direct multimodal registration was used, to 69% when synthetic images were used. For the abdomen, the two registration methods have similar performance. Therefore, it is absolutely crucial to ensure that the synthetic image has correct spatial correspondence with the source image before it is used for registration.

Many synthesis-aided registrations aim at a particular multimodality pair, such as CT-MRI, T1-T2, and T1-PD. A dedicated image synthesis method needs to be developed for intensity mapping. Recently, Hoffmann *et al.* developed a contrast-agnostic image registration network, SynthMorph [37]. This method was inspired by a contrast-agnostic MRI segmentation method [38]. Deep learning-based image registration methods can predict the registration result very quickly in a single or a few iterations. However, they only work for the image modality pair on which they are trained. A network trained on T1-T2 image pair may not work on T1-PD image registration. SynthMorph was trained not on real images but on synthetic images that were generated from random label images. First, a generative model was used to create random label maps. Based on these label maps, synthetic images were generated with arbitrary contrast, deformation, and artifacts. Whenever available, the label of a real patient can also be used to train the SynthMorph network. To show the effectiveness of this contrast-agnostic registration, they compared the results with NiftyReg, ANT, and VoxelMorph using acquired real MRI contrast pairs with gradually varying contrast. FLASH images were acquired with a varying flip angle to get images with contrast from PD-weighted to T1-weighted. On MRI brain images, they have shown that SynthMorph has the most consistent performance that is nearly independent of the image contrast to be registered. In contrast, the performance of other methods deteriorated as the image contrast deviated from the original image contrast.

To sum up, image synthesis-based registration has almost consistently outperformed traditional MI-based image registration for multimodal registration. However, it is important to note that the quality of the synthetic images plays an important role in the performance of the subsequent registration. Image synthesis can transform the more challenging multimodal registration to a relatively easy monomodal registration. But the synthetic images will inevitably introduce some biased or inaccurate information into the registration process. For image synthesis in both directions, it is crucial to verify that the synthetic image does not arbitrarily create some artificial contrast. Such fake contrast may steer the registration toward a biased result. Such phenomenon is especially apparent when synthesizing image modality with rich information e.g., MRI from image modality with less structural information, e.g., CT. To date, the majority of body sites tested for synthesis-aided registration is the brain, this is partially due to the relatively small deformation for the brain. With deep-learning-based image synthesis methods available, many body parts have been explored for image synthesis [39]. There is much potential to explore the performance and possibility of multimodal registration on other body parts. Additionally, synthesis-aided registration has been focused on using traditional iterative registration methods to register the synthetic monomodal images. With more and more deep-learning-based registration networks proposed [40], it is possible to devise a unified network that could perform image synthesis and deformation prediction at the same time.

References

[1] D. L. Collins, P. Neelin, T. Peters et al., "Automatic 3D intersubject registration of MR volumetric data in standardized talairach space," *Journal of Computer Assisted Tomography*, 18, 192–205 (1994).

[2] P. M. Thompson, D. MacDonald, M. S. Mega et al., "Detection and mapping of abnormal brain structure with a probabilistic atlas of cortical surfaces," *Journal of Computer Assisted Tomography*, 21(4), 567–581 (1997).

[3] P. M. Thompson, J. Moussai, S. Zohoori et al., "Cortical variability and asymmetry in normal aging and Alzheimer's disease," *Cerebral Cortex*, 8(6), 492–509 (1998).

[4] A. W. Toga, and P. M. Thompson, "The role of image registration in brain mapping," *Image and Vision Computing*, 19(1–2), 3–24 (2001).

[5] P. Risholm, A. J. Golby, and W. Wells, 3rd, "Multimodal image registration for preoperative planning and image-guided neurosurgical procedures," *Neurosurgery Clinics of North America*, 22(2), 197–206 (2011).

[6] D. A. Woodrum, K. R. Gorny, B. M. Greenwood et al., "MRI-guided prostate biopsy of native and recurrent prostate cancer," *Seminars in Interventional Radiology*, 33(3), 196–205 (2016).

[7] Y. Fu, Y. Lei, T. Wang et al., "Biomechanically constrained non-rigid MR-TRUS prostate registration using deep learning based 3D point cloud matching," *Medical Image Analysis*, 67, 101845 (2021).

[8] M. B. Hisham, S. N. Yaakob, R. A. A. Raof et al., "Template matching using sum of squared difference and normalized cross correlation," *2015 IEEE Student Conference on Research and Development (SCOReD)*, 13–14 December 2015, Kuala Lumpur, Malaysia, 100–104 (2015).

[9] S. C. Strother, J. R. Anderson, X. L. Xu et al., "Quantitative comparisons of image registration techniques based on high resolution MRI of the brain," *Journal of Computer Assisted Tomography*, 18, 954–962 (1994).

[10] X. Liu, D. Jiang, M. Wang et al., "Image synthesis-based multi-modal image registration framework by using deep fully convolutional networks," *Medical & Biological Engineering & Computing*, 57, 1037–1048 (2018).

[11] S. Guan, T. Wang, C. Meng et al., "A review of point feature based medical image registration," *Chinese Journal of Mechanical Engineering*, 31, 1–16 (2018).

[12] D. Rivest-Hénault, H. Sundar, and M. Cheriet, "Nonrigid 2D/3D registration of coronary artery models with live fluoroscopy for guidance of cardiac interventions," *IEEE Transactions on Medical Imaging*, 31(8), 1557–1572 (2012).

[13] R. A. McLaughlin, J. Hipwell, D. J. Hawkes et al., "A comparison of a similarity-based and a feature-based 2-D-3-D registration method for neurointerventional use," *IEEE Transactions on Medical Imaging*, 24(8), 1058–1066 (2005).

[14] S. Habert, P. Khurd, and C. Chefd'Hotel, "Registration of multiple temporally related point sets using a novel variant of the coherent point drift algorithm: Application to coronary tree matching," *Medical Imaging 2013: Image Processing* (2013).

[15] M. P. Heinrich, M. Jenkinson, M. Bhushan et al., "MIND: Modality independent neighbourhood descriptor for multi-modal deformable registration," *Medical Image Analysis*, 16(7), 1423–1435 (2012).

[16] D. Jiang, Y. Shi, X. Chen et al., "Fast and robust multimodal image registration using a local derivative pattern," *Medical Physics*, 44, 497–509 (2017).

[17] S. Klein, U. A. van der Heide, I. Lips et al., "Automatic segmentation of the prostate in 3D MR images by atlas matching using localized mutual information," *Medical Physics*, 35(4), 1407–1417 (2008).

[18] D. Loeckx, P. Slagmolen, F. Maes et al., "Nonrigid image registration using conditional mutual information," *IEEE Transactions on Medical Imaging*, 29, 19–29 (2010).

[19] J. Woo, M. L. Stone, and J. L. Prince, "Multimodal registration via mutual information incorporating geometric and spatial context," *IEEE Transactions on Image Processing*, 24, 757–769 (2015).

[20] H. Rivaz, Z. Karimaghaloo, and D. L. Collins, "Self-similarity weighted mutual information: A new nonrigid image registration metric," *Medical Image Analysis*, 18(2), 343–358 (2014).

[21] K. J. Friston, J. Ashburner, C. D. Frith et al., "Spatial registration and normalization of images," *Human Brain Mapping*, 3, 165–189 (1995).

[22] A. Guimond, A. Roche, N. Ayache et al., "Three-dimensional multimodal brain warping using the Demons algorithm and adaptive intensity corrections," *IEEE Transactions on Medical Imaging*, 20, 58–69 (2001).

[23] S. Roy, A. Carass, A. Jog et al., "MR to CT registration of brains using image synthesis," *Medical Imaging 2014: Image Processing* (2014).

[24] J. D. Lawson, E. Schreibmann, A. B. Jani et al., "Quantitative evaluation of a con... beam computed tomograph™ planning computed tomography deformable image registration method for adaptive radiation therapy," *Journal of Applied Clinical Medical Physics*, 8, 96–113 (2007).

[25] B. B. Avants, C. L. Epstein, M. Grossman et al., "Symmetric diffeomorphic image registration with cross-correlation: Evaluating automated labeling of elderly and neurodegenerative brain," *Medical Image Analysis*, 12(1), 26–41 (2008).

[26] J. E. Iglesias, E. Konukoglu, D. Zikic et al., "Is synthesizing MRI contrast useful for intermodality analysis?," *Medical Image Computing and Computer-Assisted Intervention: MICCAI... International Conference on Medical Image Computing and Computer-Assisted Intervention*, 16 (Pt 1), 631–638 (2013).

[27] F. Rousseau, P. A. Habas, and C. Studholme, "A supervised patch-based approach for human brain labeling," *IEEE Transactions on Medical Imaging*, 30, 1852–1862 (2011).

[28] M. Chen, A. Carass, A. Jog et al., "Cross contrast multi channel image registration using image synthesis for MR brain images," *Medical Image Analysis*, 36, 2–14 (2017).

[29] A. Jog, S. Roy, A. Carass et al., "Magnetic resonance image synthesis through patch regression," *2013 IEEE 10th International Symposium on Biomedical Imaging*, 350–353 (2013).

[30] D. Forsberg, Y. Rathi, S. Bouix et al., "Improving registration using multi-channel diffeomorphic demons combined with certainty maps," *Multimodal Brain Image Analysis: First International Workshop, MBIA 2011, Held in Conjunction with MICCAI 2011* (2011).

[31] X. Cao, J. Yang, Y. Gao et al., "Region-adaptive deformable registration of CT/MRI pelvic images via learning-based image synthesis," *IEEE Transactions on Image Processing*, 27, 3500–3512 (2018).

[32] E. M. McKenzie, A. P. Santhanam, D. Ruan et al., "Multimodality image registration in the head-and-neck using a deep learning derived synthetic CT as a bridge," *Medical Physics*, 47, 1094–1104 (2019).

[33] J.-Y. Zhu, T. Park, P. Isola et al., "Unpaired image-to-image translation using cycle-consistent adversarial networks," *2017 IEEE International Conference on Computer Vision (ICCV)*, 2242–2251 (2017).

[34] S. Klein, M. Staring, K. Murphy et al., "elastix: A toolbox for intensity-based medical image registration," *IEEE Transactions on Medical Imaging*, 29, 196–205 (2010).

[35] Y. Fu, Y. Lei, J. Zhou et al., "Synthetic CT-aided MRI-CT image registration for head and neck radiotherapy," *Medical Imaging 2020: Biomedical Applications in Molecular, Structural, and Functional Imaging* (2020).

[36] C. Tanner, F. Özdemir, R. Profanter et al., "Generative adversarial networks for MR-CT deformable image registration," *ArXiv*, arXiv:1807.07349 (2018).

[37] M. Hoffmann, B. Billot, D. N. Greve et al., "SynthMorph: Learning contrast-invariant registration without acquired images," *IEEE Transactions on Medical Imaging*, 41, 543–558 (2021).

[38] B. Billot, D. N. Greve, K. V. Leemput et al., "A learning strategy for contrast-agnostic mri segmentation," *arXiv preprint*, arXiv:2003.01995 (2020).

[39] Y. Lei, J. Harms, T. Wang et al., "MRI-only based synthetic CT generation using dense cycle consistent generative adversarial networks," *Medical Physics*, 46, 3565–3581 (2019).

[40] Y. Fu, Y. Lei, T. Wang et al., "Deep learning in medical image registration: A review," *Physics in Medicine and Biology*, 65, 20TR01 (2020).

16

CT Image Standardization Using Deep Image Synthesis Models

Md Selim, Jie Zhang, and Jin Chen
University of Kentucky, Lexington, KY, USA

CONTENTS

16.1 Introduction

Computed tomography (CT) is one of the most commonly used imaging modalities for diagnostics. It can capture detailed anatomical features (Beutel et al., 2000; Mahesh, 2011; Prince & Links, 2006; Webb & Kagadis, 2003). The use of customized (non-standard) protocol configuration in CT scanning provides flexibility for personalized clinical requirements and patients' diagnostics needs (Midya et al., 2018). However, this flexibility comes with a cost of variations among acquired CT images in terms of radiomic features. Even for the same patient, their image features may be significantly different from each other when non-standardized acquisition protocols are used (Chirra et al., 2018; Liang et al., 2017). Such variations can create ambiguity in sharing data across different institutions as the institutions might not be using identical protocols or scanners from different vendors (Buckler et al., 2011). It also creates a barrier to developing robust tools for downstream tasks. For example, an image segmentation tool is trained to segment images for random acquisition

DOI: 10.1201/9781003243458-20

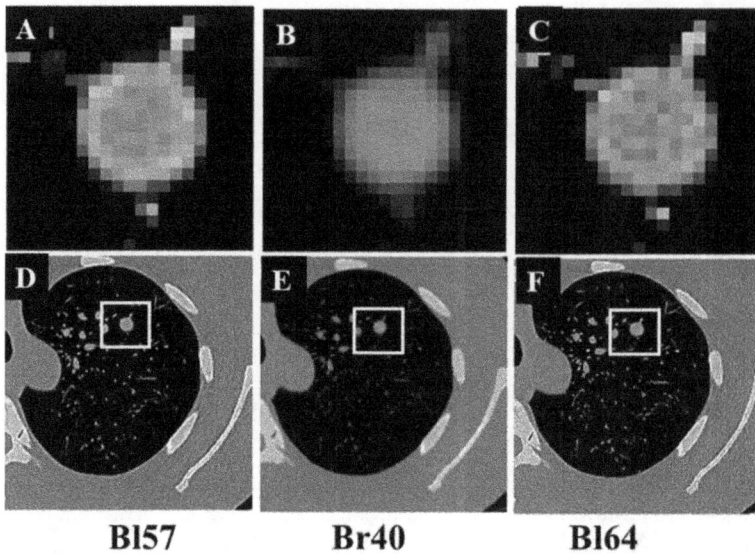

FIGURE 16.1
CT images acquired using three different image reconstruction kernels. The discrepancy in image features may significantly limit the promise of large-scale cross-center radiomic studies.

protocols A may not work accurately on the images acquired using protocol B. An example of using non-standardized acquisition protocols in CT imaging is shown in Figure 16.1. A chest phantom has been scanned using the same CT scanner, and images have been reconstructed using three different kernels (Bl57, Br40, and Bl64, Siemens Healthineers, Erlangen, Germany). The same tumors in the scanned images are significantly different from each other in terms of visual appearance as well as radiomic features.

The radiomic feature discrepancy problem in CT images could be potentially addressed by defining and using a standard image acquisition protocol. However, it is impractical to use the same image acquisition protocol in all clinical practices, not only because there are already multiple CT scanner manufacturers in the market (Paul et al., 2012) but also because of the limitations of using a fixed protocol for all patients under all situations in diagnosis, staging, therapy selection, and therapy assessment of tumor malignancies (Gierada et al., 2010).

The radiomic feature discrepancy problem can be formalized by either normalizing the radiomic features of all the non-standard images or standardizing CT images and then extracting the radiomic features from the standardized images. Image feature-based standardization models have been used to extract and standardize radiomic features successfully. For example, Andrearczyk et al. presented multilayer perceptron neural network-based radiomic feature standardization, where the network uses image features as inputs and produces its corresponding standardized features as outputs (Andrearczyk et al., 2019). Nevertheless, each individual feature in the model follows its own distribution, and the network needs the identical feature set in input as it was used during the training. This greatly limits the usage of the model. Furthermore, in a practical scenario, a radiomic model may require a problem-specific feature set that might be different from the input features. Also, the existing image feature extractors do not have a standard feature set because the radiomic features are not well-defined. For example, PyRadiomics (van

Griethuysen et al., 2017) GLCM feature class has 24 total sub-features, but IBEX (Zhang et al., 2015) GLCM has a total of 22 sub-features which can be further sub-categorized in total 594 features. Thus, standardization in the image domain provides much flexibility to the end-user radiomic study.

Image synthesis algorithms have been recently developed aiming to synthesize images with similar feature-based distributions compared to that of the target images while preserving anatomic details. In this article, we will discuss the image synthesis-based CT image standardization approaches to solve the radiomic feature discrepancy problem. The aim of the CT image standardization problem is to develop a postprocessing framework to standardize and normalize existing CT images while preserving most of the anatomic details.

16.2 Background

Understanding CT image standardization using an image synthesis approach requires a few key concepts starting from image acquisition, radiomic features, image synthesis models, etc.

16.2.1 CT Image Acquisition and Reconstruction Parameters

CT images are typically acquired by setting several parameters, such as kVp, Pitch, mAs, reconstruction Field Of View (FOV), slice thickness, reconstruction kernels, etc. Here, kVp stands for kilovoltage peak, which is the highest voltage produced by the X-ray tube during the scan. During a CT scan, the X-ray tube rotates around the table where the scanned subject is kept, and the table moves with the subject. Pitch is the distance of the table for one x-ray rotation. mAs is the amount of current passing through the scanner. FOV is the viewing scope of the scan. Slice thickness is the distance between two adjacent scans, and the reconstruction kernel is the key algorithm to reconstruct images from the X-ray captured signal. Varying the settings of CT image acquisition and reconstruction parameters and the selection of different CT scanners may affect the image quality and subsequently alter radiomic features extracted from the resulting CT images. For example, in Figure 16.1, Bl64 and Br40 are two different reconstruction kernels used to generate images from the same chest phantom. Br40 produces a smoother image, and Bl64 produces a sharper image. As the two images have different texture patterns, their calculated radiomic features will also be different.

16.2.2 Radiomic Features

Image features, commonly known as radiomic features, are critical for tumor characterization (Yip & Aerts, 2016). Mathematical and statistical models are used to extract these features from images. These features reflect the cellular and genetic levels of phenotypic patterns that are hidden from the naked eye (Basu et al., 2011; Yang & Knopp, 2011; Yip & Aerts, 2016). Thus, there is a great potential to capture tumor heterogeneity and phenotypic details with radiomic features. However, the effectiveness of radiomic features, especially for large-scale cross-institute studies, is greatly reduced due to the non-standard practice of medical image acquisition (Berenguer et al., 2018), since radiomic features are dependent on both inter and intra-scanner protocol settings (Berenguer et al., 2018; Hunter et al., 2013) or data synthesis is an active research area in computer vision, computer graphics, and natural language processing (Huang et al., 2018). By definition, image synthesis is a process of generating synthetic images using limited information (Hall & Greenberg, 1983). The given information includes text description, random noise, or any other types of information. With the recent breakthrough in deep learning, image synthesis algorithms have been applied in the areas of text-to-image generation (Reed et al., 2016), detecting lost frame in a video (Kwon & Park, 2019) image-to-image transformation (Karras et al., 2019), and medical imaging (Yu et al., 2019) successfully.

16.2.3 Deep Generative Models for Image Synthesis

16.2.3.1 U-net

U-Net is a special fully connected neural network originally proposed for medical image segmentation (Ronneberger et al., 2015). Precise localization and relatively small training data requirements are the major advantages of U-Net. A U-Net usually has three parts, down-sampling, bottleneck, and up-sampling, where the up-sampling and down-sampling are symmetric. There are also connections from down-sampling layers to the corresponding up-sampling layers to recover lost information during down-sampling. However, while U-net is effective in generating structural information, it suffers from learning and keeping texture details (Ravishankar et al., 2017). This issue can be overcome by adopting U-net in a more sophisticated generative model called Generative Adversarial Networks (GANs) (Huang et al., 2018).

16.2.3.2 Generative Adversarial Network

Generative Adversarial Networks (GAN), which are often used for data and image synthesis (Huang et al., 2018), normally consist of a generator G and a discriminator D. The generator that could be a U-Net is responsible for generating fake data from noise, and the discriminator tries to identify whether its input is drawn from the real or fake data. Among all the GAN models, cGAN can synthesize new images based on a prior distribution (Mirza & Osindero, 2014). However, since the image features of the synthesized data and that of the target data may not fall into the same distribution, cGAN may not be directly applicable for the CT image standardization problem.

16.3 CT Image Standardization Model

Deep learning methods dominate computer vision tasks and most of the available methods for CT image standardization leverage deep learning-based models for image synthesis. We have selected six models from the literature. Among these models, four deep learning models named GANai (Liang et al., 2019), STAN-CT (Selim et al., 2020), RadiomicGAN (Selim et al., 2021b), and CVH-CT (Selim et al., 2021a) are based on the GAN model, and two models use a similar CNN-based generative model named Choe et al. (Choe et al., 2019) and Jin et al. (Yoon et al., 2021). Table 16.1 summarizes the models' descriptions.

TABLE 16.1

Overview of the CT Image Standardization Models

Model Name	Neural Network Model Type	Standardization Approach	Evaluation Criteria
CNN (Choe et al.)	Fully connected convolution network	One-to-one	CCC
CNN (Jin et al.)	Fully connected convolution network	One-to-one	CCC
GANai	GAN (Pix2Pix)	One-to-one	Absolute Error
STAN-CT	GAN (Pix2Pix)	Many-to-one	Relative Absolute Error
RadiomicGAN	GAN (Pix2Pix)	Many-to-one	CCC
CVH-CT	CycleGAN	One-to-one	CCC

16.3.1 CNN-based CT Image Standardization

Choe et al. developed a Convolutional Neural Network (CNN) based approach for CT image standardization. The CNN model consists of six consecutive blocks of convolution and activation layers. Each convolutional has a 3 × 3 kernel with 64 filters. The model learns the residual representation (difference between the target and the input images) of the target images, and then the resulting residual image is combined with its source image to generate a synthesized image. Figure 16.2 illustrates the detailed architecture of the model. The model, since it trains a CNN from scratch, requires large training data. It used both B50f and B30f reconstruction kernels in the experiment, and its performance was evaluated in both directions. Pulmonary nodules or masses were semi-automatically segmented, and 702 radiomic features (tumor intensity, texture, and wavelet features) were

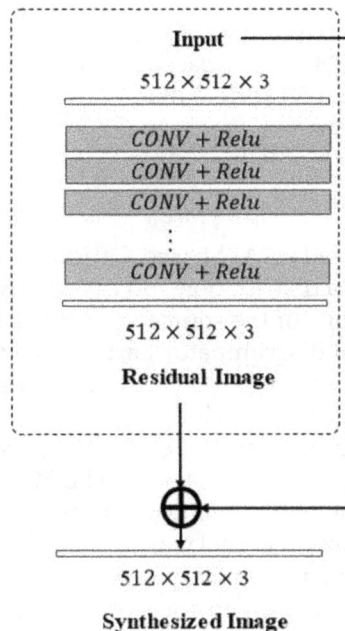

FIGURE 16.2

Network architecture for CNN-based CT image standardization used in Choe et al. and Jin et al.

extracted. Measurement variability in radiomic features was evaluated using the concordance correlation coefficient (CCC). A total of 104 patient data were studied. The conversion increased the number of stable radiomic features from 15.2% to 57.4% in terms of CCC. The result also concluded that the texture features and the wavelet features were predominantly affected by reconstruction kernels.

Jin et al. extended the model proposed by Choe et al. to standardize the CT images on a larger dataset. They investigated the impact of standardization or conversion between two groups of reconstruction kernels, namely smooth and sharp. The smooth kernels produced images with a smoother texture, and the sharp kernel had a sharp noise texture. The CNN model was trained using the data from two groups of 32 lung cancer patients. The validation was done using 223 cancer patient data from an external cohort using different CT scanners and kernels. They extracted 89 features from 23 feature groups and found that the median CCC score between two groups before the conversion was 0.504. The median CCC score for smooth to sharp conversion was 0.589, and from sharp to smooth was 0.835. This concludes that the conversion from a sharp reconstruction kernel to a smoother reconstruction can produce better stable features.

Jin et al. extended the harmonization from single scanner to multi scanner multi-institution data standardization. Unlike the other methods where the standardization was done as a many-to-one conversion approach, this article did not specify any particular reconstruction kernel as a standard one. The study concluded that sharp-to-smooth conversion produced more stable radiomic features compared with the smooth to sharp image conversion. However, recommended lung screening reconstruction kernels produce shaper texture, and higher performance is naturally required for the smooth-to-sharp conversion in cancer study.

16.3.2 GAN-based CT Image Standardization

GAN-based models are widely used in image synthesis, including image denoising, image segmentation, image denoising, etc. GAN-based models have also used attention in the CT image standardization problem. Figure 16.3 illustrates four GAN-based models designed to synthesize standard CT images.

16.3.2.1 GANai

Liang et al. proposed an improved cGAN-based CT image standardization model named GANai, which can learn the data distribution and can generate similar to the target image by preserving the radiomic details of the source image. The GAN is composed of a U-Net as a generator, and a traditional discriminator part is a feed-forward network. The training of the GANai network is done using multiple generators G and discriminators D. The multiple Gs and Ds are trying to force each other to improve progressively. This training strategy is named alternative improvement. For further improvement and stabilization of the network, ensemble learning was used. It had used Bl57 and Bl64 reconstruction kernels from Siemens CT scanner in the experiment. Eight texture features were extracted from soft tissues ROIs. Bl64 reconstruction kernel was defined as a standard target protocol, and Bl57 was converted to Bl64 reconstructed images. Absolute errors between the synthesized images and target images were computed for performance evaluation. The results indicate that the GANai model reduced the absolute error significantly compared with other models. However, the study did not report the errors before conversion, so the relative improvement using the model cannot be determined.

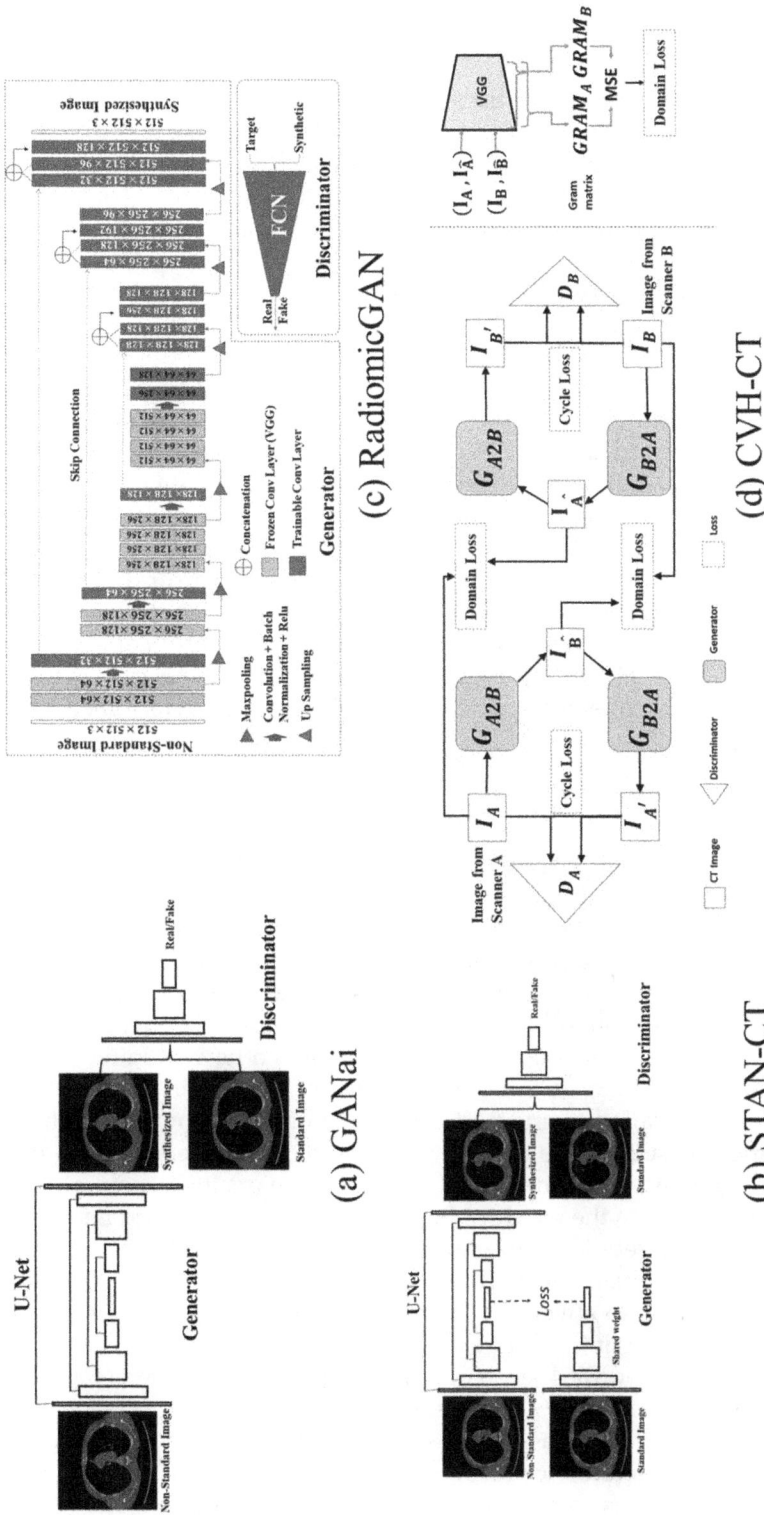

FIGURE 16.3
Network architectures for GAN-based CT image standardization used in GANai, STAN-CT, RadiomicGAN, and CVH-CT models.

16.3.2.2 STAN-CT

Selim et al. addressed the CT image standardization and normalization problem with a GAN model by minimizing the latent feature loss between target images and input images. This poses two fundamental computational challenges: 1) to effectively map between target images and synthesized images with great pixel-level details, and 2) to maintain the texture consistency among the synthesized images. In this paper, we present an end-to-end solution called STAN-CT. In STAN-CT, we introduce two new constraints in GAN loss. Specifically, the study adopted a latent-space-based loss for the generator to establish a one-to-one mapping from target images to synthesized images. Also, a feature-based loss was adopted for the discriminator to critic the texture features of the standard and the synthesized images. Furthermore, to synthesize CT images in the Digital Imaging and Communications in Medicine (DICOM) format, STAN-CT introduces a DICOM reconstruction framework that can integrate all the synthesized image patches to generate a DICOM file for clinical use. The framework ensured the quality of the synthesized DICOM by systematically identifying and pruning low-quality image patches.

It had used Bl57, Br40, and Bl64 reconstruction kernels from Siemens CT scanner in the experiment. Five texture features were extracted from soft tissue ROIs. Bl64 reconstruction kernel was defined as a standard target protocol, and images acquired using Bl57 and Br40 were converted to Bl64 reconstructed images. Relative absolute errors between the synthesized images and target images were computed for performance evaluation. The results indicate that the STAN-CT model reduced the error significantly compared with other models, including GANai. However, the result did not mention the error before standardization, so the relative improvement using the model cannot be determined.

16.3.2.3 RadiomicGAN

Selim et al. presented a GAN-based deep learning model for CT image standardization and normalization focused on harmonizing CT images acquired with non-standard reconstruction kernels as it is one of the most significant factors of feature inconsistency. RadiomicGAN employs a hybrid architecture for image texture feature extraction and embedding. Its encoder consists of multiple consecutive neural blocks, including both pre-trained and trainable convolutional layers. To address the dynamic pixel range-related problem in transfer learning, RadiomicGAN uses a new training strategy named *Dynamic Window-based Training* (DWT), which allows us to train a model using pixels within a selected range called "window." The range of a window can be automatically broadened or shrank based on the pixels where the model suffers most in the previous training iteration, allowing us to fine-tune the trainable layers in RadiomicGAN using the frequently appeared pixels in the window. In summary, given its hybrid network structure, RadiomicGAN can effectively learn the radiomic feature distributions from the standard CT images and then harmonize non-standard CT images. A dynamic window-based training approach is developed to effectively address the pixel range difference problem and thus enable transfer learning in the medical image domain.

It had used Bl57, Br40, and Bl64 reconstruction kernels from Siemens CT scanner in the experiment. 1,401 radiomic features were extracted from different types of tissues. Bl64 reconstruction kernel was defined as a standard target protocol, and images acquired using Bl57 and Br40 were converted to Bl64 reconstructed images. CCC between the synthesized images and target images was computed for performance evaluation. The results summarize that Bl57 to Bl64 conversion is relatively easy compared with the Bl40 to Bl64 conversion because the CCC scores between the Bl57 and Bl64 before conversion were relatively high. The model required less training data as it leveraged a pre-trained network.

16.3.2.4 CVH-CT

The cross-vendor CT image harmonization remains a critical bottleneck for inter-institutional data harmonization in the standardization task. This is mainly because it is difficult to obtain paired imagery data. For example, a patient is scanned using a GE scanner, and it is less likely that the patient will be scanned using a Siemens scanner in close time. This method provides a novel deep learning model called CVH-CT (cross-vendor harmonization of CT images) for cross-vendor CT image harmonization. CVH-CT relaxed the need for paired training data. Using unpaired training data, CVH-CT can synthesize images from vendor A to vendor B and vice versa. CVH-CT integrates a self-attention layer named CBAM (Convolutional Block Attention Map) to systematically learn the global features that appear in the images due to the use of different vendors. CVH-CT borrowed the CycleGAN loss and improved it with a feature-based domain loss to determine the feature gap between the synthesized images and the target images in the target domain.

Images from two different scanners named Siemens and General Electric (GE) were used to evaluate the performance of the model. Phantom images were captured and converted from GE to Siemens to GE and also from GE to Siemens. CCC scores of six feature classes were calculated and used to measure the performance. In both conversions, the model CVH-CT achieved a mean CCC score above 0.85. The performance was evaluated based on soft-tissue ROIs.

16.4 Discussion and Conclusion

The studied image synthesis-based models are divided into two main categories, CNN and GAN. Also, the discussed model can synthesize images either for images within a scanner or between scanners. Choe et al., GANai, STAN-CT, and RadiomicGAN show encouraging performance in terms of CT image standardization with paired training data which significantly limit the use of these models within a particular scanner. Jin et al. proposed a new approach by grouping all reconstruction kernels into two groups and tried to convert images from one group to another. CVH-CT model proposed a framework for standardizing images between scanners by relaxing the requirement of paired data but with some limitations. CVH-CT learns only a one-to-one mapping function from one non-standard image distribution to another fixed standard distribution. But in a real case scenario, the model needs to learn many-to-one mapping where multiple non-standardized image distributions need to map into a single standard image distribution. This many-to-one mapping can be addressed by acquiring paired training data of many different non-standard image protocols and one standard image protocol, but the paired data is hard to collect if the standard image and non-standard images are from different scanners or vendors. An alternative naive approach may be using our paired data model and unpaired data model one after another like a chain to make a bridge between different scanner, which make the framework very complex and loses usability. A single framework is required which can utilize both paired and unpaired images to train a single model for this standardization task.

All the models' performances were evaluated based on the radiomic features. However, the evaluated feature set and the evaluation matrices are not identical in different models. GANai and STAN-CT analyze basic key texture features for model evaluation. Other models used more than 100 features, and those features are sub-features driven from the major feature classes. CCC is used in most of the models for feature comparison. However, it is not practical to expect identical values from the synthesized image as the target image feature values. The literature suggested that CCC > 0.85 as reproducible or stable features. However, the 15% gap may produce different feature values, which does not ensure the reproducible downstream task. The analysis of the 85% CCC score in the downstream task also needs to be further investigated. As the models' evaluation used different feature sets and datasets, a fair comparison is also not possible by looking at the reported performance. A comparative study is required on a standard dataset that includes images from different scanners. Finally, the impact of standardization on the downstream task also needs further study.

In summary, the above models created a solid foundation for standardizing CT images from diverse sources, enabling the identification of critical radiomic features largely applicable to cancer prognosis and personalized treatments. Additionally, the discussed deep learning models in this study are not limited to the image standardization task in the CT imaging domain. These models can easily be adopted on generic image synthesis tasks, including image segmentation, image denoising, image conversion, etc.

References

Andrearczyk, V., Depeursinge, A., & Müller, H. (2019). Neural network training for cross-protocol radiomic feature standardization in computed tomography. *Journal of Medical Imaging*, 6(2), 1. https://doi.org/10.1117/1.JMI.6.2.024008

Basu, S., Kwee, T. C., Gatenby, R., Saboury, B., Torigian, D. A., & Alavi, A. (2011). *Evolving role of molecular imaging with PET in detecting and characterizing heterogeneity of cancer tissue at the primary and metastatic sites, a plausible explanation for failed attempts to cure malignant disorders*. Springer.

Berenguer, R., Pastor-Juan, M. D. R., Canales-Vázquez, J., Castro-García, M., Villas, M. V., Mansilla Legorburo, F., & Sabater, S. (2018). Radiomics of CT features may be nonreproducible and redundant: Influence of CT acquisition parameters. *Radiology*, 288, 407–415.

Beutel, J., Kundel, H. L., & van Metter, R. L. (2000). *Handbook of medical imaging: Physics and psychophysics* (Vol. 1). Spie Press.

Buckler, A. J., Bresolin, L., Dunnick, N. R., Sullivan, D. C., & Group. (2011). A collaborative enterprise for multi-stakeholder participation in the advancement of quantitative imaging. *Radiology*, 258(3), 906–914.

Chirra, P., Leo, P., Yim, M., Bloch, B. N., Rastinehad, A. R., Purysko, A., Rosen, M., Madabhushi, A., & Viswanath, S. (2018). Empirical evaluation of cross-site reproducibility in radiomic features for characterizing prostate MRI. *Medical Imaging 2018: Computer-Aided Diagnosis*, vol. 10575, 105750B.

Choe, J., Lee, S. M., Do, K.-H., Lee, G., Lee, J.-G., Lee, S. M., & Seo, J. B. (2019). Deep learning–based image conversion of CT Reconstruction kernels improves radiomics reproducibility for pulmonary nodules or masses. *Radiology*, 292(2), 365–373.

Gierada, D. S., Bierhals, A. J., Choong, C. K., Bartel, S. T., Ritter, J. H., Das, N. A., Hong, C., Pilgram, T. K., Bae, K. T., Whiting, B. R., & others. (2010). Effects of CT section thickness and reconstruction kernel on emphysema quantification: Relationship to the magnitude of the CT emphysema index. *Academic Radiology*, 17(2), 146–156.

Hall, R. A., & Greenberg, D. P. (1983). A testbed for realistic image synthesis. *IEEE Computer Graphics and Applications, 3*(8), 10–20.

Huang, H., Yu, P. S., & Wang, C. (2018). An introduction to image synthesis with generative adversarial nets. *ArXiv Preprint*, ArXiv:1803.04469.

Hunter, L. A., Krafft, S., Stingo, F., Choi, H., Martel, M. K., Kry, S. F., & Court, L. E. (2013). High quality machine-robust image features: Identification in nonsmall cell lung cancer computed tomography images. *Medical Physics, 40*(12), 121916.

Karras, T., Laine, S., & Aila, T. (2019). A style-based generator architecture for generative adversarial networks. *Proceedings of the IEEE Conference on Computer Vision and Pattern Recognition*, 4401–4410, Long Beach, CA, USA.

Kwon, Y.-H., & Park, M.-G. (2019). Predicting future frames using retrospective cycle GAN. *Proceedings of the IEEE Conference on Computer Vision and Pattern Recognition*, 1811–1820, Long Beach, CA, USA.

Liang, G., Fouladvand, S., Zhang, J., Brooks, M. A., Jacobs, N., & Chen, J. (2019). GANai: Standardizing CT images using generative adversarial network with alternative improvement. *2019 IEEE International Conference on Healthcare Informatics (ICHI)*, 1–11, Xian, China.

Liang, G., Zhang, J., Brooks, M., Howard, J., & Chen, J. (2017). Radiomic features of lung cancer and their dependency on CT image acquisition parameters. *Medical Physics, 44*(6), 3024.

Mahesh, M. (2011). Fundamentals of medical imaging. *Medical Physics, 38*(3), 1735.

Midya, A., Chakraborty, J., Gönen, M., Do, R. K. G., & Simpson, A. L. (2018). Influence of CT acquisition and reconstruction parameters on radiomic feature reproducibility. *Journal of Medical Imaging, 5*(1), 11020.

Mirza, M., & Osindero, S. (2014). Conditional generative adversarial nets. ArXiv:1411.1784v1.

Paul, J., Krauss, B., Banckwitz, R., & others. (2012). Relationships of clinical protocols and reconstruction kernels with image quality and radiation dose in a 128-slice CT scanner: Study with an anthropomorphic and water phantom. *European Journal of Radiology, 81*(5), e699–e703.

Prince, J. L., & Links, J. M. (2006). *Medical imaging signals and systems.* Pearson Prentice Hall Upper Saddle River.

Ravishankar, H., Venkataramani, R., Thiruvenkadam, S., & others. (2017). Learning and incorporating shape models for semantic segmentation. *International Conference on Medical Image Computing and Computer-Assisted Intervention*, 203–211, Quebec City, Quebec, Canada.

Reed, S., Akata, Z., Yan, X., Logeswaran, L., Schiele, B., & Lee, H. (2016). Generative adversarial text to image synthesis. *ArXiv Preprint*, ArXiv:1605.05396.

Ronneberger, O., Fischer, P., & Brox, T. (2015). U-Net: Convolutional networks for biomedical image segmentation. *Medical Image Computing and Computer-Assisted Intervention*, 234–241.

Selim, M., Zhang, J., Fei, B., Zhang, G.-Q., & Chen, J. (2020). STAN-CT: Standardizing CT image using generative adversarial network. *AMIA Annual Symposium Proceedings, 2020,* Virtual worldwide.

Selim, M., Zhang, J., Fei, B., Zhang, G.-Q., Ge, Gary, & Chen, J. (2021a). Cross-vendor CT image data harmonization using CVH-CT. *AMIA Annual Symposium Proceedings, 2021,* San Diego, CA, USA.

Selim, M., Zhang, J., Fei, B., Zhang, G.-Q., & Chen, J. (2021b). CT image harmonization for enhancing radiomics studies. *2021 IEEE International Conference on Bioinformatics and Biomedicine (BIBM)*, 1057–1062, Virtual worldwide. https://doi.org/10.1109/BIBM52615.2021.9669448

van Griethuysen, J. J. M., Fedorov, A., Parmar, C., Hosny, A., Aucoin, N., Narayan, V., Beets-Tan, R.G., Fillion-Robin, J.C., Pieper, S. & Aerts, H.J. (2017). Computational radiomics system to decode the radiographic phenotype. *Cancer Research, 77*(21), e104–e107.

Webb, A., & Kagadis, G. C. (2003). Introduction to biomedical imaging. *Medical Physics, 30*(8), 2267.

Yang, X., & Knopp, M. V. (2011). Quantifying tumor vascular heterogeneity with dynamic contrast-enhanced magnetic resonance imaging: A review. *BioMed Research International, 2011.*

Yip, S. S. F., & Aerts, H. J. W. L. (2016). Applications and limitations of radiomics. *Physics in Medicine & Biology, 61*(13), R150.

Yoon, J. H., Sun, S. H., Xiao, M., Yang, H., Lu, L., Li, Y., Schwartz, L. H., & Zhao, B. (2021). Convolutional neural network addresses the confounding impact of CT reconstruction kernels on radiomics studies. *Tomography*, 7(4), 877–892. https://doi.org/10.3390/tomography7040074

Yu, Z., Xiang, Q., & Meng, J. et al. (2019). Retinal image synthesis from multiple-landmarks input with generative adversarial networks. *Biomedical Engineering Online*, 18(1), 62.

Zhang, L., Fried, D. V., Fave, X. J., Hunter, L. A., Yang, J., & Court, L. E. (2015). IBEX: An open infrastructure software platform to facilitate collaborative work in radiomics. *Medical Physics*, 42(3), 1341–1353.

Section V

Clinic Usage of Medical Image Synthesis

17

Image-Guided Adaptive Radiotherapy

Yang Sheng and Jackie Wu

Duke University Medical Center, Durham, NC, USA

Taoran Li

University of Pennsylvania, Philadelphia, PA, USA

CONTENTS

17.1 Introduction

Modern radiation therapy aims to deliver a highly localized radiation dose to the tumor volume. The success of radiation therapy treatment relies on reproducible patient anatomy during the simulation process as well as throughout the entire treatment course (Figure 17.1). Even with a robust external immobilization technique, the patient's anatomy could still vary substantially due to multiple reasons, including internal organ motion, patient weight loss, breathing motion etc. Additional margin is commonly prescribed to account for such day-to-day anatomy variation. With the advent of online imaging components, image-guided radiation therapy (IGRT) enables more comprehensive and accurate visualization of patients' daily anatomy changes. It allows the radiation therapy plan to be adapted based on the daily anatomy to deliver a more conformal dose to the target while minimizing unnecessary dose spillage to the organs-at-risk (OARs), which is more commonly known as adaptive radiotherapy. The variety of available imaging techniques has opened up a wide opportunity to realize adaptive radiotherapy workflow for cancer patient of many treatment sites in real clinical settings. Figure 17.2 summarizes the adaptive therapy workflow from the initial simulation and planning to online plan adaptation and finally to dose accumulation and offline plan adaptation. This chapter will discuss several clinically available applications of the adaptive therapy concept in a modern-day clinic.

DOI: 10.1201/9781003243458-22

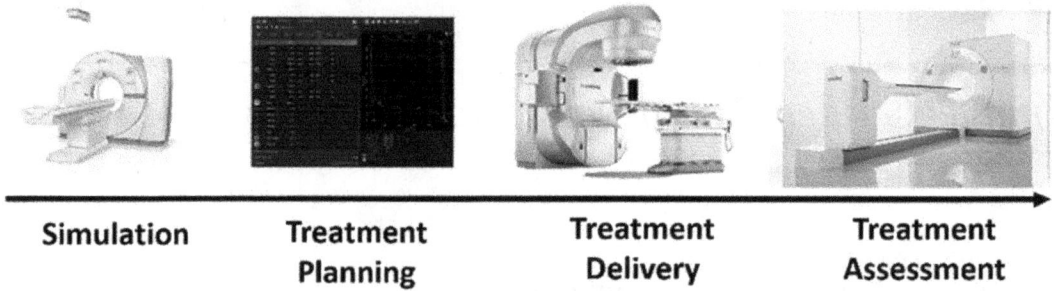

Simulation   Treatment   Treatment   Treatment
                    Planning   Delivery   Assessment

FIGURE 17.1
Radiation therapy workflow. (Image courtesy of Varian Medical Systems, Inc. and Siemens Healthineers. All rights reserved.)

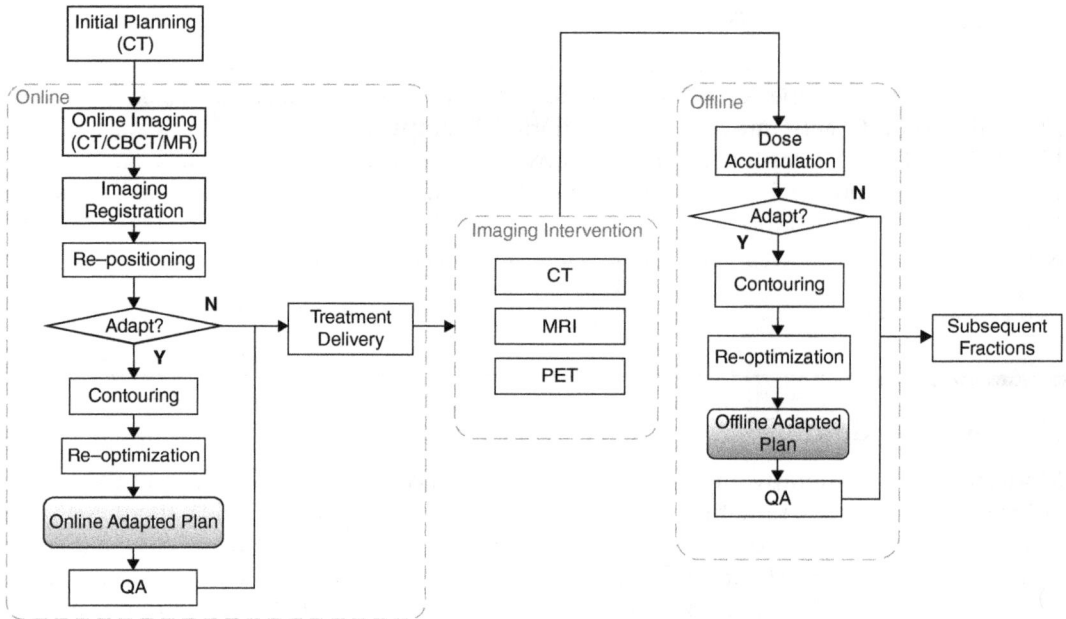

FIGURE 17.2
Summary of adaptive radiation therapy workflow including both online adaptive approach and offline adaptive approach.

17.2 CBCT-based Online Adaptive Radiation Therapy System

Ethos™ system is one of the first few commercialized adaptive radiotherapy workflow solution widely available to the public. Designed and manufacture by Varian Medical Systems (Palo Alto, California, USA), it is built upon the novel high throughput Halcyon platform, an O-ring Linac (Figure 17.3). It is designed as a self-contained system that enables online adaptive workflow composed of three steps: initial planning, on-couch adaptation, and treatment monitoring.

For initial planning, it enables template-driven creation of physician's objectives with the ability to perform real-time trade-off navigation at the time of prescription. A graphics

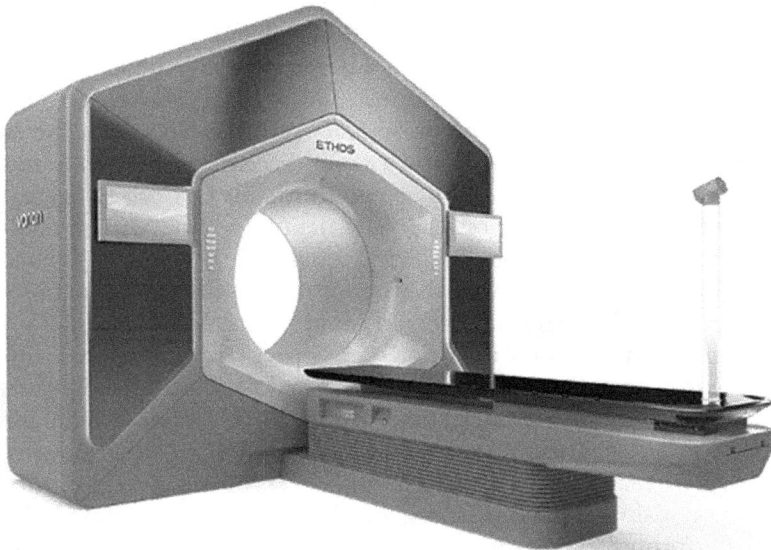

FIGURE 17.3
Varian Halcyon system equipped with Ethos™. (Image courtesy of Varian Medical Systems, Inc. All rights reserved.)

processing unit (GPU) powered optimizer also enables automated plan generation for intensity modulated radiation therapy (IMRT) and volumetric modulated arc therapy (VMAT) plans. Roover et al. reported initial clinical experience of using the automated treatment planning workflow for prostate stereotactic body radiation therapy (SBRT) and compared it with C-arm Linacs (1). They found that the automatic planning of prostate SBRT was feasible and provided plan quality similar to those manual plans on C-arm Linacs. Several other groups have published similar findings regarding the treatment plan quality generated using the Ethos™ initial planning system. Nasser et al. evaluated the initial planning strategies and planning quality of head-and-neck (HN) patients using the adaptive planning solution (2). They found that the AI optimizer was capable of generating a plan with better PTV coverage and lower OAR doses barring higher and spatially undesirable hot spots. Scheuermann et al. assessed the initial plan quality and consistency of the adapted plans over the course of the treatment for HN patients (3). They found that adapted plans met slightly fewer clinical goals than manually created clinical plans, with the majority of plans with elevated max dose to OARs. Hao et al. evaluated the performance of the same planning module for upper abdominal malignancies (4). Results showed that AI-assisted plan adaptation helped avoid OAR constraint violation while both calculation-based and measurement-based quality assurance (QA) showed a decent gamma passing rate. Bolt et al. assessed the automated plan quality of cervical cancer patients generated by the Ethos™ planning solution (5). Results showed that automated plans were clinically acceptable, with room for further improvement as compared to the manual plans.

The on-couch adaptation process is based on CBCT (auto-contouring) and synthetic CT (plan re-optimization), and introduces benefits in several folds. First, it is equipped with iterative CBCT (iCBCT), which enhances the onboard image quality by reducing noise and improving contrast-to-noise ratio. As compared to a traditional analytical technique for reconstruction, such as filtered back projection (FBP), it offers substantial advantages of reduced noise and artifacts. Cai et al. provided comprehensive benchmark performance testing on Halcyon kV system (6). They found that volumetric CBCT can be achieved within a single breathing cycle, and the noise level was reduced from the conventional reconstruction method. The study by Washio et al. discovered that dose reduction was achievable using iCBCT (7). They found that two-thirds of the dose to the patient can be reduced as compared to the conventional reconstruction method while maintaining low-contrast detectability. In another study by Jarema et al., it was found that Halcyon iCBCT images were capable of accurate dose calculation for pelvic patients based on both Gamma analysis and dose-volume histogram (DVH) endpoints for OARs (8).

Secondly, the on-couch adaption module is equipped with automated contouring for both targets and nearby organs, which are then used in the adaptive re-optimization based on daily anatomy. Auto contouring can be done within 15–30 seconds, and planning can be achieved within 2.5 minutes. The auto segmentation modules are either powered by U-Net and DenseNet, which feeds in full 3D iCBCT image (9) for select anatomical sites, or by contour propagation using deformable registration between planning CT and online CBCT. Mao et al. performed a retrospective study to evaluate the accuracy of the automatically segmented volumes and the corresponding dosimetric impact (10). They compared two approaches of adaptive radiation therapy workflow using Ethos™. The first approach was fully automated auto contouring without modification (automatic-ART), and the other approach was manual modification added to auto segmentation (supervised-ART). The result showed that automatic contouring was considered clinically acceptable. Yoon et al. also conducted comparisons between automatic contours and physician-reviewed contours using Ethos™ emulator using subjective rating scales and objective quantifiers (e.g. dice correlation coefficient, 95% Hausdorff distance) for HN patients. Results can conclude the automated contours were largely satisfactory and helpful for regaining daily plan objectives (11).

Additionally, the built-in calculation-based plan QA tool (Mobius™) and log file-based plan QA verification, avoided impeding the adaptive workflow with a conventional QA approach. The ability to import and view multi-modality images such as PET/CT or MR is also integrated into the initial planning and online adaptation workflow. Last but not least, voxel-based dose accumulation and tracking is available offline to assess the progress of the treatment.

A few groups have reported initial clinical experience of using Ethos™ for adaptive radiation therapy workflow. Ward et al. used Ethos™ system as an intervention approach for upper abdomen patient cohort to avoid substantial daily anatomical changes (12). They found that the process could be done in 30 minutes and was well tolerated by the patients. Byrne et al. performed a systematic assessment of Ethos™ workflow for prostate cancer adaptive therapy in terms of contouring accuracy, treatment plan quality, and treatment time (13). A total of 184 adaptive fractions were delivered, and results showed that 95% of the fractions favored the adapted plan. Average online adaptive contouring and replanning took 19 minutes, which was a clinically efficient time frame. Kim et al. developed a visually guided respiratory motion management workflow that was compatible with Ethos™ adaptive workflow (14). They used an air pressure sensor to detect patient respiratory motion in real time for CBCT acquisition. Moazzezi et al. systematically assessed the workflow of using Ethos™ to treat prostate adaptive therapy with nodal involvement (15). They adopted auto-segmented contours without manual edits, and the results showed that clinical target volume (CTV) coverage and normal tissue dose were both improved.

However, they also alert that 1 out of 25 patients experienced substantial contour variation, which required manual revision for the auto segmentation. This study suggested that special attention/mechanism is needed in place to ensure safe plan adaptation for all patients. Pokharel et al. assessed the performance of the intelligent optimization engine (IOE) component of Ethos™ in a prostate adaptive planning setting, retrospectively (16). Results showed that Ethos™ is capable of producing high-clinical-quality plans for both IMRT and VMAT in a timely fashion. On average, 2-full arc VMAT plan can be generated in 13 minutes and 12-field IMRT plan will only need 5 minutes. Multiple clinical trials are being conducted by the vendor-organized Adaptive Intelligence Consortium, which aims at providing clinical evidence of the benefit of CBCT-based online adaptive radiation therapy.

17.3 MR-Guided Real-Time Adaptive Radiation Therapy

Recent efforts saw tremendous advancement in the technology of integrated IGRT units. Magnetic resonance imaging (MRI) offers superior soft tissue contrast, which is unparalleled by computed tomography (CT)/CBCT treatment planning and online image guidance. MR is routinely used in the treatment planning process for sites such as cranial and abdominal lesions. Prior to the introduction of the MR-guided radiation therapy unit, online treatment verification for aforementioned treatment sites relies heavily on bony anatomy or anatomical landmarks, such as surgical clips. However, such anatomical landmarks may not perfectly reflect the true position and morphology of the soft tissue, which is prone to deformation and translation. This nature of the malleability of soft tissue creates tremendous challenges for accurate treatment delivery and limits potential dose escalation, which could improve treatment outcomes. Society calls for more advanced online MR-based imaging techniques to mitigate this issue.

The initial effort of implementing MR-guided radiation therapy was carried out by ViewRay Inc. (Oakwood, OH, USA). They introduced ViewRay System (VRS), a 0.35-T whole-body MRI integrated Co-60 radiation therapy unit (17). Such low-field MR offers several advantages, including low magnetic susceptibility, low absorption rate, and minimal perturbation of the dose distribution (17). The integration of a radioactive decay source with the MR component offers the advantage of limited interference between the two, which is commonly seen between MR and Linac duo thanks to the electron return effect. However, the radioactive material radiation unit is commonly known for larger penumbra, less penetration, and higher surface dose. It motivates the researchers to continue to look into MR-Linac philosophy.

UMC Utrecht in the Netherlands later introduced the prototype concept of MR-Linac using a 1.5-T Philips MRI and an Elekta 6-MV accelerator (18, 19). The RF interference between the radiation generating component and the MR imaging component was mitigated by redesigning the Faraday cage via positioning the accelerator outside the cage (18). This prototype enabled multi-leaf collimator (MLC) and, therefore, IMRT delivery. The electron return effect was mitigated by IMRT delivery. First 1.5-T MR-Linac patient treatment was reported in 2017 by Raaymakers et al. (20). Four patients with spinal metastases were treated with an IMRT plan, which was created online while the patient was on the treatment couch. Results demonstrated high accuracy of less than 1.7% dose deviation (20). The MR component geometry accuracy is critical for accurate dose delivery. Tijssen et al. (21) provided thorough MR component commissioning guidelines. On the other hand,

ViewRay Inc. introduced the MRIdian Linac system, which is a 0.35-T MR-guided radiotherapy (MRgRT) equipped with a 6 MV flattening-filter-free linear accelerator. They introduced six shielding compartments (a.k.a. buckets) mounted on the gantry to shield the magnetic field (22). It also provides RF shielding, so the MR component is not perturbed by the Linac operation. The system is capable of gating the photon beam using cine MRI images tracking anatomical structures.

For those two commercially available MRgRT solutions, one key component of adaptive workflow is using online MR images for accurate dose calculation. One of the popular approaches is to generate synthetic CT from online MR images. In Elekta Unity adaptive workflow solution, the planning CT images are deformably registered to the daily MR images. The pre-treatment contours are subsequently propagated to the MR images, together with the average electron density (ED) within a certain contour to create the synthetic CT for dose calculation (23). ViewRay MRIdian stereotactic MR-guided adaptive radiation therapy (SMART) solution adopted similar approach of propagating electron density from planning CT to the MR of the day using deformable image registration. It also offers the flexibility of editing electron density map before optimization if there is discrepancy in air pockets or filling of OARs (24).

Several studies have introduced MR-guided adaptive radiation therapy workflow since the introduction of MR-guided radiation unit. Werensteijn-Honingh et al. reported the clinically treated oligometastatic lymph nodes using the Elekta Unity system for online adaptation (Figure 17.4) (25). Treatment sessions with online adaptation can be achieved in 60 minutes (25). Winkel et al. further explained the plan adaptation mechanism in the Elekta Unity system (23). It utilized the concept of "adapt to shape", which uses deformable image registration (DIR) between pre-treatment CT and daily MRI, and subsequently deforms the structure to guide re-optimization on adapted contours (23). Finazzi et al. reported the SMART protocol treatment for peripheral lung tumors using the MRIdian system (26). The daily plan adaptation workflow starts with acquiring daily breath-hold 3DMR. A "predicted plan" is generated using anatomy-of-the-day to recalculate the dose

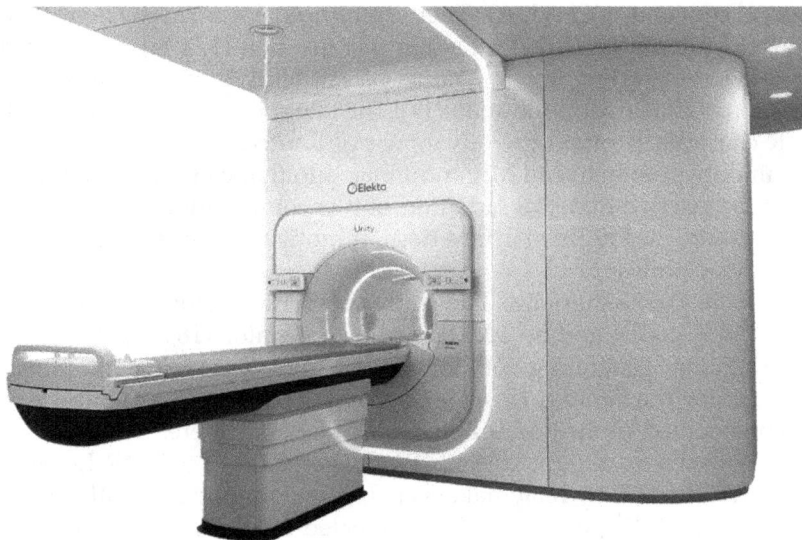

FIGURE 17.4
Elekta MR Linac unit with adaptive therapy solution Unity. (Image courtesy of Elekta.)

after image registration and adaptation of contours. Then the daily gross tumor volume (GTV) is rigidly registered to the baseline MR followed by an online couch shift. OAR contours are deformable and adapted before re-optimization is initiated. The attending physician makes a discretionary decision about whether to use the "predicted plan" or the adapted plan. Results showed that online adaptation ensured accurate dose delivery to peripheral lung tumors (26). However, they also found that on-table plan adaptation may not be the priority for most peripheral lung tumors as it can save up to 10 minutes from plan adaptation (26). Intven et al. reported the online adaptive radiation therapy workflow using the Elekta Unity system treating rectal cancer patients (27). They reported an average of 48 minutes in room time with adequate CTV coverage, even with margin reduction (27). de Leon et al. reported 6-month clinical experience of implementing the Unity system treating a total of 37 patients, including treatment sites of prostate, colorectal, breast, melanoma, lung, bladder, and duodenal (28). They reported an average of 50 minutes and 38 minutes for stereotactic and non-stereotactic treatment (28). Bernchou et al. reported a phantom-based end-to-end dose geometry accuracy benchmark for the Elekta Unity system (29). It was found that the adaptive system is capable of achieving 0.1–0.9 mm accuracy (29). Paulson et al. reported using 4DMRI to guide abdominal adaptive radiotherapy using the Unity system (30). Both the "adapt-to-position" and "adapt-to-shape" workflow produced high dose accuracy. Dunlop et al. analyzed the delivered dose accuracy between MR-Linac and conventional C-arm Linac (31). They found both workflows deliver satisfying outcomes, with MR-Linac slightly better than C-arm Linac (86% vs 80% constraints met) (31). Axford et al. developed and implemented an end-to-end assessment procedure for MR-Linac adaptive radiotherapy workflow using a pelvic phantom (32).

17.4 CBCT-Guided Adaptive Proton Therapy

Proton radiation therapy is capable of offering superior dosimetric benefits for select disease sites thanks to the physical nature of proton-tissue interaction and corresponding depth dose curve. The focused dose deposit towards the tail end of the penetration trajectory, which is also known as the Bragg peak, makes proton radiation therapy one of the preferred modalities for treatment where the close proximity of OAR is observed, as well as for pediatric patients.

Successful proton radiation therapy relies on both treatment planning and patient position verification prior to the treatment. Unlike photon radiation therapy, where the dose is mostly shift-invariant, proton dose calculation is much more sensitive to changes in water-equivalent depth (WED) in the beam path and has the potential risk of under-dosing the target if daily WED is higher or over-dosing OARs if the OAR is directly behind the distal edge of Bragg peak. Therefore, proton therapy is, in general, much more susceptible to anatomical variations and, therefore can benefit more from rigorous IGRT and online adaptive radiation therapy. Daily IGRT typically uses in-room setup verification primarily involving kilovoltage (kV) techniques (33). Two-dimension kV images have been routinely used with the angle corresponding to the treatment field being used. Bony anatomy positioning needs special attention as it would severely perturb the proton range. In the past decade, volumetric imaging using kV system has been introduced to proton delivery systems and offers a substantial improvement in IGRT (Figure 17.5). It also enabled offline adaptive proton therapy (34), and potentially daily adaptive proton therapy (DAPT) workflow (35).

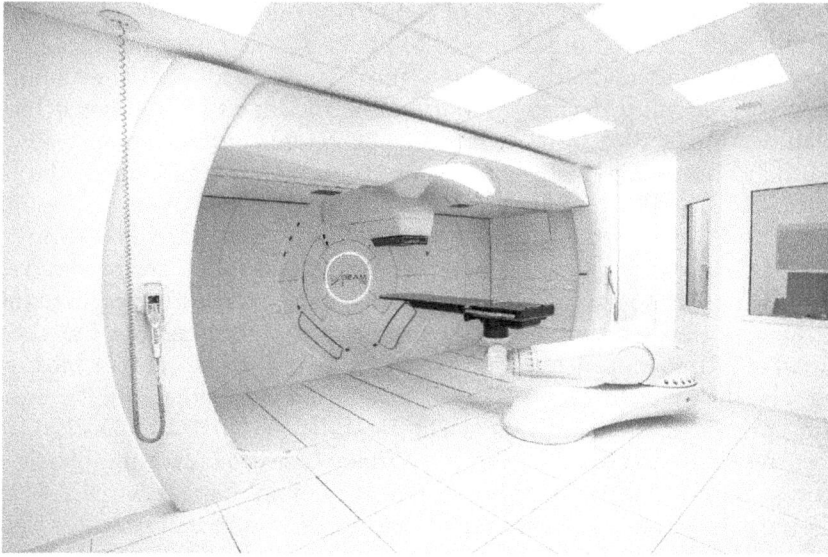

FIGURE 17.5
Varian ProBeam 360° proton therapy system with CBCT capability. (Image courtesy of Varian Medical Systems, Inc. All rights reserved.)

Hua et al. introduced a ceiling-mounted C-arm CBCT image system for a proton vault (36). It was integrated with a discrete scanning beam nozzle (PROBEAT V, Hitachi, Tokyo, Japan). The system is capable of high mechanical precision and offers high image quality. Cho et al. developed a CBCT and cone beam digital tomosynthesis (CBDT) capable imaging system using Varian orthogonal kV image panels integrated with the proton unit (37). They used Feldkamp's cone-beam algorithm for CBCT reconstruction and the filtered back projection (FBP) approach for digital tomosynthesis. Reconstructed images using quantitative AAPM phantom showed decent geometric robustness and image quality. CBCT-based volumetric imaging has become daily image guidance routine for photon radiation therapy. However, CBCT image guidance for proton therapy faces challenges in accurately representing HU to verify the proton range, as well as an elevated level of scatter noises and artifacts. Therefore, proper correction of CBCT image is needed. Park et al. demonstrated the feasibility of phantom based *a priori* CT-based scatter correction approach for CBCT-based proton dose calculation (38). They found that water equivalent path length error was substantially reduced with correction. Andersen et al. evaluated an *a priori* scatter correction method for CBCT-based range and dose calculation (39). They used Catphan and Alderson phantom to acquire onboard CBCT. Planning CT of the phantom was projected to each acquisition angle's projection, which was subtracted from the onboard acquired cone-beam projections to estimate the scatter and beam hardening component. Nomura et al. developed a combined scatter correction approach for CBCT-based proton therapy dose calculation (40). Two existing correction methods were explored: point-spread function (PSF)-based scatter correction and fast adaptive scatter kernel superposition (fASKS) model. They were able to improve the mean absolute relative error of the proton range in the pelvic phantom from 5.03% to 2.57% (40). Arai et al. adopted a different approach to using CBCT for proton daily dose calculation (41). They studied a histogram-matching algorithm on HN phantom, pelvic phantom, and HN patients. Both rigid and DIR were applied to match CBCT to CT. This approach was capable of achieving a 94.1%

gamma passing rate using modified CBCT (41). Landry et al. evaluated the CT to CBCT registration approach for proton therapy dose calculation using a deformable phantom (42). A Morphons DIR algorithm was analyzed. The result showed 2-3 mm deformation accuracy, which translated to 2% accuracy of the proton range (42). Moreover, no distortion in CT number intensity in the uniform region was observed.

More recently, with the help of a deep neural network, a novel scatter correction approach has been made available. Lalonde et al. developed a U-net-based projection scatter correction approach (43). They were capable of achieving 98.7% gamma passing rate using the proposed approach as compared to the reference dose calculation. Uh et al. trained a deep neural network combining training data of both abdomen and pelvis from children and young adults for accurate CBCT correction (44). They used cycle-consistent generative adversarial network (GAN) architecture. Results showed that a combined dataset with age-normalized body size improved the HU accuracy of CBCT and proton dose estimation. Harms et al. developed a cycle-GAN model to derive a relative stopping power map using daily onboard CBCT for dose calculation for HN patients (45). This approach was capable of passing the gamma index test (3%, 3mm) with 94% accuracy as compared to CT-based dose calculation.

Due to the sensitivity of proton range uncertainty or variation as a result of the anatomical change, plan adaptation is often needed for proton therapy to account for reflect daily anatomy. Figure 17.6 summarizes the workflow of CBCT-guided adaptive proton therapy implemented with scatter correction. One popular approach is registering daily CBCT to planning CT. DIR is often applied. The warped planning CT towards daily CBCT can be used to estimate daily dose to the patient, which can be used subsequently to adapt the plan. The University of Pennsylvania pioneered the application of this adaptive approach to the IBA system using synthetic CTs from online CBCTs. Veiga et al. performed a comprehensive evaluation of the DIR approach to warp planning dose toward daily CBCT anatomy (46). The result validated that this approach was capable of providing similar water equivalent

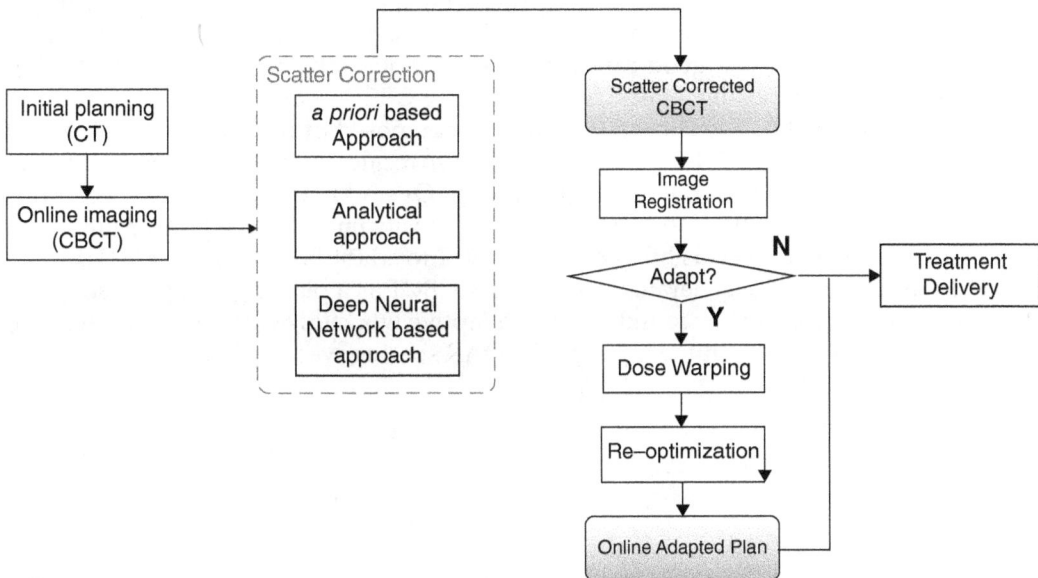

FIGURE 17.6
Workflow of CBCT-guided adaptive proton therapy.

thickness and dosimetric information as compared to acquiring a daily diagnostic level CT. Bondesson et al. validated 4D-CBCT image correction for adaptive proton radiation therapy with anthropomorphic lung phantom (47). DIR was used to create 4D virtual CT to aid proton dose calculation and accumulation. Good agreement was observed in dose calculation between 4D-CT-based calculation and 4D virtual CT-based calculation. Thummerer et el. performed a comparison among three CBCT correction approaches for HN proton adaptive therapy (48). They compared the deep convolutional neural network-based approach, DIR approach, and image-based analytical correction approach. Results showed accurate dose calculation from images generated with deep neural network and DIR approach, while the analytical correction approach resulted in compromised dose calculation accuracy. Thummerer et al. further compared CBCT guided synthetic CT approach with MRI-based synthetic CT for HN proton adaptive therapy (49). It was shown that a significant image quality difference was observed between the two but no significant dose calculation accuracy was observed, indicating that readily available orthogonal kV imaging components can provide a direct solution without additional hardware/software resource requirements. Wang et al. quantitatively assessed the HN patient anatomical change during the course of the treatment using virtual proton depth radiographs (PDRs) (50). DIR was used to register daily CBCT to planning CT, which was subsequently used to project cumulative water equivalent thickness to create PDRs. The PDRs were used to identify large anatomical variation areas and were used to establish the relation with dose distribution changes.

A few pre-clinical studies have demonstrated the feasibility of CBCT-based adaptive proton therapy in a clinical setting for various treatment sites. Xu et al. assessed the daily dose accumulation for prostate proton radiation therapy under a robust optimization setting (51). They registered daily CBCT to plan CT using DIR and accumulated daily dose thereafter. They found that ±3mm (patient setup)/±3% (proton range) was a sufficient uncertainty estimate for robust optimization. Ornelas et al. assessed an adaptive workflow for HN patients using daily CBCT for guidance (52). Patients in this study were retrospectively planned with robust optimization intensity modulated proton therapy (RO-IMPT) with ± 3 mm/± 3% uncertainty. The result demonstrated adequate coverage of CTV for the majority of treatment fractions. In addition, they found synthetic CT generated from deformed CBCT serves as a good candidate for decision-making in an adaptive setting. Veiga et al. performed the first clinical investigation of CBCT-guided proton adaptive therapy for lung cancer (34). The clinical workflow is composed of fast range-corrected dose distribution using virtual CT, followed by offline dose recalculation on the virtual CT scan, which is used as a trigger for the decision to rescan. Overall performance was decent, but the authors also warned that subtle anatomical or density change needs special attention. Kurz et al. investigated the feasibility of using CBCT to adapt IMPT plans for HN patients (53). They used virtual CT generated from weekly CBCT as a planning image to adapt the plan in an offline setting. They found that this adaptive workflow was capable of reducing D2 in the planning target volume (PTV) while OAR sparing was partially improved.

17.5 Online Adaptive RT Using Plan-of-the-Day

Patient undergoing radiation therapy is prone to internal anatomy change due to multiple reasons, including tumor shrinkage, day-to-day organ motion, organ filling, weight loss, etc. (54, 55). This poses tremendous challenges for accurate treatment delivery on daily basis. Most sensitive treatment sites include the abdominal region and pelvic region. One of the popular approaches to addressing daily anatomy change is contour adaptation (56). Contour is considered an additional layer of images overlaid on top of the diagnostic and planning images to guide dose distribution for therapeutic purposes. Therefore, closely monitoring the variation of contour, including deformation and translation, can help improve overall treatment accuracy. With accurate information on daily contour changes, a.k.a. "anatomy-of-the-day", the physician is capable of assessing the plan quality on current anatomy. Visualizing daily anatomy is of key importance in providing such information.

Heijkoop et al. developed and implemented an online adaptive plan-of-the-day protocol using nonrigid motion management for cervical cancer (57). They scanned two CT images, one with a full bladder and one with an empty bladder. A library consisting of one IMRT plan based on predicted internal target volume (ITV) volume using two CT scans and a 3D conformal plan was constructed. Daily CBCT was used to visualize and verify daily anatomy, which is subsequently used to determine the optimal plan for the day. Bondar et al. developed a similar approach for adaptive cervical cancer radiation therapy (58). Nine to ten variable bladder filling CT scans were acquired for each patient at pretreatment and after 40 Gy. A library of plans constructed based on variable bladder-filling anatomy was used to pick the "plan of the day." Daily bladder volume measured online was used to identify the plan to use for the day. Bondar et al. also developed a symmetric nonrigid registration method to address the issue of large organ deformation in cervical cancer patients (59). A unidirectional thin plate spline robust point matching algorithm was developed. A total of five CT scans with large variability in organ shape, volume, and deformation for each patient were acquired to train and validate the model. Results showed that the proposed method substantially reduced the residual error of registration from between 5.8 to 70.1 mm to between 1.9 to 8.5 mm.

A similar adaptive approach is employed for other pelvic sites, such as prostate cancer. Chen et al. described an adaptive workflow for managing interfractional anatomical variations for prostate cancer (60). They generated a series of plans based on planning images and selected the "plan-of-the-day" based on anatomy similarity. Xia et al. built a library of prostate plans based on a few potential prostate locations, which were referred to as multiple adaptive plans. A best-matched plan was chosen based on "prostate position of the day," which is either determined by image registration of implant markers or aligned to the pelvic bones.

17.6 Dose Gradient Adaptation

The previous section discussed the adaptive therapy approach of using contour as a surrogate to ensure conformal tight dose gradient delivery to the target. Although using this approach could ensure adequate target coverage, it comes with the cost of potential excessive margin to account for potential contour variation, and therefore less optimal. Another commonly adopted approach of adaptive radiation therapy is dose gradient adaptation. Instead of contour adaptation and subsequent plan selection based on a library of preexisting plans, one can adapt the dose gradient directly to conform the radiation dose to the "anatomy of the day." This approach could ensure conformal dose gradient shaping around the "anatomy-of-the-day," and therefore reducing excessive toxicity. Several studies have reported this approach (61–78).

Lim et al. explored two adaptive strategies for cervical cancer patients (61). Their patient cohort underwent weekly MR scans over a 5-week treatment course. DIR was used to assess the accumulated dose, which will be used to trigger either anatomy-driven mid-treatment replan or dosimetry-driven replan. Stewart et al. adopted a similar weekly imaging approach for cervical cancer patients (62). They fused weekly MRI images with planning MRI and compared the efficacy of two planning approaches: one using a 3-mm margin with no replanning, and the other using a 3-mm margin with automated replan on weekly anatomy. Results showed that replanning was capable of providing more consistent target coverage. Kerkhof et al. simulated online MRI guidance adaptive approach for cervical cancer (63). They replanned based on online MRI images with a 4 mm margin and compared the plan based on the pre-treatment IMRT plan with a 15 mm margin to PTV and a 10 mm margin to nodal PTV, respectively. Results showed that a significant reduction in healthy tissue dose was observed. Oh et al. assessed the cost and benefit of adaptive replanning schemes for cervical cancer patients based on online weekly MRI (64). They found online MRI is a critical part of adaptive therapy and offline replanning enhanced adaptation quality.

The previous section mentioned one stream of prostate adaptive therapy using rigid registration and adopting "plan-of-the-day" approach. Another stream of approach is to adapt the gradient of the plan to tailor towards the daily anatomy. The overall goal of gradient manipulation is to conform the radiation dose to the daily PTV shape and/or to navigate away from daily OAR shape. Sheng et al. reported a dose gradient adaptation method for prostate radiation therapy (65). The workflow of dose gradient adaptation is shown in Figure 17.7. They established a prostate anatomy atlas with corresponding isodose lines. The anatomy similarity was defined, and the query case can identify the most similar atlas case. The atlas case's isodose will then be warped towards the query case to serve as the "goal dose." This approach is demonstrated in (Figure 17.7) allows better dose conformity and gradient relative to the prostate anatomy than using a library plan. Wu et al. developed a dose gradient adaptation approach based on daily anatomy using DIR (66). The goal dose was created to guide voxel-based re-optimization online using linear programming. Li et al. developed an online plan re-optimization approach with RTOG objective, patient-specific objectives, and daily CBCT anatomy-adapted objectives, which was

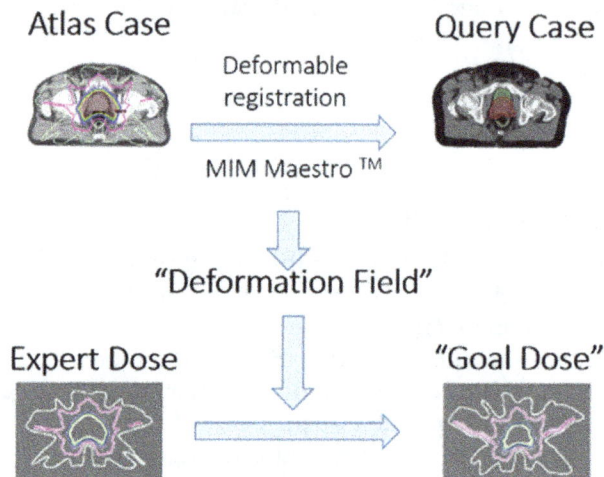

FIGURE 17.7
Workflow for dose gradient adaptation to create goal dose.

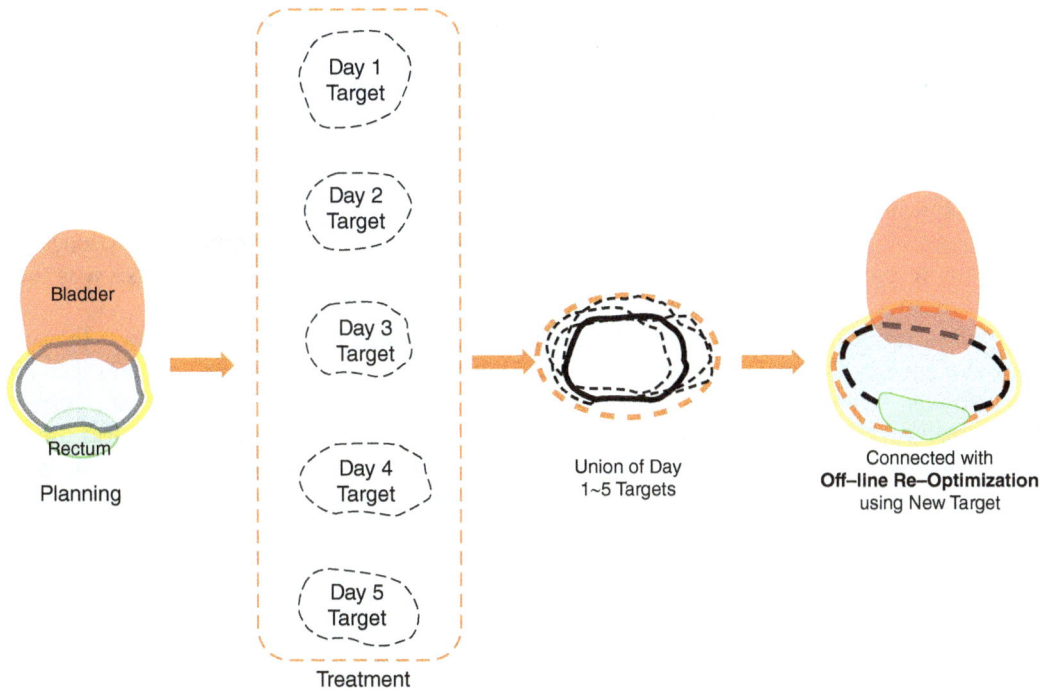

FIGURE 17.8
Offline re-optimization approach to adapt dose gradient to accommodate anatomy variation.

achievable within 2 min (67, 68). Online re-optimization can be combined with offline re-optimization, which is based on the union of PTVs for the first few fractions (Figure 17.8). Mohan et al. developed an approach of deforming the intensity distribution to guide transforming the leaf sequencing to adapt to daily anatomy for IMRT treatment (70). Fu et al. addressed the issue of interfractional prostate motion using online adaptive replanning based on CBCT images (71). Instead of re-optimizing the fluence intensity, they directly modify the MLC leaf position for each subfield to account for the shape change in the beam's eye view. Feng et al. developed a direct aperture deformation (DAD) approach which is capable of morphing the treatment aperture as a potential online correction approach using the deformation field acquired between the planning image and the daily image (72). Direct aperture optimization was further developed and enhanced by Men et al. using GPU acceleration, which significantly improves online adaptive therapy efficiency (73). Ahunbay et al. proposed a "gradient maintenance" (GM) approach to adapt the plan for daily prostate anatomy (74). They generated a series of partial concentric rings around the target toward the OAR on planning and daily images, which was subsequently used to determine inverse optimization objective functions. Thongphiew et al. compared the repositioning approach and the online adaptive approach for prostate IMRT treatment (75). They found that all IGRT techniques are sufficient to address simple geometric variation while large variation would favor the online re-optimization approach, which is achievable with less than 2 min.

The aforementioned gradient adaptation approach is more or less dependent on the input of daily anatomy as well as the deformation field related to the planning image. This dose gradient adaptation approach can be generalized to a broader scope of adaption upon user request, which is also known as interactive planning. Otto initially introduced this concept in 2014 (79). He developed a novel algorithm of Achievable Dose Estimate (ADE), which is an efficient algorithm to update the dose in real time considering physical delivery limitations. Ziegenhein et al. developed an interactive dose-shaping approach for IMRT treatment planning (80). This approach is composed of two modules: the dose modification module, where the planner can request local dose modification patches, which are used to guide fluence adaptation in the backend, as well as the recovery module, which aims to recover lost dose coverage elsewhere. They validated the proposed approach on the prostate patient cohort and discovered that the interactively planned dose was well under Quantec constraints (81).

Another effort in dose gradient adaptation has been seen in adaptive radiation therapy. Bock et al. combined robust treatment planning with adaptive re-optimization (82). The initial plan was generated with the assumption of anticipating systematic and random errors. The re-optimization was triggered when the estimated dose is not satisfactory, and the re-optimized plan will be used for subsequent fractions. Ferrara et al. addressed the same issue in prostate radiation therapy using the combination of adaptive strategy and robust optimization (83). They adopted an alternative approach of employing robust optimization during re-optimization process using the deformation field between daily CBCT and planning CT.

References

1. De Roover R, Crijns W, Poels K, Dewit B, Draulans C, Haustermans K, et al. Automated treatment planning of prostate stereotactic body radiotherapy with focal boosting on a fast-rotating O-ring linac: Plan quality comparison with C-arm linacs. *Journal of Applied Clinical Medical Physics/American College of Medical Physics* 2021;22(9):59–72.

2. Nasser N, Caudell JJ, Moros EG, Feygelman V, Redler G. Initial plan quality evaluation using a novel ai-driven planning system and paradigm for adaptive head and neck patients. *International Journal of Radiation Oncology, Biology, Physics* 2021;111(3):e97.

3. Scheuermann RM, Marcel J, Anderson N, Apinorasethkul O, Cooper K, Kempsey B, et al. Evaluation of dosimetric quality of auto-generated plans by a novel online adaptive system for head and neck radiotherapy. *International Journal of Radiation Oncology, Biology, Physics* 2020;108(3):e355–e6.

4. Hao Y, Cai B, Laugeman E, Kim T, Jiang Z, Green OL, et al. Plan quality and fidelity evaluation of CBCT-guided stereotactic online adaptive radiotherapy (CT-STAR) with an artificial intelligence (AI) aided plan re-optimization. *International Journal of Radiation Oncology, Biology, Physics* 2020;108(3):e786–e7.

5. Bolt M, Shelley C, Hollingdale R, Chadwick S, Barnard A, Leverton A, et al. PO-1574 Evaluation of automated plan quality for cervical cancer using the Ethos TPS. *Radiotherapy and Oncology* 2021;161:S1299–300.

6. Cai B, Laugeman E, Mazur TR, Park JC, Henke LE, Kim H, et al. Characterization of a prototype rapid kilovoltage x-ray image guidance system designed for a ring shape radiation therapy unit. *Medical Physics* 2019;46(3):1355–70.

7. Washio H, Ohira S, Funama Y, Ueda Y, Morimoto M, Kanayama N, et al. Dose reduction and low-contrast detectability using iterative CBCT reconstruction algorithm for radiotherapy. *Technology in Cancer Research & Treatment* 2022;21:15330338211067312.

8. Jarema T, Aland T. Using the iterative kV CBCT reconstruction on the Varian Halcyon linear accelerator for radiation therapy planning for pelvis patients. *Physica Medica: PM: An International Journal Devoted to the Applications of Physics to Medicine and Biology: Official Journal of the Italian Association of Biomedical Physics* 2019;68:112–6.

9. Archambault Y, Boylan C, Bullock D, Morgas T, Peltola J, Ruokokoski E, et al. Making on-line adaptive radiotherapy possible using artificial intelligence and machine learning for efficient daily re-planning. *Medical Physics International Journal* 2020;8(2): 77–86.

10. Mao W, Riess J, Kim J, Vance S, Chetty IJ, Movsas B, et al. Evaluation of auto-contouring and dose distributions for online adaptive radiation therapy of patients with locally advanced lung cancers. *Practical Radiation Oncology* 2022;12(4):e329–38.

11. Yoon SW, Lin H, Alonso-Basanta M, Anderson N, Apinorasethkul O, Cooper K, et al. Initial evaluation of a novel cone-beam ct-based semi-automated online adaptive radiotherapy system for head and neck cancer treatment - A timing and automation quality study. *Cureus* 2020;12(8):e9660.

12. Ward A, Martinou M, Kidane G, Graham S. Daily adaptive radiotherapy using the varian ETHOS system to improve dose distribution during treatment to the upper abdomen. *Clinical Oncology* 2022;34:e10.

13. Byrne M, Archibald-Heeren B, Hu Y, Teh A, Beserminji R, Cai E, et al. Varian ethos online adaptive radiotherapy for prostate cancer: Early results of contouring accuracy, treatment plan quality, and treatment time. *Journal of Applied Clinical Medical Physics* 2022;23(1):e13479.

14. Kim T, Ji Z, Lewis B, Laugeman E, Price A, Hao Y, et al. Visually guided respiratory motion management for Ethos adaptive radiotherapy. *Journal of Applied Clinical Medical Physics* 2022;23(1):e13441.

15. Moazzezi M, Rose B, Kisling K, Moore KL, Ray X. Prospects for daily online adaptive radiotherapy via ethos for prostate cancer patients without nodal involvement using unedited CBCT auto-segmentation. *Journal of Applied Clinical Medical Physics* 2021;22(10):82–93.

16. Pokharel S, Pacheco A, Tanner S. Assessment of efficacy in automated plan generation for Varian Ethos intelligent optimization engine. *Journal of Applied Clinical Medical Physics* 2022;23(4):e13539.

17. Mutic S, Dempsey JF. The ViewRay system: Magnetic resonance–Guided and controlled radiotherapy. *Seminars in Radiation Oncology* 2014;24(3):196–9.

18. Lagendijk JJW, Raaymakers BW, van Vulpen M. The magnetic resonance imaging–linac system. *Seminars in Radiation Oncology* 2014;24(3):207–9.

19. Raaymakers BW, Lagendijk JJW, Overweg J, Kok JGM, Raaijmakers AJE, Kerkhof EM, et al. Integrating a 1.5 T MRI scanner with a 6 MV accelerator: Proof of concept. *Physics in Medicine and Biology* 2009;54(12):N229–37.

20. Raaymakers BW, Jürgenliemk-Schulz IM, Bol GH, Glitzner M, Kotte ANTJ, van Asselen B, et al. First patients treated with a 1.5 T MRI-Linac: Clinical proof of concept of a high-precision, high-field MRI guided radiotherapy treatment. *Physics in Medicine & Biology* 2017;62(23): L41–L50.

21. Tijssen RHN, Philippens MEP, Paulson ES, Glitzner M, Chugh B, Wetscherek A, et al. MRI commissioning of 1.5T MR-linac systems – A multi-institutional study. *Radiotherapy and Oncology* 2019;132:114–20.

22. Klüter S. Technical design and concept of a 0.35 T MR-Linac. *Clinical and Translational Radiation Oncology* 2019;18:98–101.

23. Winkel D, Bol GH, Kroon PS, van Asselen B, Hackett SS, Werensteijn-Honingh AM, et al. Adaptive radiotherapy: The Elekta Unity MR-linac concept. *Clinical and Translational Radiation Oncology* 2019;18:54–9.

24. Bohoudi O, Bruynzeel AME, Senan S, Cuijpers JP, Slotman BJ, Lagerwaard FJ, et al. Fast and robust online adaptive planning in stereotactic MR-guided adaptive radiation therapy (SMART) for pancreatic cancer. *Radiotherapy and Oncology: Journal of the European Society for Therapeutic Radiology and Oncology* 2017;125(3):439–44.

25. Werensteijn-Honingh AM, Kroon PS, Winkel D, Aalbers EM, van Asselen B, Bol GH, et al. Feasibility of stereotactic radiotherapy using a 1.5 T MR-linac: Multi-fraction treatment of pelvic lymph node oligometastases. *Radiotherapy and Oncology* 2019;134:50–4.

26. Finazzi T, Palacios MA, Haasbeek CJA, Admiraal MA, Spoelstra FOB, Bruynzeel AME, et al. Stereotactic MR-guided adaptive radiation therapy for peripheral lung tumors. *Radiotherapy and Oncology* 2020;144:46–52.

27. Intven MPW, de Mol van Otterloo SR, Mook S, Doornaert PAH, de Groot-van Breugel EN, Sikkes GG, et al. Online adaptive MR-guided radiotherapy for rectal cancer; feasibility of the workflow on a 1.5T MR-linac: Clinical implementation and initial experience. *Radiotherapy and Oncology* 2021;154:172–8.

28. de Leon J, Crawford D, Moutrie Z, Alvares S, Hogan L, Pagulayan C, et al. Early experience with MR-guided adaptive radiotherapy using a 1.5 T MR-Linac: First 6 months of operation using adapt to shape workflow. *Journal of Medical Imaging and Radiation Oncology* 2022;66(1):138–45.

29. Bernchou U, Christiansen RL, Bertelsen A, Tilly D, Riis HL, Jensen HR, et al. End-to-end validation of the geometric dose delivery performance of MR linac adaptive radiotherapy. *Physics in Medicine & Biology* 2021;66(4):045034.

30. Paulson ES, Ahunbay E, Chen X, Mickevicius NJ, Chen G-P, Schultz C, et al. 4D-MRI driven MR-guided online adaptive radiotherapy for abdominal stereotactic body radiation therapy on a high field MR-Linac: Implementation and initial clinical experience. *Clinical and Translational Radiation Oncology* 2020;23:72–9.

31. Dunlop A, Mitchell A, Tree A, Barnes H, Bower L, Chick J, et al. Daily adaptive radiotherapy for patients with prostate cancer using a high field MR-linac: Initial clinical experiences and assessment of delivered doses compared to a C-arm linac. *Clinical and Translational Radiation Oncology* 2020;23:35–42.

32. Axford A, Dikaios N, Roberts DA, Clark CH, Evans PM. An end-to-end assessment on the accuracy of adaptive radiotherapy in an MR-linac. *Physics in Medicine & Biology* 2021;66(5):055021.

33. MacKay RI. Image guidance for proton therapy. *Clinical Oncology* 2018;30(5):293–8.

34. Veiga C, Janssens G, Teng C-L, Baudier T, Hotoiu L, McClelland JR, et al. First clinical investigation of cone beam computed tomography and deformable registration for adaptive proton therapy for lung cancer. *International Journal of Radiation Oncology, Biology, Physics* 2016;95(1):549–59.

35. Albertini F, Matter M, Nenoff L, Zhang Y, Lomax A. Online daily adaptive proton therapy. *The British Journal of Radiology* 2020;93(1107):20190594.

36. Hua C, Yao W, Kidani T, Tomida K, Ozawa S, Nishimura T, et al. A robotic C-arm cone beam CT system for image-guided proton therapy: Design and performance. *The British Journal of Radiology* 2017;90(1079):20170266.

37. Cho MK, Kim JS, Cho Y-B, Youn H, Park SY, Cho S, et al. *CBCT/CBDT equipped with the x-ray projection system for image-guided proton therapy.* Proceedings Volume 7258, Medical Imaging 2009: Physics of Medical Imaging; 72582V (2009) https://doi.org/10.1117/12.811525. Event: SPIE Medical Imaging, Lake Buena Vista (Orlando Area), Florida, United States; 2009.

38. Park Y-K, Sharp GC, Phillips J, Winey BA. Proton dose calculation on scatter-corrected CBCT image: Feasibility study for adaptive proton therapy. *Medical Physics* 2015;42(8):4449–59.

39. Andersen AG, Park Y-K, Elstrøm UV, Petersen JBB, Sharp GC, Winey B, et al. Evaluation of an a priori scatter correction algorithm for cone-beam computed tomography based range and dose calculations in proton therapy. *Physics and Imaging in Radiation Oncology* 2020;16:89–94.

40. Nomura Y, Xu Q, Peng H, Takao S, Shimizu S, Xing L, et al. Modified fast adaptive scatter kernel superposition (mfASKS) correction and its dosimetric impact on CBCT-based proton therapy dose calculation. *Medical Physics* 2020;47(1):190–200.

41. Arai K, Kadoya N, Kato T, Endo H, Komori S, Abe Y, et al. Feasibility of CBCT-based proton dose calculation using a histogram-matching algorithm in proton beam therapy. *Physica Medica* 2017;33:68–76.

42. Landry G, Dedes G, Zöllner C, Handrack J, Janssens G, Orban de Xivry J, et al. Phantom based evaluation of CT to CBCT image registration for proton therapy dose recalculation. *Physics in Medicine and Biology* 2014;60(2):595–613.

43. Lalonde A, Winey B, Verburg J, Paganetti H, Sharp GC. Evaluation of CBCT scatter correction using deep convolutional neural networks for head and neck adaptive proton therapy. *Physics in Medicine & Biology* 2020;65(24):245022.

44. Uh J, Wang C, Acharya S, Krasin MJ, Hua C-H. Training a deep neural network coping with diversities in abdominal and pelvic images of children and young adults for CBCT-based adaptive proton therapy. *Radiotherapy and Oncology* 2021;160:250–8.

45. Harms J, Lei Y, Wang T, McDonald M, Ghavidel B, Stokes W, et al. Cone-beam CT-derived relative stopping power map generation via deep learning for proton radiotherapy. *Medical Physics* 2020;47(9):4416–27.

46. Veiga C, Janssens G, Baudier T, Hotoiu L, Brousmiche S, McClelland J, et al. A comprehensive evaluation of the accuracy of CBCT and deformable registration based dose calculation in lung proton therapy. *Biomedical Physics & Engineering Express*. 2017;3(1):015003.

47. Bondesson D, Meijers A, Janssens G, Rit S, Rabe M, Kamp F, et al. Anthropomorphic lung phantom based validation of in-room proton therapy 4D-CBCT image correction for dose calculation. *Zeitschrift für Medizinische Physik* 2022;32(1):74–84.

48. Thummerer A, Zaffino P, Meijers A, Marmitt GG, Seco J, Steenbakkers RJHM, et al. Comparison of CBCT based synthetic CT methods suitable for proton dose calculations in adaptive proton therapy. *Physics in Medicine & Biology* 2020;65(9):095002.

49. Thummerer A, de Jong BA, Zaffino P, Meijers A, Marmitt GG, Seco J, et al. Comparison of the suitability of CBCT- and MR-based synthetic CTs for daily adaptive proton therapy in head and neck patients. *Physics in Medicine & Biology* 2020;65(23):235036.

50. Wang P, Yin L, Zhang Y, Kirk M, Song G, Ahn PH, et al. Quantitative assessment of anatomical change using a virtual proton depth radiograph for adaptive head and neck proton therapy. *Journal of Applied Clinical Medical Physics* 2016;17(2):427–40.

51. Xu Y, Diwanji T, Brovold N, Butkus M, Padgett KR, Schmidt RM, et al. Assessment of daily dose accumulation for robustly optimized intensity modulated proton therapy treatment of prostate cancer. *Physica Medica* 2021;81:77–85.

52. De Ornelas M, Xu Y, Padgett K, Schmidt RM, Butkus M, Diwanji T, et al. CBCT-based adaptive assessment workflow for intensity modulated proton therapy for head and neck cancer. *International Journal of Particle Therapy* 2021;7(4):29–41.

53. Kurz C, Nijhuis R, Reiner M, Ganswindt U, Thieke C, Belka C, et al. Feasibility of automated proton therapy plan adaptation for head and neck tumors using cone beam CT images. *Radiation Oncology* 2016;11(1):64.

54. Yan D, Vicini F, Wong J, Martinez A. Adaptive radiation therapy. *Physics in Medicine and Biology* 1997;42:123–32.

55. Glide-Hurst CK, Lee P, Yock AD, Olsen JR, Cao M, Siddiqui F, et al. Adaptive radiation therapy (ART) strategies and technical considerations: A state of the ART review from NRG oncology. *International Journal of Radiation Oncology, Biology, Physics* 2021;109(4):1054–75.

56. Lim-Reinders S, Keller BM, Al-Ward S, Sahgal A, Kim A. Online adaptive radiation therapy. *International Journal of Radiation Oncology, Biology, Physics* 2017;99(4):994–1003.

57. Heijkoop ST, Langerak TR, Quint S, Bondar L, Mens JWM, Heijmen BJM, et al. Clinical implementation of an online adaptive plan-of-the-day protocol for nonrigid motion management in locally advanced cervical cancer IMRT. *International Journal of Radiation Oncology, Biology, Physics* 2014;90(3):673–9.

58. Bondar ML, Hoogeman MS, Mens JW, Quint S, Ahmad R, Dhawtal G, et al. Individualized nonadaptive and online-adaptive intensity-modulated radiotherapy treatment strategies for cervical cancer patients based on pretreatment acquired variable bladder filling computed tomography scans. *International Journal of Radiation Oncology, Biology, Physics* 2012;83(5):1617–23.

59. Bondar L, Hoogeman MS, Vásquez Osorio EM, Heijmen BJM. A symmetric nonrigid registration method to handle large organ deformations in cervical cancer patients. *Medical Physics* 2010;37(7Part1):3760–72.

60. Chen W, Gemmel A, Rietzel E. A patient-specific planning target volume used in 'plan of the day' adaptation for interfractional motion mitigation. *Journal of Radiation Research* 2013;54(suppl_1):i82–i90.

61. Lim K, Stewart J, Kelly V, Xie J, Brock KK, Moseley J, et al. Dosimetrically triggered adaptive intensity modulated radiation therapy for cervical cancer. *International Journal of Radiation Oncology, Biology, Physics* 2014;90(1):147–54.

62. Stewart J, Lim K, Kelly V, Xie J, Brock KK, Moseley J, et al. Automated weekly replanning for intensity-modulated radiotherapy of cervix cancer. *International Journal of Radiation Oncology, Biology, Physics* 2010;78(2):350–8.

63. Kerkhof EM, Raaymakers BW, van der Heide UA, van de Bunt L, Jürgenliemk-Schulz IM, Lagendijk JJW. Online MRI guidance for healthy tissue sparing in patients with cervical cancer: An IMRT planning study. *Radiotherapy and Oncology* 2008;88(2):241–9.

64. Oh S, Stewart J, Moseley J, Kelly V, Lim K, Xie J, et al. Hybrid adaptive radiotherapy with on-line MRI in cervix cancer IMRT. *Radiotherapy and Oncology* 2014;110(2):323–8.

65. Sheng Y, Li T, Zhang Y, Lee WR, Yin FF, Ge Y, et al. Atlas-guided prostate intensity modulated radiation therapy (IMRT) planning. *Physics in Medicine and Biology* 2015;60(18):7277–91.

66. Wu QJ, Thongphiew D, Wang Z, Mathayomchan B, Chankong V, Yoo S, et al. On-line re-optimization of prostate IMRT plans for adaptive radiation therapy. *Physics in Medicine and Biology* 2008;53(3):673–91.

67. Li T, Wu Q, Zhang Y, Vergalasova I, Lee WR, Yin FF, et al. Strategies for automatic online treatment plan reoptimization using clinical treatment planning system: A planning parameters study. *Medical Physics* 2013;40(11):111711.

68. Wu QJ, Li T, Wu Q, Yin FF. Adaptive radiation therapy: Technical components and clinical applications. *Cancer Journal* 2011;17(3):182–9.

69. Siciarz P, McCurdy B, Hanumanthappa N, Van Uytven E. Adaptive radiation therapy strategies in the treatment of prostate cancer patients using hypofractionated VMAT. *Journal of Applied Clinical Medical Physics/American College of Medical Physics* 2021;22(12):7–26.

70. Mohan R, Zhang X, Wang H, Kang Y, Wang X, Liu H, et al. Use of deformed intensity distributions for on-line modification of image-guided IMRT to account for interfractional anatomic changes. *International Journal of Radiation Oncology, Biology, Physics* 2005;61(4):1258–66.

71. Fu W, Yang Y, Yue NJ, Heron DE, Huq MS. A cone beam CT-guided online plan modification technique to correct interfractional anatomic changes for prostate cancer IMRT treatment. *Physics in Medicine and Biology* 2009;54(6):1691–703.

72. Feng Y, Castro-Pareja C, Shekhar R, Yu C. Direct aperture deformation: An interfraction image guidance strategy. *Medical Physics* 2006;33(12):4490–8.

73. Men C, Jia X, Jiang SB. GPU-based ultra-fast direct aperture optimization for online adaptive radiation therapy. *Physics in Medicine and Biology* 2010;55(15):4309–19.

74. Ahunbay EE, Li XA. Gradient maintenance: A new algorithm for fast online replanning. *Medical Physics* 2015;42(6Part1):2863–76.

75. Thongphiew D, Wu QJ, Lee WR, Chankong V, Yoo S, McMahon R, et al. Comparison of online IGRT techniques for prostate IMRT treatment: Adaptive vs repositioning correction. *Medical Physics* 2009;36(5):1651–62.

76. Ghilezan M, Yan D, Martinez A. Adaptive radiation therapy for prostate cancer. *Seminars in Radiation Oncology* 2010;20(2):130–7.

77. Hurkmans CW, Dijckmans I, Reijnen M, van der Leer J, van Vliet-Vroegindeweij C, van der Sangen M. Adaptive radiation therapy for breast IMRT-simultaneously integrated boost: Three-year clinical experience. *Radiotherapy and Oncology: Journal of the European Society for Therapeutic Radiology and Oncology* 2012;103(2):183–7.

78. Ghose S, Holloway L, Lim K, Chan P, Veera J, Vinod SK, et al. A review of segmentation and deformable registration methods applied to adaptive cervical cancer radiation therapy treatment planning. *Artificial Intelligence in Medicine* 2015;64(2):75–87.

79. Otto K. Real-time interactive treatment planning. *Physics in Medicine and Biology* 2014;59(17):4845–59.

80. Ziegenhein P, Ph Kamerling C, Oelfke U. Interactive dose shaping part 1: A new paradigm for IMRT treatment planning. *Physics in Medicine and Biology* 2016;61(6):2457–70.

81. Ph Kamerling C, Ziegenhein P, Sterzing F, Oelfke U. Interactive dose shaping part 2: Proof of concept study for six prostate patients. *Physics in Medicine and Biology* 2016;61(6):2471–84.

82. Böck M, Eriksson K, Forsgren A, Hårdemark B. Toward robust adaptive radiation therapy strategies. *Medical Physics* 2017;44(6):2054–65.

83. Ferrara E, Beldì D, Yin J, Vigna L, Loi G, Krengli M. Adaptive strategy for external beam radiation therapy in prostate cancer: Management of the geometrical uncertainties with robust optimization. *Practical Radiation Oncology* 2020;10(6):e521–e8.

Section VI

Perspectives

18

Validation and Evaluation Metrics

Jing Wang and Xiaofeng Yang
Emory University, Atlanta, GA, USA

CONTENTS

18.1 Introduction

Image synthesis applications have shown increasing clinical values to potentially simplify the imaging process, reduce the dose exposure, or avoid unwanted contrast injection to patients. However, the clinical adoption of such new techniques will strongly depend on their robustness and quality. Thus, extensive validation and evaluation need to be performed to assess the models before incorporation into the clinical workflow [1].

Several types of validation in research or clinical assessment have been proposed and used in previous studies. There exists qualitative validation, which is mainly a visual comparison between synthetic and real images, as well as quantitative metrics to evaluate the models with specifically calculated numbers. The quantitative metrics include dosimetric agreement and similarity measures. Both qualitative and quantitative validations are vital, and they are usually combined when used in research reports.

18.2 Overview of Qualitative Validation

An intuitive way of qualitative validation is simply by visual comparison. By comparing the synthetic images with the ground truth images side by side, it is very straightforward to evaluate the accuracy of synthetic images and check the overall quality by observing possible unusual local effects or artifacts. Often better visual results are correlated to better

DOI: 10.1201/9781003243458-24

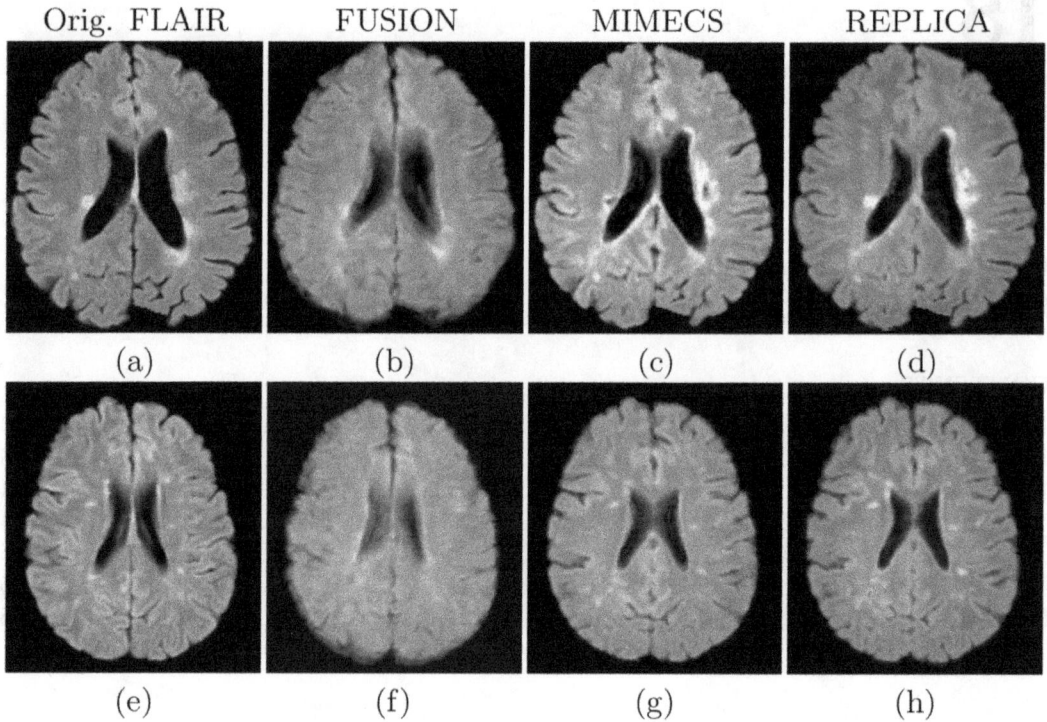

FIGURE 18.1
Visual comparison of FUSION, MIMECS, and REPLICA for the FLAIR synthesis task. The two rows show images from different subjects. (a) and (e) real FLAIRs, (b) and (f) FUSION results, (c) and (g) MIMECS results, (d) and (h) REPLICA results. (Reprint permission of [2].)

numerically calculated metrics. For example, Figure 18.1 is a visual comparison between original (real) FLAIR images and synthetic FLAIR images [2]. Looking at Figure 18.1, we can see the clearly favored results by the REPLICA method compared to the other two methods, with clearer structures and fewer blurring artifacts.

18.3 Overview of Quantitative Validation

One of the quantitative validations of medical imaging synthesis (e.g., CT or MRI synthesis) includes the structural similarities between synthetic and real images. Second, if the image synthesis process is embedding segmentation tasks, then the dice similarity will also be considered. The second main validation is the dosimetric agreement, since the most applicable scenario of sCT is for MRI-only radiotherapy [3].

18.3.1 Similarity Measures

Several numerical metrics have been developed to quantify the local or overall similarities between the output and ground truth images. The mean absolute error (MAE) is a straightforward metric to calculate the pixel difference between images:

$$\text{MAE} = \frac{\sum_{i=1}^{i=N} |R_i - S_i|}{N}, \tag{18.1}$$

where N is the total number of pixels in the images, R_i and S_i are the intensities of ith pixel on real CT and sCT images, respectively.

The peak signal-to-noise ratio (PSNR) measures the peak error between two images, and is defined as:

$$\text{PSNR} = 10 \times \log_{10} \left(\frac{\text{MAX}^2}{\text{MSE}} \right), \tag{18.2}$$

where MSE is the mean square error between the two images and MAX is the maximum dynamic range of signal.

The normalized cross-correlation (NCC) measures the similarity of image structures [4, 5] and is defined as:

$$\text{NCC} = \frac{1}{N} \sum_{i=1}^{i=N} \frac{1}{\sigma_r \sigma_s} (R_i \times S_i), \tag{18.3}$$

where σ_r and σ_s are the standard deviations of real CT and sCT images, respectively.

Apart from the metrics calculated from pixel-to-pixel intensity correlations, the structural similarity metric (SSIM) has been among the most popular metrics since its introduction. It is calculated based on the similarity of statistics (e.g., mean, variance, and covariance) inside small patches between the real and synthetic images, which is a product of luminance, contrast, and structure. However, the SSIM was designed to work on all positive values, which could sometimes be violated for CT HU numbers or z-normalization [1].

18.3.2 Dice Measures

In some situations, the bone structures and organs need to be segmented during or after image synthesis, and the dice similarity is thus an important metric to measure the overlapping between contoured organs in real images and segmented organs in synthetic images (e.g., sCT). The dice similarity coefficient (DSC) is defined as:

$$Dice_{\text{organ}} = \frac{2(V_{CT} \cap V_{sCT})}{V_{CT} + V_{sCT}} \tag{18.4}$$

Where V_{CT} and V_{sCT} are the volumes of segmented organs on real CT and sCT images, respectively. Besides DSC, sensitivity and specificity are two standard metrics to quantify the detection accuracy of segmented targets, depending on how to set the threshold of CT HU values for each organ.

18.3.3 Dosimetric Agreement

The dosimetric agreement refers to the dose differences when calculating the doses based on sCTs compared to the real CTs, which can be quantified with dose volume histogram

(DVH) points. The general equivalent uniform dose (gEUD) can be used to quantify biologically relevant differences of the entire DVH. Other metrics, such as gamma index [6], can reveal spatial correlation of dose deviance.

References

1. Gourdeau, D., S. Duchesne, and L. Archambault, On the proper use of structural similarity for the robust evaluation of medical image synthesis models. *Medical Physics*, 2022. **49**(4): pp. 2462–2474.
2. Jog, A., et al., Random forest regression for magnetic resonance image synthesis. *Medical Image Analysis*, 2017. **35**: pp. 475–488.
3. Johnstone, E., et al., Systematic review of synthetic computed tomography generation methodologies for use in magnetic resonance imaging – only radiation therapy. *International Journal of Radiation Oncology* Biology* Physics*, 2018. **100**(1): pp. 199–217.
4. Yoo, J.-C. and T.H. Han, Fast normalized cross-correlation. *Circuits, Systems and Signal Processing*, 2009. **28**(6): pp. 819–843.
5. Briechle, K. and U.D. Hanebeck. Template matching using fast normalized cross correlation. in *Optical Pattern Recognition XII*. 2001. SPIE 4387, Orlando, FL.
6. Low, D.A., et al., A technique for the quantitative evaluation of dose distributions. *Medical Physics*, 1998. **25**(5): pp. 656–661.

19

Limitations and Future Trends

Xiaofeng Yang

Emory University, Atlanta, GA, USA

CONTENT

Recent years have witnessed the trend of deep learning being increasingly used in the application of medical imaging. The latest networks and techniques have been borrowed from the field of computer vision and adapted to specific clinical tasks in radiology and radiation oncology. As reviewed in this paper, learning-based image synthesis is an emerging and active field - all the reviewed studies were published within the last three years. With further development in both artificial intelligence and computing hardware, more learning-based methods are expected to facilitate the clinical workflow with novel applications. Although the reviewed literature shows the success of deep learning-based image synthesis in various applications, there remain some open questions to be answered in future studies.

Due to the limitations of graphics processing unit (GPU) memory, some of the deep learning approaches examined were trained on two-dimensional (2D) slices. Since the loss functions of 2D models do not account for continuity in the third dimension, slice discontinuities can be observed. Some studies trained models on three-dimensional (3D) patches to exploit 3D spatial information with even less memory burden [1], while a potential drawback is that the larger-scale image features may be hard to extract [2]. Training on 3D image stacks is expected to achieve a more homogeneous conversion result. Fu *et al.* compared the performance of 2D and 3D models using the same U-net implementation [3]. They found that 3D-generated synthetic CT exhibited smaller mean absolute error (MAE) and more accurate bone regions. However, to achieve robust performance, the 3D model needs more training data to learn more parameters. A compromise is to use multiple adjacent slices that allow the model to capture more 3D context or to train different networks for all three combinations of orthogonal 2D planes to produce pseudo-3D information [4].

The previous studies illustrate the advantages of learning-based methods over conventional methods in performance as well as clinical application. Learning-based methods generally outperform conventional methods in generating more realistic synthetic images with higher similarity to real images and better quantitative metrics. Depending on the hardware, training a model in development usually takes hours to days for learning-based methods. However, once the model is trained, it can be applied to new patients to generate synthetic images in seconds to minutes. Conventional methods vary widely in specific methodologies and implementations, resulting in a wide range of run times. Iterative

DOI: 10.1201/9781003243458-25

methods such as compressed sensing (CS) were shown to be unfavorable due to significant costs in time and compute power.

Unlike conventional methods, learning-based methods require large training datasets. The size of training sets has been shown to affect the performance of machine learning in many challenging computer vision problems as well as medical imaging tasks [5–8]. Generally, a larger training set size with greater data variation can reduce overfitting of the model and enable better performance. Compared with studies in some medical imaging applications where it is common to see thousands of patients enrolled, studies in medical image synthesis involve far fewer patients. A training size of dozens of patients is more common in many studies, while hundreds of patients per set are rare and can be considered as a relatively "large" study. Moreover, it is very common to see the leave-X-out or N-fold cross validation strategy used in evaluating methods. The lack of an independent test set unseen by the model may complicate the generalization of results for broad clinical applications. The current small sample norm arises from circumstances which vary from application to application. In radiation oncology, clinical patient volume is inherently lower than other specialties such as radiology, so that fewer eligible patients are available for study. In addition to limitations in data collection, data cleaning further eliminates a portion of data that are low in quality or represent outliers, such as image pairs with suboptimal registration. In order to address the problem posed by limited training data, novel techniques have been proposed, such as transfer learning [9], self-supervised/weakly supervised/unsupervised learning [10, 11], and data augmentation [12]. These methods either diminish or completely eliminate dependence on training data sample size; although they may not be applicable to all medical image synthesis applications.

In the training stage, most of the reviewed studies require paired datasets, i.e., the source image and target image must exhibit pixel-to-pixel correspondence. This requirement poses difficulties in collecting sufficient eligible datasets and demands high accuracy in image registration. Some networks, such as CycleGAN, can relax the requirement for paired image datasets, which can be beneficial for clinical applications enrolling large numbers of patients for training.

Although the advantages of learning-based methods have been demonstrated, it should be noted that their performance can be unpredictable when input images during production differ significantly from training images. In most of the reviewed studies, unusual cases are excluded. However, unusual cases may be realistically observed in the clinical setting, and in these cases, the application of learning-based methods should be approached with diligence and caution. For example, hip prostheses create severe artifacts on both CT and MR images; thus, it is of clinical and practical interest to understand the effect of their inclusion in training or testing datasets for learning-based models, but this effect has not yet been studied. Similar unusual cases may also be encountered in other forms in other imaging modalities and are worthy of investigation, such as medical implants that introduce artifacts, obesity resulting in greater image noise, and anatomic deformities or abnormalities.

Due to the limitation in the number of available datasets, most studies used N-fold cross-validation or the leave-N-out strategy. The small to intermediate number of patients in training and testing datasets are appropriate for feasibility studies, but are not sufficient for evaluating clinical utility. Moreover, the representativeness of training/testing datasets relative to a particular clinic's population requires special attention in a clinical study. Suboptimal demographic diversity may reduce the robustness and generalizability of any model. Most studies reviewed here trained models using data from a single institution with a single scanner. Model performance across hardware of several models or manufacturers, wherein image characteristics cannot be exactly matched, is an important consideration due to frequent hardware replacement and upgrade in the modern clinical setting. Boni *et al.* recently presented a proof-of-concept study that predicted synthetic images of one clinical site using a model trained on data from two other sites and demonstrated clinically acceptable results [13]. Further studies could include datasets from multiple centers and adopt a leave-one-center-out training and/or test strategy in order to validate the consistency and robustness of the network.

Before being deployed into clinical workflow, there are still several challenges to be addressed. To account for potentially unpredictable synthetic images that can result from non-compliance with imaging protocols in training data or unexpected anatomic variation, additional quality assurance (QA) steps would be essential in clinical practice. QA procedures would aim to check the consistency of model performance routinely or after upgrade by re-training the network with additional patient datasets and verify synthetic image quality on specific cases.

References

1. Lei, Y., et al., MRI-only based synthetic CT generation using dense cycle consistent generative adversarial networks. *Medical Physics*, 2019. **46**(8): pp. 3565–3581.
2. Dinkla, A.M., et al., Dosimetric evaluation of synthetic CT for head and neck radiotherapy generated by a patch-based three-dimensional convolutional neural network. *Medical Physics*, 2019. **46**(9): pp. 4095–4104.
3. Fu, J., et al., Deep learning approaches using 2D and 3D convolutional neural networks for generating male pelvic synthetic computed tomography from magnetic resonance imaging. *Medical Physics*, 2019. **46**(9): pp. 3788–3798.
4. Schilling, K.G., et al., Synthesized b0 for diffusion distortion correction (Synb0-DisCo). *Magnetic Resonance Imaging*, 2019. **64**: pp. 62–70.
5. Azizi, S., et al., Transfer learning from RF to B-mode temporal enhanced ultrasound features for prostate cancer detection. *International Journal of Computer Assisted Radiology and Surgery*, 2017. **12**(7): pp. 1111–1121.
6. Gulshan, V., et al., Development and validation of a deep learning algorithm for detection of diabetic retinopathy in retinal fundus photographs. *JAMA*, 2016. **316**(22): pp. 2402–2410.
7. Mohamed, A.A., et al., A deep learning method for classifying mammographic breast density categories. *Medical Physics*, 2018. **45**(1): pp. 314–321.
8. Sun, C., et al., Revisiting unreasonable effectiveness of data in deep learning era. 2017. arXiv:1707.02968.
9. Shin, H., et al., Deep convolutional neural networks for computer-aided detection: CNN architectures, dataset characteristics and transfer learning. *IEEE Transactions on Medical Imaging*, 2016. **35**(5): pp. 1285–1298.
10. Chen, H., et al., Reblur2Deblur: Deblurring videos via self-supervised learning. in *2018 IEEE International Conference on Computational Photography (ICCP)*. Pittsburgh, PA. 2018.
11. Ahn, J. and S. Kwak, Learning pixel-level semantic affinity with image-level supervision for weakly supervised semantic segmentation. 2018. arXiv:1803.10464.
12. Shin, H.-C., et al., *Medical Image Synthesis for Data Augmentation and Anonymization Using Generative Adversarial Networks*. 2018. Cham: Springer International Publishing.
13. Brou Boni, K.N.D., et al., MR to CT synthesis with multicenter data in the pelvic area using a conditional generative adversarial network. *Physics in Medicine & Biology*, 2020. **65**(7): p. 075002.

Index

For Product Safety Concerns and Information please contact our EU
representative GPSR@taylorandfrancis.com
Taylor & Francis Verlag GmbH, Kaufingerstraße 24, 80331 München, Germany

www.ingramcontent.com/pod-product-compliance
Lightning Source LLC
Chambersburg PA
CBHW080930220326
41598CB00034B/5738